NORMAL AND MALIGNANT HEMATOPOIESIS

New Advances

PEZCOLLER FOUNDATION SYMPOSIA

NORMAL AND MALIGNANT HEMATOPOIESIS

New Advances

Edited by

Enrico Mihich

Roswell Park Cancer Institute
Buffalo, New York

and

Donald Metcalf

The Walter and Eliza Hall Institute of Medical Research
Royal Melbourne Hospital
Victoria, Australia

SPRINGER SCIENCE+BUSINESS MEDIA, LLC

Library of Congress Cataloging-in-Publication Data

Normal and malignant hematopoiesis : new advances / edited by Enrico
Mihich and Donald Metcalf.
     p.   cm. -- (Pezcoller foundation symposia ; 6)
     Proceedings of the Sixth Pezcoller Symposium on Normal and
Malignant Hematopoiesis: New Advances, held June 29-July 1, 1994, in
Rovereto, Italy--T.p. verso.
   Includes bibliographical references and index.
   ISBN 978-1-4613-5789-6   ISBN 978-1-4615-1927-0 (eBook)
   DOI 10.1007/978-1-4615-1927-0
   1. Lymphoproliferative disorders--Congresses. 2. Hematopoiesis-
-Congresses. 3. Hematopoiesis--Regulation--Congresses.   I. Mihich,
Enrico. II. Metcalf, Donald. III. Pezcoller Symposium on Normal
and Malignant Hematopoiesis: New Advances (1994 : Revereto, Trento,
Italy) IV. Series.
   [DNLM: 1. Hematopoiesis--genetics-- congresses . 2. Gene
Expression Regulation--congresses. 3. Leukemia, Myeloid--genetics-
-congresses. 4. Gene Expression Regulation, Neoplastic--congresses.
WH 140 N8413 1995]
RC646.2.N67 1995
616.4'1--dc20
DNLM/DLC
for Library of Congress   95-37107
                                                               CIP

Proceedings of the Sixth Pezcoller Symposium on Normal and Malignant Hematopoiesis: New Advances,
held June 29-July 1, 1994, in Rovereto, Italy

ISBN 978-1-4613-5789-6

© 1995 Springer Science+Business Media New York
Originally published by Plenum Press, New York in 1995
Softcover reprint of the hardcover 1st edition 1995

10 9 8 7 6 5 4 3 2 1

THE PEZCOLLER FOUNDATION

The Pezcoller Foundation was created in 1979 by Professor Alessio Pezcoller (1896–1993), who was the chief surgeon of the S. Chiara Hospital in Trento from 1937 to 1966 and who gave a substantial portion of his estate to support its activities; the Foundation also benefits from the cooperation of the Savings Bank Cassa di Risparmio di Trento e Rovereto.

The main goal of this nonprofit foundation is to provide and recognize scientific progress on life-threatening diseases, currently focusing on cancer. Toward this goal, the Pezcoller Foundation awards, every two years, the Pezcoller Prize, recognizing highly meritorious contributions to medical research; it also sponsors a series of annual symposia promoting interactions among scientists working at the cutting edge of basic oncological sciences.

The award selection process is managed by the European School of Oncology in Milan, Italy, with the aid of an international committee of experts chaired by Professor U. Veronesi.

The symposia are held in the Trentino Region of Northern Italy and their scientific focus is selected by Enrico Mihich with the collaboration of an international Standing Symposia Committee. A Program Committee determines the content of each symposium.

The first symposium focused on *Drug Resistance: Mechanisms and Reversal* (E. Mihich, Chairman, 1989); the second on *The Therapeutic Implications of the Molecular Biology of Breast Cancer* (M.E. Lippman and E. Mihich, Co-Chairmen, 1990); the third on *Tumor Suppressor Genes* (D.M. Livingston and E. Mihich, Co-Chairmen, 1991); the fourth on *Cell Adhesion Molecules: Cellular Recognition Mechanisms* (M.E. Hemler and E. Mihich, Co-Chairmen, 1992); and the fifth on *Apoptosis* (E. Mihich and R.T. Schimke, Co-Chairmen, 1993). The seventh symposium (1995) was focused on *Cancer Genes: Functional Aspects* (E. Mihich and D. Hausman, Co-Chairmen).

PREFACE

The Sixth Pezcoller Symposium, entitled Normal and Malignant Hematopoiesis: New Advances, was held in Rovereto, Italy, June 29-July 1, 1994, and was focused on the genetic basis of the regulation of hematopoiesis and the molecular mechanisms of this regulation. The function of such genes as ALL1, RAS, Abl, PML, and Cyclin D, the mechanisms of transcriptional control and the pathways of signal transduction, the role of certain cytokines, the involvement of growth versus differentiation programs were all discussed in comparing normal and leukemic hematopoiesis. Approaches to cancer treatment based on the stimulation of differentiation with retinoids and the modulation of certain cytokines were also considered.

Leukemias are clonal neoplasms of one or another type of hematopoietic cell with the leukemic clone becoming dominant over pre-existing normal hematopoietic populations. Both types of population share many biological features in common—the need for continuous cell division, the necessity to make decisions regarding self-renewal versus differentiation commitment, the need to undergo often striking morphological and functional changes during maturation. Although it is likely that each of these processes is aberrant in leukemic populations, there is evidence that similar regulatory controls are active in both types of population. It is pertinent, therefore, to review the rapidly expanding body of information on the molecular control of normal hematopoiesis in the specific context of identifying the abnormal response patterns exhibited by leukemic cells. In particular, there is need to link the now extensive data on specific chromosomal translocations in leukemia and the expanding list of oncogenes known to be involved in leukemogenesis with the extensive data on normal molecular control mechanisms. The Sixth Annual Pezcoller Symposium brought together experts in each of the fields considered in an attempt to develop a coherent molecular explanation of the leukemic state.

We wish to thank the participants in the Symposium for their substantial contributions and their participation in the spirited discussions which followed. We would also like to thank Drs. David M. Livingston, Owen Witte, Pier Giuseppe Pelicci, Tadatsugu Taniguchi, and Carlo Croce for their essential input as members of the Program Committee, and Ms. A. Toscani for her invaluable assistance. The aid of the Bank Cassa di Risparmio di Trento and Rovereto, and the Municipal, Provincial, and Regional Administrations in supporting this Symposium through the Pezcoller Foundation are also acknowledged with deep appreciation. Finally, we wish to thank the staff of Plenum Publishing Corporation for their efficient cooperation in the production of these Proceedings.

Enrico Mihich
Donald Metcalf

CONTENTS

REGULATION OF NORMAL HEMOPOIESIS

Donald Metcalf

The Walter and Eliza Hall Institute of Medical Research
P.O. Royal Melbourne Hospital
3050 Victoria, Australia

INTRODUCTION

The following is a summary of an oral presentation delivered because of the unavoidable absence of Dr. Ogawa from the meeting. No detailed referencing will be included as the subject has been reviewed extensively elsewhere (1,2,3).

MULTIPLICITY OF HEMOPOIETIC REGULATORS

The process by which multipotential stem cells form lineage-restricted committed progenitor cells that then generate maturing progeny in the eight major lineages of blood cells involves not only the controlled formation of very large numbers of cells from each stem cell but also a complex series of cellular events including differentiation commitment, maturation induction, controlled release of mature cells and often their functional activation in the tissues.

A logical method for regulating such events might have been the use of a large series of regulators, each controlling a single aspect of this complex biology. However, what has emerged from two decades of in vitro analysis of these events is a likely situation that is both more complex and simpler. Multiple hemopoietic regulators, now numbering more than 25, have been documented as controlling hemopoiesis, with the likelihood that an equal number have yet to be discovered. Each of these regulators has actions on cells of more than one lineage and, as a consequence, typically six to eight regulators are already known to have actions on cells in a single lineage. However, each regulator exhibits polyfunctionality and is able to control more than one aspect of the biology of the responding hemopoietic cells.

It is already evident therefore that the regulatory system is complex and has elements of redundancy that at the very least provide a fail-safe system with the potential to ensure maintenance of regulation.

Despite the existence of these systemic regulators, hemopoiesis in the adult is in fact restricted to the marrow and spleen. This is because specialized stromal cells in these organs play a vital role in sustaining and regulating hemopoiesis either by cell contact processes or by the local production of regulators acting over a short distance. In some cases, such as for

Normal and Malignant Hematopoiesis, Edited by Enrico Mihich and Donald Metcalf
Plenum Press, New York, 1995

the stem cell factor (SCF) or flk ligand, stromal cells can display these regulatory molecules on their membranes. In addition, there are a large number of adhesion molecules that are displayed on the surface of stromal cells and these may also play an indirect or direct regulatory role. However, none of the regulatory molecules so far characterized is unique to stromal cells and it must be presumed that our knowledge of the role played by stromal cells is still very incomplete.

Information is more complete on the functional actions of various secreted hemopoietic regulators on stem and progenitor cells.

CONTROL OF HEMOPOIETIC STEM CELLS

Most stem cells have the morphology of small lymphocyte-like cells and number only one per 100,000 marrow cells. They are capable of extensive self-renewal but the mechanisms controlling this are unknown. They also have an extensive capacity to generate progenitor cell progeny that are committed to a restricted range of lineages - most often one or two. Characteristically, stem cells cannot be stimulated to proliferate by single hemopoietic regulators but respond only to combinations of such regulators (4). Certain regulators such as stem cell factor or flk ligand are needed in such combinations not so much as direct proliferative stimuli but because they are able to activate stem cells to become responsive to other regulators such as the colony stimulating factors, IL-6, IL-11 and IL-7.

Combination of two or more members of these two classes of regulators greatly amplifies progenitor cell production, with limited evidence that particular colony stimulating factors (CSFs) such as Multi-CSF can influence the type of progenitor cells being generated (5). In the case of combinations such as SCF plus G-CSF or SCF plus IL-6, the combined action leads to the generation of many progenitor cells that are not able to be stimulated to further cell division by this combination (6). Effective exploitation of such progenitor cells therefore requires the action of other regulatory molecules.

The curious regulatory requirements of stem cells may represent a protective device to avoid the possibility that excess levels of any one regulator might be able to stimulate stem cells to expend themselves by forming progenitor cell progeny.

CONTROL OF PROGENITOR CELLS

Progenitor cells are large to medium blast-like cells and number about two to five per 1000 marrow cells. At least when stimulated by single regulators, progenitor cells have little or no capacity for self-generation and are therefore transit populations that expend themselves by each generating up to 10^5 maturing progeny. This proliferative capacity can be readily monitored by colony formation in semisolid cultures and such proliferation can be effectively stimulated by low concentrations of single regulatory factors (1,2). Thus for committed granulocyte-macrophage progenitor cells, cell proliferation can be stimulated by GM-CSF, Multi-CSF, M-CSF or G-CSF at concentrations of 50 to 5,000 pg/ml. Other regulators such as stem cell factor, IL-6 or IL-11 have a weak capacity to stimulate granulocyte colony formation but much higher concentrations are required.

These regulators have a positive action in stimulating cell division with a concentration-dependent action shortening the cell cycle. For M-CSF, this action is based on induction of transcription of cyclin D1 and possibly D2.

POLYFUNCTIONAL ACTIONS OF HEMOPOIETIC REGULATORS

As best documented for the CSFs, hemopoietic regulators are not simply stimuli for cell proliferation but also have actions on differentiation commitment, maturation induction, cell survival and functional activation of mature cells. The various CSFs have an ability to commit bipotential granulocyte-macrophage progenitor cells to the more restricted formation of either granulocytes or macrophages and, when acting on certain leukemic cells, to suppress self-generation, with commitment of the cells to a restricted proliferative capacity, with or without maturation. Maturation is clearly a highly complex process involving a multiplicity of inductive events as the maturing cells develop a highly specialized morphology and set of functions. While there is some evidence for a certain intrinsic capacity of these cells to initiate maturation, the weight of the evidence indicates that maturation can be actively initiated or made much more efficient by the action of the CSFs.

CSFs have the ability to maintain the transport integrity of the membrane of cells ranging from progenitor cells to mature granulocytes and macrophages. Withdrawal of CSF in vitro leads to failure of this transport system with death by apoptosis. Superimposed on this basic function is the ability of the CSFs to enhance the functional activity of mature granulocytes and macrophages in a wide variety of functions.

The polyfunctionality of these hemopoietic regulators in part compensates for the complexity of a regulatory system apparently comprised of an unnecessarily large number of regulators with overlapping functions. Each regulator is able to control multiple aspects of the biology of cells in a particular lineage and usually is active on cells in more than one lineage.

This polyfunctionality of necessity involves the activation of a number of quite distinct signaling pathways in a responding cell and has raised the obvious problem of how this is accomplished, because each regulator has only a single type of membrane receptor. Information is accumulating rapidly on this question with the cloning of cDNAs encoding the various receptors. All receptors, when in high-affinity form, are composed of at least two chains, either in the form of homodimers or heterodimers. Analysis using mutagenized receptors has identified distinct cytoplasmic domains signaling proliferative, maturation induction or membrane transport actions and in principle therefore multiple signaling is accomplished by selective interactions with discrete portions of the activated receptor (7,8). It remains unclear how the maturation stage of a responding cell influences the type of response able to be elicited by an activated receptor. Are the same set of signals initiated in all cells but fail to elicit certain responses because of gene shutdown or, for example, do receptors in a postmitotic cell no longer activate mitotic signaling intermediates?

THE BIOLOGY OF REGULATOR PRODUCTION AND DEGRADATION IN VIVO

Hemopoietic regulators differ from classical hormones in typically being the products of multiple cell types in multiple locations in the body. In the case of the CSFs, a wide variety of cell types is known to be able to produce one or more CSF, often simultaneously. These cells include endothelial cells, stromal cells, fibroblasts, macrophages and lymphocytes.

Under basal conditions, regulators such as the CSFs are produced in very low concentrations but the CSF-producing cells respond promptly to inductive signaling by greatly amplifiying transcription of CSF mRNA and the production of mature glycoprotein. The system is also highly labile following withdrawal of the inducing signals because of the short half-life of the mRNAs.

The regulatory mechanisms controlling basal levels of regulator production are not known but inducing signals for CSF production are in general products of microorganisms, such as lipopolysaccharide, that either act directly on CSF-producing cells or elicit molecules such as IL-1 that serve as the direct inducing signal.

In the main therefore, under emergency conditions, regulators of the CSF type are demand-generated with the CSF-producing cells being widely distributed to make early and effective contact with the products of invading microorganisms.

The biology of CSF-producing tissues is such that CSF production can occur simultaneously in different locations but for different purposes (1). Under basal conditions, most CSF production may occur locally in the bone marrow and achieve the necessary level of granulocyte and macrophage formation. Conversely, if a minor focus of infection is present in some tissue, local CSF production at that site could attract preexisting cells to the location and achieve functional activation of these cells to eliminate the microorganisms.

The mechanisms by which hemopoietic regulators are degraded are less well understood but, for M-CSF and G-CSF, the likelihood is that receptor-mediated endocytosis by target hemopoietic cells could represent a major pathway of degradation. This is the likely situation under basal conditions where regulator concentrations are low and production likely to be mainly in the hemopoietic tissues containing vast numbers of receptor-bearing cells. Despite this, the evidence is not strong in the case of the CSFs that simple depletion of mature cells is an effective signal either for inducing regulator production or for resulting in higher than normal concentrations of regulators. The situation probably differs from one regulator to another because low red cell levels certainly are associated with high erythropoietin levels and low platelet levels with high levels of thrombopoietin (mpl ligand).

REDUNDANCY OF HEMOPOIETIC REGULATORS

The existence of multiple regulators able to exert superficially common overlapping actions on cells of various lineages has prompted many to conclude that hemopoietic regulation is a highly redundant process. This question can only be resolved by gene inactivation experiments in which the genes encoding particular regulators or their receptors have been inactivated.

In every case so far analyzed, deletion of the regulator has resulted in an identifiably abnormal phenotype, so in no case has a completely redundant regulator been documented. In the case of the CSFs, inactivation of the G-CSF gene leads to a profound fall in circulating neutrophils (9) while inactivation of GM-CSF has no obvious effects on the numbers of hemopoietic cells but does result in an abnormal accumulation of surfactant in lung alveoli often with associated lung infection (10). Inactivation of M-CSF leads to a selective depletion of certain macrophage populations including osteoclasts which results in the development of osteopetrosis (11). Thus while these three CSFs have some actions in common, each plays a unique role in certain aspects of the biology of granulocyte-macrophage populations and this pattern seems likely to hold true for all hemopoietic regulators.

It is also becoming evident that regulators, when acting in combination, can achieve a more efficient production of mature cells than is achievable by the use of a single regulator acting alone. Such combinations can also ensure responses involving broader ranges of cells and achieve more subtlety in the location of the mature cells being generated.

CONCLUSIONS

Hemopoiesis is regulated by a combination of local stromal cell control and a large series of glycopotein regulatory factors able to act either locally or systemically. These regulatory factors control multiple aspects of the biology of hemopoietic cells and their maturing progeny by complex signaling pathways able to be initiated following binding of the regulators to their membrane receptors. Gene inactivation studies are indicating that, while many of these regulators have overlapping actions, each has unique functions. The control system therefore exhibits subtlety and flexibility rather than redundancy.

REFERENCES

1. Metcalf D, Nicola NA: The Colony Stimulating Factors: Molecular Biology to Clinical Aspects. Cambridge University Press, NY 1994 (in press).
2. Metcalf D: The colony-stimulating factors: discovery to clinical use. Phil Trans R Soc Lond B *333*: 147-173, 1991.
3. Metcalf D: Hemopoietic regulators: Redundancy or subtlety? Blood *82*: 3515-3523, 1993.
4. Li CL, Johnson GR: Rhodamine 123 reveals heterogeneity with murine Lin⁻, Sca-1⁺ hemopoietic stem cells. J Exp Med *175*: 1443-1447, 1992.
5. Metcalf D: Lineage commitment of hemopoietic progenitor cells in developing blast cell colonies: Influence of colony stimulating factors. Proc Natl Acad Sci USA *88*: 11310-11314, 1991.
6. Metcalf D: The cellular basis for enhancement interactions between stem cell factor and the colony stimulating factors. Stem Cells *11*: Suppl 2:1-11, 1993.
7. Fukunaga R, Ishizaka-Ikeda E, Nagata S: Growth and differentiation signals mediated by different regions in the cytoplasmic domain of granulocyte-stimulating factor receptor. Cell *74* 1079-1087. 1993.
8. Sakamaki K, Miyajima I, Kitamura T, Miyajima A: Critical cytoplasmic domains of the common beta subunit of the human GM-CSF, IL-3 and IL-5 receptors for growth signal transduction and tyrosine phosphorylation. EMBO J *11*: 3541-3549, 1993.
9. Lieschke GJ, Grail D, Hodgson G, Metcalf D, Stanley E, Cheers C, Fowler, KJ. Basu S, Zhan YF, Dunn AR: Mice lacking granulocyte colony-stimulating factor have chronic neutropenia, granulocyte and macrophage progenitor cell deficiency and impaired neutrophil mobilization. Blood 1994 (in press).
10. Stanley E, Lieschke GJ, Grail D, Metcalf D, Hodgson G, Gall JAM, Maher DW, Cebon J, Sinickas V, Dunn AR: Granulocyte/macrophage colony-stimulating factor-deficient mice show no major perturbation of hematopoiesis but develop a characteristic pulmonary pathology. Proc Natl Acad Sci USA *91*: 5592-5596, 1994.
11. Wiktor-Jedrzejczak W, Urbanowska E, Aukerman SK, Pollard JW, Stanley ER, Ralph P, Ansari AA, Sell KW, Szperl M. Correction by CSF-1 of defects in the osteopetrotic *op/op* mouse suggests local, developmental and humoral requirements for this growth factor. Exp Hematol *19*: 1049-1054, 1991.

DISCUSSION

H. Beug

I am wondering about the capacity of normal cells for self-renewal. Most think that only the totipotent stem cell self-renews and other progenitors go through a more or less strict program of division until they end up in the final stage. However B-cell progenitors are one example, and chicken erythroid precursor cells, may be another cell type capable of self-renewal. What are the present ideas on this question?

D. Metcalf

It is remarkable how dogmas change. The more you learn, the fewer dogmas you retain. One of the dogmas twenty years ago was that progenitor cells could not self-renew. However, these experiments were performed under a very defined set of culture conditions with the use of a single regulator. We may have underestimated the true ability of those cells. We have recently been looking at human progenitor cells and their behavior when stimulated by quite complex combinations of regulators. They seem to exhibit a certain level of self-generation. So it may be that, if you use a complex combination of regulators, even a relatively mature cell may have a broader range of biological behavior than is presently believed. Some hold an even more extreme view, that progenitor cells are clearly capable of self-generation. I would not go so far but I think we may have underestimated their capacity for some self-renewal.

H. Beug

A restricted capacity for self-renewal would make sense with respect to the behavior of human leukemia cells, which appear not to be immortalized. If you grow human leukemia cells, they will undergo a crisis after 60 generations. So this may be more typical of the restricted self-renewal behavior of progenitor cells in those cases where you do not have a stem cell leukemia.

D. Metcalf

That is true, but I think that some of the current difficulties we are having growing leukemic cells are technical. Nobody has yet been able to grow primary T-cell lymphomas from a mouse most likely because we do not have the correct regulators. So when we say a leukemic cell culture will die out after x number of generations, is this a true index of the behavior of such cells or is it simply the way we are growing them?

H. Beug

Well, from what I recall about those experiments, the leukemic cells were actually growing nicely for 50 divisions, or enough to die early, as would have been expected if incorrect regulators had been used. The results look more like an example of the Hayflick phenomenon.

D. Metcalf

Yes, but in the same context you cannot keep long-term cultures of human marrow cells beyond about eight weeks, with our current techniques.

H. Beug

That may be more a question of the use of inadequate regulators.

D. Metcalf

It is a puzzle really, how an immortalized cell line like the FDC-P1 can persist in very active cell division for more than 15 years.

H. Beug

It may be that animals with short life spans of their fibroblasts, immortalize easily, because the cells undergo the mutations necessary for immortalization readily. Organisms like humans with a long life span of their fibroblasts will almost never undergo the necessary genetic changes leading to immortalization. This may be why there are not human equivalents of the murine FDC-P1 line.

I. Weissman

There are immortalized T-cell lines.

H. Beug

Yes, that is another exception. Lymphocytes like stem cells do not seem to show the behavior expected from the Hayflick phenomenon.

L. Nadler

I am very interested by the GM-CSF knockout mouse where presumably G-CSF gives it the potential to form granulocytes. However, I would like to deal with the dendritic cells because these cells are going to be very important in tumor immunity. The so-called peripheral blood dendritic cells or inter-digitating reticulum cell is really the first line antigen-presenting cell and really is responsible for most of the subsequent immune response. Although dendritic cells can be induced to proliferate by stem cell factor, they require GM-CSF for functional activity. Does the GM-CSF knockout mouse have these cells? Can you see in the normal interfollicular areas, normal numbers of inter-digitating reticulum cells in this mouse?

D. Metcalf

That has not been looked at in the GM-CSF knockout mice. What we have done is made cell suspensions and sorted them using a battery of monoclonal antibodies believed to be selective for dendritic cells with high-class II expression. The answers have been a little variable. Initial experiments did show a depletion of dendritic cells and this agreed nicely with our GM-CSF transgenic mice which have a hundred-fold elevation of dendritic cells. Later experiments have produced more equivocal data. However, I think the answer will be that there is a subset of dendritic cells, whose functional activation is crucially dependent on stimulation by GM-CSF.

L. Nadler

You would predict in those mice that two things would happen that would tell us something of the immunology of tumors. First that it appears both in immunogenic and non-immunogenic tumors, these cells infiltrate but that even when they infiltrate, sometimes their class I and class II molecules are sufficiently down regulated that the tumor antigens are not presented, and, second, that one could not amplify an immune response sufficiently so that although maybe macrophages could get in and deliver an immune response, if you gave these mice a challenge of bacteria, or an immunogen, you would not get an immune response at a normal time. You would see a significant delay in the immune response. Is that known at all?

D. Metcalf

We have not looked yet in the knockout mice. We did this in the GM-CSF transgenic mice hoping that we would have superimmune responses for dendritic cell-dependent antigens and that a GM-CSF transgenic mouse with its hundred-fold excess of dendritic cells would be more resistant to tumor development than normal. The immune responses that I have measured have been normal in those transgenic mice and cross-breeding of these mice with a strain susceptible to T-lymphoma development has really not prevented T-lymphoma development. So at present there are two sets of observations. There are the Dranoff-Mulligan experiments where you get spectacular preimmunization by putting the GM-CSF retrovirus into tumor cells and using these to immunize against subsequent tumor cell challenge. On the other hand, there is the failure of the GM-CSF transgenic mice to suppress thymic lymphoma development. There are a variety of experiments that show that GM-CSF coupled to various antigens makes them superantigens.

R. Perlmutter

Following up on Lee Nadler's question, do you have any insight into the pathogenesis of the broncho-pneumonia in the GM knockout animals? Is it that the alveolar macrophages are especially sensitive to the absence of GM-CSF and hence cannot clear material satisfactorily after which a pneumonia then develops, or does the alveolar abnormality have something to do with the alveolar cells themselves?

D. Metcalf

I cannot tell you that. The in vivo models are very complex and slow to analyze. The explanation I favor is that in GM-CSF knockout mice there is failure of the alveolar macrophages to clear the material, whether because it is produced in an abnormal manner or it accumulates because it is not being cleared in the normal manner. The accumulated material then provides a favorable nidus for various microorganisms. These animals were kept in a conventional animal house which has many disadvantages but in this instance provided a good model of natural infection. The studies of Dranoff et al. (Science 294:713, 1994) reported no infections in their GM-CSF knockout mice and I think maybe that is just reflecting the difference in availability of microorganisms for the two sets of animals.

R. Perlmutter

Are there defects in the phagocytotic behavior of other macrophage populations that you can demonstrate?

D. Metcalf

Yes, they cannot handle challenge, Listeria monocytogens infections normally.

R. Perlmutter

And that is clearly because of phagocytotic defect?

D. Metcalf

They cannot handle the infection. It is an intra cellular organism.

R. Perlmutter

No, that is why I was asking the question.

D. Metcalf

Our initial studies with Listeria infections were in GM-CSF transgenic mice and the was quite complex. The cells from these animals had an increased ability to phagocytose the organism but a decreased ability to kill it.

R. Perlmutter

I am trying to understand the general requirement for GM and mature behavior of macrophages.

D. Metcalf

Yes, but the model is intrinsically complex because the organism actually resides in the macrophages.

A. D'andrea

Does the G-CSF receptor actually drive differentiation of granulocytes or does the G-CSF receptor actually enable granulocytes to remain viable and for this reason to complete differentiation? The reason I think this is an important question is because there is an alleged differentiation domain of the G-CSF receptor. But one could imagine that mutations in that alleged domain weaken the receptor, make it less mitogenic, if you will, and make the cells accumulate more in the G-1 phase of the cell cycle, where they are more inclined to stochastically initiate maturation. I believe this is an important point because there is a lot of confusion as to whether or not these receptors are actively driving maturation or merely weakly mitogenic and more inclined to promote differentiation indirectly.

D. Metcalf

I think that the data from Nagata's group (Fukunaga et al., Cell 74:1079-1087, 1993) are pretty convincing. If you delete the C-terminal region of the receptor, the cells remain capable of proliferating, but are unable to initiate the earliest steps in maturation. What has been controversial are the reports on naturally-occurring examples of this in congenital agranulocytosis. Don et al. (Proc. Natl. Acad. Sci. 91:4480-4484, 1994) have described a patient where there was deletion of that region of the G-CSF receptor and in this disease the failure to produce mature granulocytes is accompanied by a maturation block at the promyelocyte stage. Now, Jim Ihle's group, on the other hand, have looked at a number of patients without observing this particular defect so the disease may be heterogeneous. Having said all that, there are experiments that make you wonder whether differentiation cannot be a default system for proliferation so, I would not want to take an extreme view. I think in part what you are saying could be true.

E. Mihich

You mentioned in passing, if I got it right, that these GM-CSF knockout mice, have abortive tendencies. Is that what you said?

D. Metcalf

No, there are certain mouse strains that are characterized by having a very high frequency of spontaneous abortion and there has been one study reporting that the injection of GM-CSF reduces the frequency of those abortions. The litter size and fertility of these knockout mice appear to be normal.

E. Mihich

In your complex scheme of growth factors you did not mention the Flk/3 ligand. Perhaps you would like to comment on that?

D. Metcalf

Well, I did not mention several ligands that have been reported recently. The Flk/3 ligand seems to behave rather like stem cell factor and has an ability to stimulate the formation of blast colonies containing progenitor cells. In line with this, expression of the receptor appears to be restricted to early hemopoietic cells. The ligand for mpl has recently been purified and cloned and is the factor earlier termed thrombopoietin. It seems to act in a fairly selective manner on megakaryocyte precursors. Having said all that, I think there are obviously other, probably membrane-displayed ligands or regulators yet to be discovered.

2

CYTOKINE REGULATION OF LYMPHOHEMOPOIETIC PROGENITORS

Makio Ogawa and Fumiya Hirayama

Ralph H. Johnson Department of Veterans Affairs Medical Center and
Department of Medicine, Medical University of South Carolina
Charleston South Carolina 29401-5799

INTRODUCTION

For many years the mechanisms regulating proliferation of lymphohemopoietic progenitors remained unknown. Studies from our laboratory and others indicated that proliferation of primitive hemopoietic progenitors is regulated by interacting groups of cytokines (1). Interleukin (IL)-6, granulocyte colony-stimulating factor (G-CSF), IL-11, leukemia inhibitory factor (LIF) and IL-12 form a group of cytokines that work synergistically with IL-3, IL-4, and granulocyte/macrophage colony-stimulating factor (GM-CSF) to support the proliferation of multipotential progenitors from cell cycle dormant progenitors. However, the identification of the cytokine combinations that would permit proliferation of lymphohemopoietic progenitors remained elusive. Definitive evidence for the existence of lymphohemopoietic stem cells was provided by studies using retroviral labeling of individual stem cells of mouse (2). Despite this evidence using *in vivo* studies, it has not been possible to identify and quantitate lymphohemopoietic progenitors in culture until recently. Following establishment of semi-solid culture assays for human and murine multilineage progenitors, a number of investigators have reported the presence of lymphocytes or precursors of lymphocytes in human (3, 4, 5) and murine (6) multilineage colonies. However, clonal origin of these colonies was never established in these studies.

Recently investigators in three laboratories independently reported cell culture assays for lymphohemopoietic progenitors capable of yielding myeloid and B-cell progenies. Baum et al (7) described a co-culture assay with a murine stromal cell line for human fetal marrow lymphohemopoietic progenitors that are capable of producing myeloid cells and pre-B cells. Cumano et al (8) described a co-culture assay using a murine stromal cell line for murine fetal lymphohemopoietic progenitors capable of generating macrophages and pre-B cells. In our laboratory we developed a two-step methylcellulose culture assay for adult murine marrow lymphohemopoietic progenitors that are capable of producing cells in all myeloid lineages as well as pre-B cells (9). Using this assay we were able to demonstrate both stimulatory and inhibitory effects of lymphohemopoietic cytokines on the early stages of B-cell development.

Normal and Malignant Hematopoiesis, Edited by Enrico Mihich and Donald Metcalf
Plenum Press, New York, 1995

RESULTS AND DISCUSSION

In order to establish clonal origin of the lymphohemopoietic progenitors, we used micromanipulation of highly enriched mouse marrow cells. Density-separated, lineage negative Ly-6 A/E (Sca-1) positive cells harvested from 5-fluorouracil (5-FU)-treated mice were individually plated in primary methylcellulose culture and incubated for 11 days. Aliquots of resulting colonies were examined for myeloid lineage expression and analyzed for B-cell potential by plating into secondary methylcellulose culture containing c-*kit* ligand (steel factor, SF) and IL-7. In the original assay the primary culture was supported by a combination of medium conditioned by pokeweed mitogen-stimulated spleen cells (PWM-SCM), SF, IL-7, and erythropoietin (Ep). Subsequently, we observed that IL-7 and Ep are not needed in the primary culture.

After establishing the primary culture conditions with PWM-SCM plus SF, we determined if combinations of early-acting cytokines can replace PWM-SCM in support of the B-lymphoid potential of primary colonies. A representative study of the early-acting cytokines is shown in Table 1. Two-factor combinations consisting of SF, such as SF plus IL-6, SF plus G-CSF and SF plus IL-11 supported proliferation of lymphohemopoietic progenitors. Earlier we had reported that IL-4 like IL-3 works synergistically with IL-6 (10) and IL-11 (11) in support of colony formation from bone marrow cells of 5-FU treated mice. Therefore, we also tested the effects of IL-4 in combinations for their support of lympho-hemopoietic progenitors. As shown in Table 1, cytokine combinations based on IL-4 such as IL-4 plus IL-6, and IL-4 plus IL-11 maintained the B-lymphoid potential of the primary colonies although less effectively than SF-containing combinations. More recently we observed that SF plus IL-12 (12) can support the proliferation of B-cell progenitors in the primary colonies. In contrast, primary colonies supported by a combination of cytokines

Table 1. Analysis of B cell potentials in primary colonies supported by combinations of recombinant cytokines

Test cytokines in 1° culture						No of 1° colonies	Mean cell no. of 1° colonies ($\times 10^{-4}$)	Average no. of pre-B cell colonies
SF	IL-3	IL-4	IL-6	IL-11	G-CSF			
+	+					20 ± 4	6.0	0
+						2 ± 2	0.2	0	+
		+						
			+			19 ± 3	9.4	38 ± 4
+				+		19 ± 3	5.6	71 ± 9
+					+	18 ± 3	6.4	54 ± 1
	+		+			16 ± 2	5.6	0
	+			+		19 ± 3	4.9	0
	+				+	15 ± 6	3.7	0
		+	+			15 ± 6	2.8	5 ± 1
		+		+		18 ± 3	2.4	9 ± 2
		+			+	0		

50 enriched marrow cells were cultured in the presence of 2 <F128>m<F255>/ml Ep,, 500 <F128>m<F255>/ml IL-7 and designated cytokine combinations. Single factors failed to support 1° colony formation except for IL-3 on day 11 of incubation. No pre-B colonies were derived from these colonies. The number of pre-B cell colonies was derived from one-twentieth aliquot of 20 pooled 1° colonies.

Table 2. Inhibitory effects of IL-3 and IL-1α on the B cell potential of primary colonies

Test cytokines in 1° culture	No. of 1° colonies	Mean cell no. of 1° colonies ($\times 10^{-4}$)	No. of pre-B cell colonies
SF, IL-11	14 ± 1	5.2	59 ± 1
SF, IL-11, IL-3	13 ± 1	8.8	0
SF, IL-11, IL-1α	22 ± 1	2.6	0

50 enriched marrow cells were cultured in the presence of designated combinations of cytokines. The number of pre-B cell colonies was derived from 1/40 of 20 pooled 1° colonies.

consisting of IL-3 such as IL-3 plus IL-6, IL-3 plus IL-11, IL-3 plus G-CSF, and IL-3 plus SF did not support the B-cell potential of primary colonies.

The observation that IL-3 neither synergized nor replaced SF in support of B-lymphoid potential of the primary colonies was very surprising since several investigators have earlier established murine pre-B cell lines (13, 14) and pro-B cell lines (15, 16, 17) that are dependent on IL-3. We therefore studied addition of IL-3 and other cytokines to permissive cytokine combinations to test if IL-3 and/or other factors can inhibit the B-lymphoid potential of the primary colonies. We have found that IL-3 and IL-1 (α & β) can independently abrogate the B-lymphoid potential of the primary colonies when added to permissive cytokine combinations such as SF plus IL-11 (18). A typical experiment showing the inhibition by these factors is shown in Table 2. Following these observations, we pooled 20 primary colonies grown under a variety of cytokine combinations and injected the cells into scid mice. Serum samples analyzed from the mice receiving 20 or the equivalent of 2 pooled primary colonies grown in the presence of SF and IL-11 contained IgM of BDF1 origin beginning 4 weeks after cell injections. In contrast, samples from the scid mice injected with 20 pooled colonies grown in the presence of either IL-3 or IL-1 did not contain donor-origin IgM. We are not certain of the physiological implication of these findings. Independently, 2 observations may corroborate our observations on IL-3. Transgenic mice expressing antisense IL-3 died in the state of B-cell hyperproliferation and/or neurological complications (19). When human CD34+ cells were transplanted into immunodeficient mice together with human marrow that had been engineered to produce human IL-3, multilineage grafting, except B-cells, was seen (20). It is possible that IL-3 and IL-1 possess negative regulatory roles in the physiology of B-lymphopoiesis.

CONCLUSIONS

In this brief chapter we have discussed our recent development of two-step methylcellulose culture system for murine lymphohemopoietic progenitors that are capable of expressing both B-cell and myeloid lineages. This culture system will be useful in the study of the very early process of murine B-lymphopoiesis. We have also presented evidence that IL-3 and IL-1 independently and negatively regulate the early stages of B-cell development in culture. Recently, *in vitro* manipulation of stem cells became the subject of intense research interest in relation to *in vitro* expansion of stem cells and "gene therapy" for genetic and malignant disorders. Use of IL-3 and IL-1 *in vitro* may need to be approached cautiously because of its possible inhibitory effects on B-cell potential of human stem cells.

REFERENCES

1. Ogawa M: Differentiation and proliferation of hematopoietic stem cells. Blood *81*: 2844-2853, 1993.
2. Lemischka IR, Raulet DH, Mulligan RC: Developmental potential and dynamic behavior of hematopoietic stem cells. Cell *45*: 917-927, 1986.
3. Messner H, Izaguirre CA, Jamal N: Identification of T lymphocytes in human mixed hemopoietic colonies. Blood *58*: 402-405, 1981.
4. Fauser AA, Neumann HA, Bross KG, Kanz L, Lohr GW: Cytotoxic T-cell clones derived from pluripotent stem cells (CFU-GEMM) of patients with Hodgkins lymphoma. Blood *60*: 1317-1320, 1982.
5. Fauser AA, Kanz L, Lohr GW: Identification of B cells in multilineage hematopoietic colonies derived from cells of patients with lymphocytic lymphoma. Proc Natl Acad Sci USA *82*: 883-885, 1985.
6. Hara H: Presence of cells in B-cell lineage in mixed (GEMM) colonies from murine marrow cells. Intl J Cell Cloning *1*: 171-818, 1983.
7. Baum CM, Weissman IL, Tsukamoto AS, Buckle A-M, Peault B: Isolation of a candidate human hematopoietic stem-cell population. Proc Natl Acad Sci USA *89*: 2804-2808, 1992.
8. Cumano A, Paige CJ, Iscove NN, Brady G: Bipotential precursors of B cells and macrophages in murine fetal liver. Nature *356*: 612-615, 1992.
9. Hirayama F, Shih JP, Awgulewitsch A, Warr GW, Clark SC, Ogawa M: Clonal proliferation of murine lymphohemopoietic progenitors in culture. Proc Natl Acad Sci USA *89*: 5907-5911, 1992.
10. Kishi K, Ihle JN, Urdal DL, Ogawa M: Murine B-cell stimulatory factor-l (BSF l)/interleukin-4 (IL-4) is a multi-CSF which acts directly on primitive hemopoietic progenitors. J Cell Physiol *139*: 463-468, 1989.
11. Musashi M, Clark SC, Sudo T, Urdal DL, Ogawa M: Synergistic interactions between interleukin-11 and interleukin-4 in support of proliferation of primitive hemopoietic progenitors of mice. Blood *78*: 1448-1451, 1991.
12. Hirayama F, Katayama N, Neben S, Donaldson D, Nickbarg EB, Clark SC, Ogawa M: Synergistic interaction between interleukin-12 and steel factor in support of proliferation of murine lymphohemopoietic progenitors in culture. Blood *83*: 92-98, 1994.
13. Palacios R, Henson G, Steinmetz M, McKearn JP: Interleukin-3 supports growth of mouse pre-B cell clones in vitro. Nature *309*: 126-131, 1984.
14. Rennick D, Jackson J, Moulds C, Lee F, Yang G: IL-3 and stromal cell-derived factor synergistically stimulate the growth of pre-B cell lines cloned from long-term lymphoid bone marrow cultures. J Immunol *142*: 161-166, 1989.
15. Palacios R, Steinmetz M: IL-3-dependent mouse clones that express B-220 surface antigen, contain Ig genes in germ-line configuration, and generate B lymphocytes in vivo. Cell *41*: 727-734, 1985.
16. McKearn JP, McCubrey J, Fagg B: Enrichment of hematopoietic precursor cells and cloning of multipotential B-lymphocyte precursors. Proc Natl Acad Sci USA *82* 7414-7418, 1985.
17. Palacios R, Karasuyama H, Rolink A: Ly1⁺ Pro-B lymphocyte clones. Phenotype, growth requirements and differentiation in vitro and in vivo. EMBO J *12*: 3687-3693, 1987.
18. Hirayama F, Clark SC, Ogawa M: Negative regulation of early B-lymphopoiesis by interleukin-3 and interleukin-1α. Proc Natl Acad Sci USA *91*: 469-473, 1994.
19. Cockayne DA, Bodine DM, Cline A, Nienhuis AW, Dunbar CE: Transgenic mice expressing interleukin-3 antisense RNA develop B-cell lymphoproliferation or neurological dysfunction. Blood *80*: Suppl 1, 349a (abstr), 1992.
20. Nolta JA, Hanley MB, Kohn DB: Sustained human hematopoiesis in immunodeficient mice by cotransplantation of marrow stroma expressing human interleukin-3: Analysis of gene transduction of long-lived productions. Blood *83*: 3041-3051, 1994.

Editor's Note

This paper was not discussed due to the unavoidable absence of Dr. Ogawa from the Symposium. However, as this paper was prepared for presentation at the Symposium, it is published as part of these proceedings.

3

HEMATOPOIETIC STEM CELLS

Irving Weissman

Department of Pathology
Stanford University School of Medicine
Palo Alto, California 94305-5324

The definition of any stem cell is that it is a cell type which is capable of multi-lineage differentiation, and self-renewal; self-renewal means regeneration of more stem cells, as measured by phenotype and function.

Several years ago, based on work by a number of laboratories, and in my laboratory by Cheryl Whitlock (1) and Christa Müller-Sicburg (2), Jerry Spangrude, Shelly Heimfeld and I identified a candidate stem cell population in mouse bone marrow that represents one in 2000 cells; these cells expressed low levels of Thy-1, Sca-1 at high levels and was negative or low for markers of the B-cell, myelomonocytic, T-cell, and erythroid lineages (3). That population of cells was highly enriched for clonogenic thymic precursors. It contained all of the cells responsible for initiation of both Whitlock-Witte (2) and Dexter (4) long term bone marrow stromal cultures. It also contained a majority of cells that were involved in myelo-erythroid differentiation in the Till-McCulloch day 12 spleen colony assay. That population of cells was two thousand-fold enriched in radioprotective capacity: about 200,000 unfractionated bone marrow cells are required to radioprotect 95-100% of the lethally irradiated mice, and 40,000 bone marrow cells to radioprotect half the animals. The Thy-1^{lo} Lin-1^{lo} Sca$^+$ population takes only 100 cells to save 95-100% of the animals, and about 30 cells to save half the animals (3). Nobuko Uchida and I demonstrated that only Thy$^+$, only Sca$^+$, only Lin$^{-/lo}$ cells, within mouse bone marrow, had long term radioprotective activity, and long term multi-lineage reconstitutive activity (5). Therefore, this is arguably the only hematopoietic stem cell (HSC) or pre-HSC in mouse bone marrow, at least in the mouse strains examined. Laurie Smith, and independently Nobuko Uchida, tested the activity of single cells by limit dilution injection of these cells into lethally irradiated animals in a competitive re-population assay (6,7). In these experiments 100 stem cells of the host genotype were injected along with 1 to 20 cells of the donor genotype. The donor and the hosts were distinguished by the Ly5 allelec markers. At the single cell level, we could expect HSCs each to be multipotent. In most cases, these putative HSCs gave rise to B cells, T cells and myelomonocytic cells. To test for self-renewal, several hosts were sacrificed at intervals up to 18 months. The best of these animals had 150,000 cells of the stem cell phenotype, derived from the single engrafting cell; much lower levels of reconstitution was the norm. When 200-500 of these regenerated Thy-1^{lo} Lin$^{-/lo}$ Sca$^+$ cells were injected, secondary hosts achieved donor-derived myelomonocyte T cell and B cell reconstitution for at least 60 weeks.

Normal and Malignant Hematopoiesis, Edited by Enrico Mihich and Donald Metcalf
Plenum Press, New York, 1995

At the single cell level, therefore, Thy 1.1^{lo} Lin $^{-/lo}$ Sca-1$^+$ BM cells are functionally and phenotypically cells which are capable of multi-lineage differentiation and self-renewal.

We wished to test the relative effectiveness of these HSCs for bone marrow transplantation". In experiments carried out by Nobuko Uchida, (8) we tested the ability of 500 or 200 Thy^{lo} Lin$^-$ Sca$^+$ cells to give rise to donor derived white blood cells, platelets, red blood cells in the bloodstream. These HSCs were compared to whole bone marrow that contained 500 or 200 stem cells; 500 HSCs were compared with 1,000,000 bone marrow cells, and 200 HSCs with 400,000 bone marrow cells. Stem cells gave rise to white blood cells, platelets and, by indirect experiment, red blood cells, at virtually the same rate as whole bone marrow that contained 500 stem cells as well as the 1,000,000 other BM cells that contained all other progenitors. In the context of bone marrow transplantation, the most important cell type is the stem cell; it's very difficult to demonstrate significant regeneration capacity (in competition with stem cells) by other progenitors or more mature cells. For example in the above described experiment, all of the cells of the myelomonocytic type found in irradiated hosts 9 to 10 days after irradiation are derived from 200 donor stem cells. In competitive repopulation, 200 stem cells from Ly5.1 donor mixed together with 400,000 Sca$^-$1$^-$ cells from Ly5.1/Ly5.2 hybrid donors gives 95, 92, 90 and 97% Ly5.1 myelomonocytic cell reconstitution at 14 days (8). Even if radioprotection is provided by stem cells, the Sca-1 negative cell types don't contribute to myelomonocyopoiesis. Surprisingly, in these same animals, the Sca$^-$ population contains plentiful B cell progenitors. At 14 days the Sca-1$^-$ cells give rise to most blood B cells but by 29 days, the great majority of B cells are derived from the stem cells (8) . Thus one can study a single wave of B cells derived from Sca-1$^-$ BM cells.

One can divide hematopoietic cells and stem cells by their ability to exclude Rhodamine 123, which is a mitochondrial dye. Chaudary and Roninson (9) showed that Rh123 is transported by the MDR gene product, PGP1, which shepherds hydrophobic low MW molecules out of the cell. We have shown that Rho^{lo} subset of HSCs gives rise to long term reconstituting cells, while the Rho^{hi} subset of stem cells, which also gives rise to multi-lineage differentiation, is only productive for 2-6 weeks (8). This provides a clear indication that HSCs are heterogenous.

The Thy-1^{lo} Lin$^-$ Sca-1$^+$ population contains all stem cells; they are multi-potent, but they are heterogeneous in terms of radioprotection and heterogenous in terms of the productive life span of a single stem cell. Stem cells can be divided into subsets that have S/G-2/M amount of DNA versus those that have GO-G1amount of DNA. 100 G0-G1 stem cells gave rise to 100% radioprotection while the dividing stem cells only radioprotect 25-30% of irradiated mice at the 100 cell level (10). At the 300 cell level S/G-2/M stem cells gave about 60% radioprotection. But the level of reconstitution of survivors with the dividing cells at the 300 cell level equals that of G0/G1 HSCs (10). This is a complicated way to say that stem cells probably include populations of resting cells which are long term reconstituting cells, that have high degrees of self-renewal, as well as populations of dividing cells some of which are on their way out of the stem cell pool with a limited lifespan and an important subset of the cells that are in a self-renewal division.

When Laurie Smith and Nobuko Uchida carried out limit dilution (single cell) reconstitutions (6,7) there were 2-3 outcomes in terms of the lifespan of these stem cells. The most reliable indicator for the lifespan of a stem cell is its continued production of myelomonocytic cells, which have a very short lifespan *in vivo*. About 1/4 to 1/3 of the clones give long-term reconstitution *in vivo*. The other 2/3 have either intermediate or very short lifespans (6,7). All are multi-lineage reconstitutors; they all include lymphoid and myelomonocytic cells. Is this difference in the productive lifespan of stem cells **stochastic**, or are they **determined**? Is there some predetermination of long-term versus mid-term, versus a short-term reconstituting HSCs? Sean Morrison found that the Thy-1^{lo} Sca-1^{hi} population of stem cells in fact can be subdivided according to c-kit and lineage markers (11). In the lineage marker low to negative population, only 1/5 of the cells are actually lineage negative. 4/5 of these HSCs express low levels of Mac-1 and 1/5 of the total population, also express CD4 at low levels. The Mac-1^{lo} subsets all express high levels of c-kit, but the lineage negative population has some c-kit negative cells that are not stem cells. At limit dilution Thy-1^{lo} Sca-1^{hi} Lin$^-$ cells give long-term outcomes in 80-90% of the cases. Both of the Mac-1^{lo} populations give a majority of transient multi-lineage repopulations the Mac-1$^{/lo}$ CD4^{lo} subset also contains populations that predominantly give rise to 2 but not 3 of the lineages. Thus in terms of productive lifespan, stem cells are predetermined (11). It's not stochastic. The Mac-1^{lo} population contains the cells that are moving out of the stem cell pool into lineage-committed differentiation pathways.

In humans also, Thy-1, expressed at low levels, is a marker of stem cells (12). CD34 is a marker that is also present on stem cells. Stem cells are negative or low for B-lineage, myelomonocytic lineage, T cell lineage and erythroid lineage markers (12). Using an *in vivo* assay of repopulation of human fetal thymus with HLA different candidate stem cell populations placed underneath the kidney capsule of a SCID mouse (that cannot reject it), the graft produces donor-derived T cells (12,13). Candidate stem cells can also be injected into human fetal bones placed in SCID mice. The human stromal micro-environment permits maturation of myelo-erythroid and B lymphoid cells. Thy$^+$ CD34$^+$ Lin$^-$ cells of donor HLA type give rise, in the long term, to B lymphoid, myeloid and CD34$^+$ cells (12). In the SCID-hu thymus, they give rise to T cells, and the T cells include the immature and more mature populations. The Thy-1 negative subsets of the CD34$^+$ Lin$^-$ population do not give rise to donor derived cells in the bone assay. Rhodamine123^{lo} CD34$^+$ Thy-1$^+$ Lin$^-$ cells are HSCs with long-term reconstituting activity in the bone (12,14). Esmael Zanjani, Ron Hoffman (et. al) have recently reported that these human stem cells can be injected into fetal sheep, and 5-8 months later the lambs contain human cells of the T cell, the B cell, myelo-erythroid lineages, as well as abundant CD34$^+$ Thy1^{lo} Lin$^-$ cells. Thus by numerous *in vivo* assays, the Thy$^+$ but not the Thy$^-$, the Rhodamine low but not the Rhodamine high, the 34$^+$ but not the 34$^-$, the Lin$^-$ but not the Lin$^+$ population contains human hematopoietic stem cell activity.

REFERENCES

1. Whitlock, C.A., Tidmarsh, G.F., Muller-Sieburg, C., and Weissman, I.L. Bone marrow stromal cells with lymphopoietic activity express high levels of a pre-B neoplasia-associated molecule. *Cell 48*:1009-1021, 1987.
2. Muller-Sieburg, C., Whitlock, C.A., and Weissman, I.L. Isolation of two early B lymphocyte progenitors from mouse marrow: a committed pre-pre-B cell and a clonogenic Thy-1^{lo} hematopoietic stem cell. *Cell 44*:653-662, 1986.
3. Spanrude, G.J., Heimfeld, S., and Weissman, I.L. Purification and characterization of mouse hematopoietic stem cells. *Science 241*:58-62, 1988.
4. Weilbaecher, K., Weissman, I.L., Blume, K., and Heimfeld, S. Culture of phenotypically defined hematopoietic stem cells and other progenitors at limiting dilution on Dexter monolayer. *Blood 78*:945-952, 1991.
5. Uchida, N. and Weissman, I.L. Searching for hematopoietic stem cells: Evidence that Thy-1.1^{lo} Lin$^-$ Sca-1$^+$ cells are the only stem cells in C57BL/Ka-Thy-1.1 bone marrow. *J. Exp. Med. 175*:175-184, 1992.
6. Smith, L., Weissman, I.L., and Heimfeld, S. Clonal analysis of hematopoietic stem-cell differentiation *in vivo*. *Proc. Natl. Acad. Sci. USA 88*:2788-2792, 1991.
7. Uchida, N. Characteristics of Mouse Hematopoietic Stem Cells. Ph.D. Thesis, 1992.
8. Uchida, N., Aguila, H.L., Fleming, W.H., Jerabek, L., and Weissman, I.L. Rapid and Sustained Hematopoietic Recovery in Lethally Irradiated Mice Transplanted with Purified Thy-1.1^{lo} Lin$^-$ Sca-1$^+$ Hematopoietic Stem Cells. *Blood 83(12)*:3758-3779, 1994.
9. Chaudhary, P.M. and Ronison, I.B. Expression and activity of P-glycoprotein, a multidrug efflux pump, in human hematopoietic stem cells. Cell 66:85-94, 1991.

10. Fleming, W.H., Alpern, E.J., Uchida, N., Ikuta, K., Spangrude, G.J., and Weissman, I.L. Functional heterogeneity is associated with the cell cycle status of murine hematopoietic stem cells. *J. Cell Biology* *122*:897-902, 1993.

11. Morrison, S.J. and Weissman, I.L. The long term repopulating subset of hematopoietic stem cells is deterministic and isolatable by phenotype. *Immunity*, in press.

12. Baum, C.M., Weissman, I.L., Tsukamoto, A.S., Buckle, A., and Peault, B. Isolation of a dandidate human hematopoietic stem cell population. *Proc. Natl. Acad. Sci. USA 89*:2804-2808, 1992

13. Peault, B., Weissman, I.L., Bau, C., McCune, J.M., and Tsukamoto, A. lymphoid reconstitution of the human fetal thymus in SCID mice with CD34$^+$ precursor cells. *J. Exp. Med.* 174:1283-1286, 1991.

14. Uchida, N. et al. Persistance of human hematopoiesis in sheep transplanted *in utero* with purified human CD34$^+$ Thy$^+$ Lin$^-$ Hematopoietic Stem Cells. Abstract accepted for publication, American Society of Hematology, Nashville, Tennessee, December 3-6, 1994.

DISCUSSION

C. Sawyers

I was curious in your bcl-2 transgenics, have they lived long enough for you to decide if there is any evidence of malignancy developing in the myeloid series?

I. Weissman

Yes, we have two kinds of founders: those that are low copy number and those that are high copy number. Neither of them died of malignancy yet and they are two years old, so they should have gotten accelerated leukemias. There is a very interesting pathology in the high copy number animals. They have an appearance in the blood stream and in the bone marrow, of a cell type that may turn out to be a myelomonocytic progenitor. It is not usually present in the blood. And, in collaboration with your old lab, we are getting all of the vectors for the genes to see if they will have accelerated leukemic genesis along the myelomonocytic pathway. But we do not know yet.

C. Peschle

Could you tell us about the membrane phenotype of the human primitive progenitors (the Thy-1 low cells which give rise to long-term reconstitution), particularly with respect to Kit, FLT3, CD45 and CD48?

I. Weissman

the CD34 high, Thy low, Lyn-2 low population, all the activities in the kit+ population, in the assay that was most quantitative, that is the long term culture initiating assay, it is in the kit low but positive, better than in the kit high.

C. Peschle

We had the same results.

I. Weissman

And I think Uchi and Suda have the same results in Japan. CD38 is a very interesting case. In fetal bone marrow, there is a true CD38 negative population, and that does contain stem cells. It looks that it is not negative, but low in adult bone marrow and adult immobilized peripheral blood.

H. Beug

You said it did not make a big difference whether you transplant stem cells alone or stem cells together with all kinds of other bone marrow progenitors. But did you try to look for short term reconstitution in some or all of the lineages?

I. Weissman

There is no doubt that the Sca- population, in fact the Thy low Lyn- Sca- population, contains a very significant fraction of cells that form myelo-erythroid colonies in the spleen and they are predominantly day 8, but there are some day 12 colonies there. Despite that, 400,000 of those cells, as I showed you, as early as day 11 and day 14, only gave minor re-population if they were included with 200 Thy low Lyn- Sca-1+ cells. So, within the context of bone marrow transplantation. Because it could be that these cells fails to home, but are normally highly functional, highly self renewing proliferative, if they got into the bone marrow. But remember, we are injecting them intravenously into a lethally irradiated animal. A lot of them get screened out in the lung and the liver on the way.

H. Beug

I was wondering whether multipotent progenitors may activate a self renewing function, as a sort of emergency pathway, when there are no true pluripotent stem cells present. If you have stem cells, you may not even have to activate them.

I. Weissman

Yes, they all die by 10 to 15 days, so it is not good enough.

R. Tepper

Do you have any evidence that the ced-3 homolog, ICE, is playing a role in inducing apoptosis in the neutrophils or in any of the other hemopoetic lineages?

I. Weissman

I have no idea. I have no experiments, but maybe Jerry will tell us.

J. Adams

No, we have no information on that, but I do have a question. I was wondering whether, in the course of the experiments in which you have produced essentially a limiting dilution (i.e., clonal) reconstitution, you have formed any decided opinion as to whether there are such entities as a T-B stem cell or perhaps a B-myelomonocytic type of stem cell?

I. Weissman

In the experiments where we did the mac low CD4 low, we had the highest frequency of non full lineage reconstitutions. Most of those were B or B in myeloid. We had one or two animals which looked like T-B but I want to caution you that since these are short term reconstituting cells, and in every animal that we sacrificed early, 10 days to two weeks, they had myeloid cells. It could have been a short term myeloid reconstitution but successful seeding of the B and T population and then perpetuation of those by whatever perpetuates them, like antigens. So, we have no evidence, at early time points, of T and B but not myeloid. There is, to my mind, no evidence at all, for a common lymphocyte progenitor. None that I know in the literature, and I would be happy to hear if somebody has any evidence.

R. Perlmutter

You are excluding Ken Shortman's bi-potential primitive cell type?

I. Weissman

Ken Shortman's is not restricted to T and B, it includes NK cells and a few myeloid cells in the thymus when it responds, so I do not think that is a T-B. And he would say that now also. That is not to say that somebody will not find it some day, but I am saying, with all of the kinds of experiments we have done, with all the markers, we can never find just T-B that was in the beginning and in the end just T-B.

E. Mihich

These limiting dilution experiments were *in vivo* obviously. But can you engineer a way to keep these particular stem cells alive for a couple of weeks *in vitro* and, if so, can you dissect what are the factors that might direct the differentiation in one direction, as compared to another one.

I. Weissman

There is no condition *in vitro* that you get massive expansion of these stem cells by phenotype and/or by function. I have not found one. We have, in Dexter cultures and in some mouse stromal lines, small increases in the level of stem cells: 2, 3, 5, tenfold, over time. Yet, as I said in every case of a long term reconstituting cell *in vivo*, and even with the short term reconstituting cells, you have massive expansion of the stem cell phenotype and function. So I think, as Don Metcalf said, it is worth one's continued search and I would agree with Don that I would be looking for stromal cell associated ones at this point. A little tougher cloning strategy.

R. Perlmutter

I like the interpretation of the *in vivo* depletion of neutrophils by phagocytosis, that the cells are being eaten alive. But the alternative interpretation, it seems to me, would be that the apoptotic death that you observed *in vitro*, occurs *in vivo* via different signals. So, did you look directly to see whether apoptosis is going on in the neutrophil population in bcl-2 transgenics *in vivo* system after thioglycolate treatment?

I. Weissman

In the non transgenic you have apoptosis occurring in the peritoneal cavity by the usual DNA ladder. And in the animals that have bcl-2 you fail to see it at that level. It may be there at a tiny level but you fail to see it.

R. Perlmutter

You did it by sorting?

I. Weissman

Yes, and the question then becomes, "Is phagocytosis by the cells that are there the only way that they are being moved?" and I do not know that. But we need to have animals where you delete cells at the activated macrophage step, to see whether that is the essential element. Just to repeat it, although I may have passed by, the failure to show negative selection of T-cells in response to super antigen, in animals that have bcl-2, might be explained, though we do not know if it is so, by their removal by phagocytes. If the pathway that says "you have an auto immune T-cell receptor complex", signals the cell to have both apoptosis and putting on the surface whatever it is that macrophages like to eat, then it could be the failure to show the bcl-2 pathway still will allow phagocytosis and removal of those cells. That is why I think it is really important to rethink each of the old experiments.

J. Adams

I wonder if you could reiterate what you said about the relationship of long term repopulating ability and the cell cycle. Does all the long term reconstituting activity reside in the resting cells or just a proportion?

I. Weissman

No, a proportion. So that 100 of the resting cells or less, that is, small cells, no or little amount of SG2 or DNA, will give rise to radio protection and long term reconstitution at that cell level. In order to get equivalent amounts in the dividing subset, you have to go above 300 cells of the same phenotype G2S & M positive. So, it is in there. Obviously that must be the population that you see in retro viral marking studies and I cannot see how it could be otherwise. So, David appropriately says that the retro viral marking studies involve 5FU depletion *in vitro* culture with IL-3 plus a stromal cell plus, in many cases, neo-selection with G4-1A and that is not exactly parallel but cells are virtually of the same phenotype: Thy low, Lyn-, Sca+ are the active cells. I should add Thy+ because they increase their Thy 1 level after 5FU.

4

TRANSCRIPTIONAL CONTROL OF HEMATOPOIETIC DEVELOPMENT

Roles of GATA-Factors

Stuart H. Orkin, Mitchell Weiss, Gordon Keller,[+] and Fong-Ying Tsai

Division of Hematology-Oncology
Children's Hospital and the Dana Farber Cancer Institute
Department of Pediatrics
Harvard Medical School
Howard Hughes Medical Institute
Boston, Massachusetts
[+] National Jewish Center
Denver, Colorado

INTRODUCTION

The production of mature blood cells from pluripotent hematopoietic stem cells reflects two interrelated and overlapping processes. First, developmental potential becomes progressively restricted as stem cells give rise to multipotential progenitors and ultimately single lineage committed precursors. Second, a cell-specific program of transcription is established which leads to the expression of markers characteristic of mature cell types. In addition, within the embryo the site of hematopoiesis shifts from its initial position within the yolk sac blood islands (primitive hematopoiesis) to the fetal liver, and ultimately to the bone marrow (definitive hematopoiesis). Although it has been thought that definitive stem cells probably arise in the yolk sac and migrate to the fetal liver, more recent evidence hints at an intraembryonic origin.

In an effort to identify proteins that might participate in some of the critical events in hematopoietic development we have characterized DNA-binding proteins that are involved in cell-specific transcription, notably that in erythroid cells. This approach is based on the premise that such nuclear regulators may also play important roles in earlier stages of either erythroid or hematopoietic development. Pursuing this reasoning, we first identified the protein GATA-1 as a central erythroid transcription factor and a related protein, GATA-2, as a critical factor for very early progenitors (or stem cells).

GATA-1, the first member of the GATA-family of zinc-finger transcription factors to be identified, recognizes target sequences in DNA with a core GATA-motif. GATA-1 is expressed at high level in erythroid cells, mast cells, and megakaryocytes, and at low level

Normal and Malignant Hematopoiesis, Edited by Enrico Mihich and Donald Metcalf
Plenum Press, New York, 1995

in hematopoietic progenitors. Gene targeting in mouse embryonic stem (ES) cells has shown that GATA-1 is necessary for normal primitive and definitive erythropoiesis. Among other vertebrate GATA-factors, two (designated GATA-2 and GATA-3) are expressed in selected hematopoietic populations, as well as some cell types outside the hematopoietic system. Whereas GATA-3 is largely restricted to T-lymphoid cells, GATA-2 is present in an expression pattern that suggests a more intimate role in hematopoiesis. It is expressed highly in progenitors, early erythroid cells, mast cells, and megakaryocytes, as well as in endothelial and nervous system cells. In erythropoiesis, expression of GATA-1 is maintained at high level throughout cell maturation, whereas expression of GATA-2 progressively declines.

Here we review recent findings with gene targeting at the GATA-1 and GATA-2 loci which focus on the specific roles of these regulators in erythroid development and hematopoiesis. We have shown by analysis of differentiation of gene targeted cells *in vitro* and generation of homozygous GATA-2 embryos that these proteins exhibit overlapping but distinct roles in development. Such a complex, laryered pattern of regulation appears to be chacteristic of mammalian developmental programs.

RESULTS AND DISCUSSION

Consequences of GATA-1 Loss for Erythroid Development

Previously we demonstrated that targeted disruption of the X-linked GATA-1 gene in ES cells prevents the formation of ES-derived erythroid cells in chimeric mice or in embryoid body (EB) differentiation *in vitro*. These findings revealed an essential role for GATA-1 but did not provide specific information regarding the stage at which a block to development occurs. To approach this problem we utilized a two-step *in vitro* ES hematopoietic differentiation assay that permits identification and isolation of pure erythroid colonies of either primitive (EryP) or definitive (EryD) origins. In this system EBs are disaggregated at selected days of culture and the recovered cells plated in erythropoietin (Epo) or Epo plus kit-ligand (KL) to generate EryP and EryD precursor-derived colonies. As shown in Figure 1, GATA-1⁻ ES cells generate no EryP colonies. These findings support a requirement for

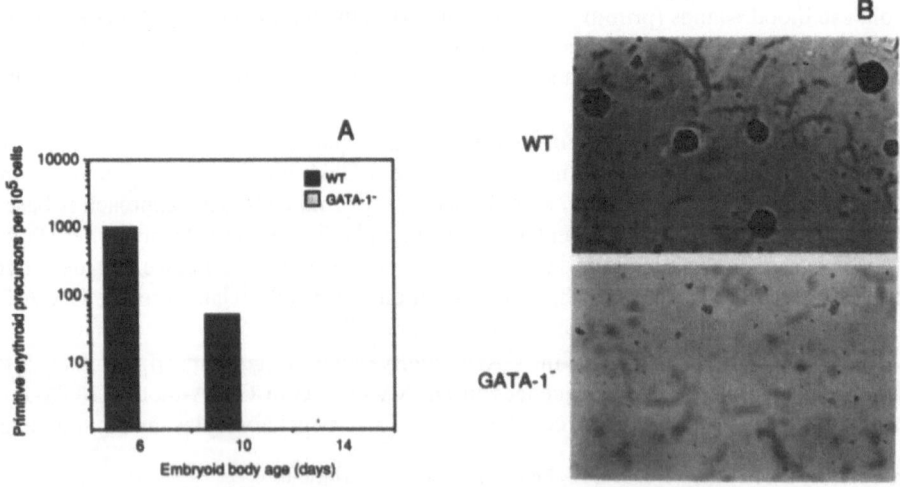

Figure 1. Primitive erythropoiesis is blocked in GATA-1⁻ ES cells. Modified from [4].

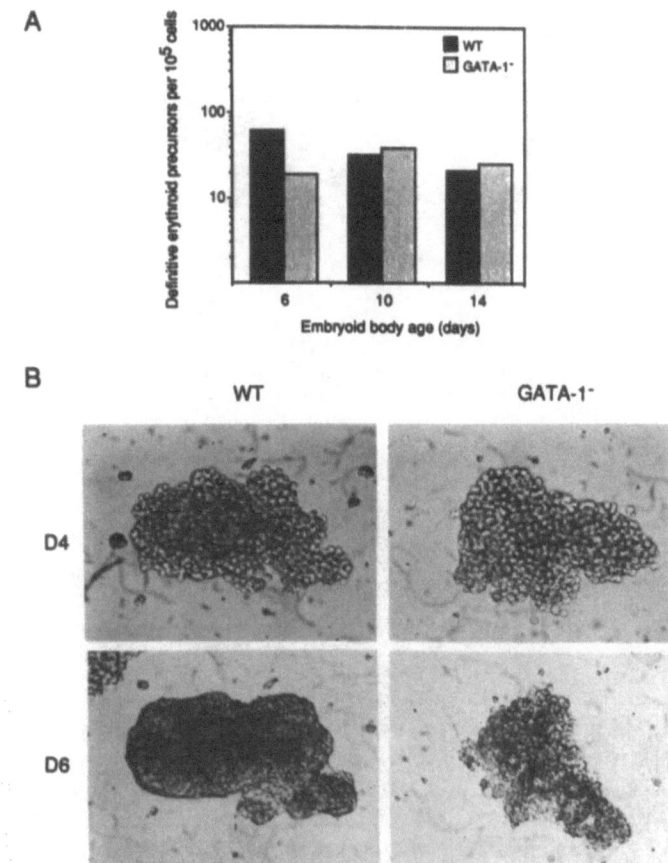

Figure 2. EryD precursors in GATA-1⁻ embryoid bodies are blocked in the final stages of development. D4 and D6: day 4 and day 6 colonies following replating. Modified from [4].

GATA-1 at the very earliest stage of primitive erythropoiesis that can be defined by current methods. In contrast, EryD colonies form at wild-type level from GATA-1⁻ ES cells (Figure 2). The cells of these colonies are arrested in their maturation at the proerythroblast stage and undergo death by day 5-6 following replating (Figure 3). Thus, erythroid commitment and partial maturation occur in EryD precursors in the absence of GATA-1, but their subsequent development is blocked.

The capacity to generate GATA-1⁻ EryD proerythroblast colonies devoid of ES cells offered a unique opportunity to discriminate between two simple models illustrated in Figure 4. On one hand, loss of GATA-1 might prevent expression of presumptive GATA-target genes, including Epo-receptor (EpoR), globins, and putative downstream transcription factors such as SCL and EKLF. Without expression of these genes, particularly the EpoR and downstream factors, erythroid development might be blocked. Alternatively, known GATA-target genes might be expressed, but the failure to express one or several unknown genes coud be responsible for the arrest and subsequent cell death. Using quantitative RT-PCR we showed that the known GATA-target genes (EpoR, SCL, EKLF, and globins) are expressed in GATA-1⁻ arrested proerythroblasts at approximately normal levels compared with wild-type proerythroblasts isolated in parallel cultures. To account for expression of these presumptive targets in the absence of GATA-1, we pursued the possibility that another member of the GATA-family of factors might compensate for GATA-1 deficiency *in vivo*. As summarized in Figure 4, we observed that GATA-2 RNA is expressed at about

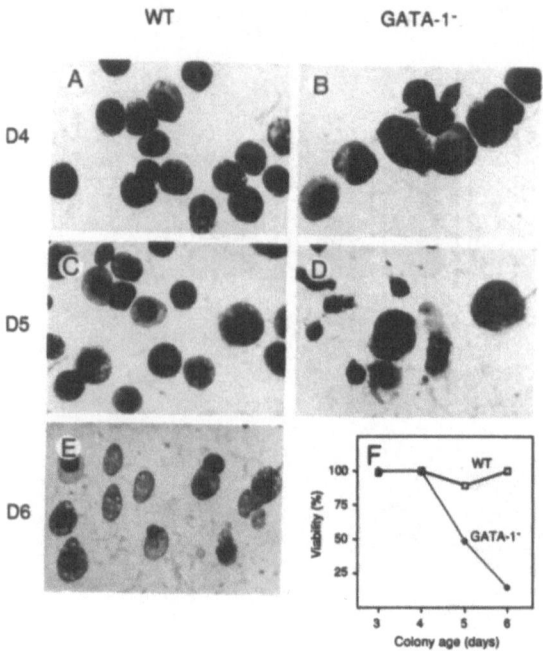

Figure 3. Maturation arrest and premature cell death in GATA-1⁻ EryD cells. D4, D5, and D6: day 4, 5, and 6 cells following replating. Modified from [4].

50-fold greater level in GATA-1⁻ as compared with wild-type cells. In contrast, GATA-3 RNA is expressed at low levels, which are similar in mutant and wild-type cells. Thus, GATA-2 expression is selectively up-regulated to a remarkable extent in the absence of GATA-1.

Our studies of *in vitro* differentiated GATA-1⁻ ES cells provide several novel insights into erythroid development. First, we are led to conclude that GATA-1 is required exceedingly early in primitive erythropoiesis, as no abortive EryP colony development is evident. This conclusion is consistent with the proposed role of GATA-1 in erythroid commitment within the embryo. The partial compensation for GATA-1 loss by over-expression of GATA-2 has several important implications. First, we postulate that GATA-2 participates in transcription of presumptive target genes, including EpoR, SCL, EKLF, and globins, in GATA-1⁻

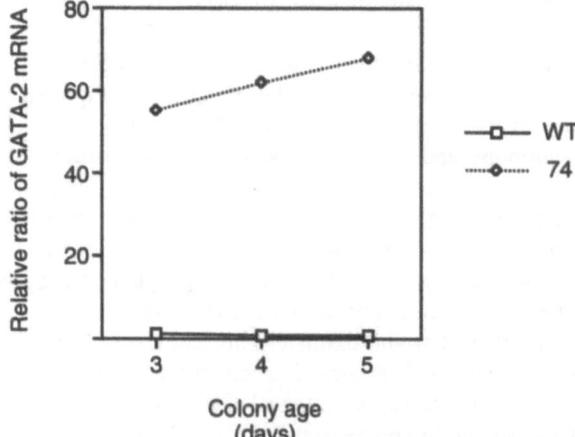

Figure 4. Marked increase in GATA-2 RNA expression in GATA-1⁻ proerythroblasts. RNA levels were quantitated by RT-PCR for wild-type (open squares) and mutant (closed) proerythroblasts harvested on days 3, 4, and 5 following replating. Modified from [4].

cells. Therefore, GATA-1 and GATA-2 proteins appear to share many functions, despite virtual nonhomology outside the conserved DNA-binding domains. This conclusion is generaly consistent with experiments with cotransfection transactivation assays that have not yet identified promoter or cell specificty in their action. Nonetheless, current findings do not exclude the likely possibility that subtle differences in the functions of GATA-factors exist.

Second, the expression of globin genes in the absence of GATA-1 implies that productive interactions proposed between locus control elements and globin gene promoters can occur under the aegis of GATA-2. A slight shift to a more embryonic pattern of globin expression is evident in the GATA-1⁻ arrested proerythroblasts, perhaps suggesting subtle differences resulting from the substitution of GATA-2 for GATA-1. Thus, the interactions of GATA-1 with other proteins within LCR-globin promoter complexes are unlikely to be entirely specific for GATA-1. Third, since GATA-2 is normally at high level in hematopoietic progenitor cells, the realization that GATA-2 may function in transcription of globin and nonglobin erythroid-expressed genes raises the intriguing possibility that a succession of GATA-factors acting at particular targets may occur during cellular differentiation. This mechanism might provide for finer regulation of target genes than achievable by a single factor. For example, in globin loci, GATA-2 might prepare LCR elements in early progenitors to be acted upon later by GATA-1 during later stages of erythroid maturation. LCRs may be maintained in a poised state prior to maturation along a single lineage, a proposal consistent with the observation that LCR elements are DNase I hypersensitive in multipotential cells. Finally, the extraordinary up-regulation of GATA-2 in the absence of GATA-1 provides the first evidence for cross-regulation between members of the GATA-family. Our findings implicate GATA-1 in the repression of GATA-2 that normally accompanies erythroid maturation. This repression could be achieved either by a direct effect of GATA-1 on GATA-2 regulatory elements or through an indirect mechanism.

The cross-regulation and potential redundancy of GATA-factors in the early stages of definitive erythropoiesis are reminiscent of the relationship of myogenic bHLH factors in muscle development. In both erythroid cells and muscle, potential redundancy is imposed by related transcription factors; yet, terminal maturation requires a specific member of the family. This complex, layered pattern of regulation is likely to be characteristic of mammalian developmental progams. The mechanism(s) by which GATA-1⁻ arrested EryD cells die is unknown. The inability of these cells to complete erythroid matuation may be due either to different properties of GATA-1 and GATA-2 and/or insufficient expression of GATA-2 (despite its marked up-regulation). Though GATA-1 and GATA-2 may overlap in many of their functions, dedicated roles for these proteins probably reflect subtle differences in their capacity to regulate target genes *in vivo*.

Consequences of GATA-2 Loss for Hematopoiesis

Several observations initially provided indirect evidence for a role for GATA-2 in hematopoietic development. First, GATA-2 is highly expressed in hematopoietic progenitors, mast cells, and megakaryocytes. Expression in erythropoiesis is limited to less mature cells. In addition, forced expression of GATA-2, but not GATA-1 or GATA-3, in chicken erythroid progenitors promotes proliferation at the expense of differentiation. Overall, these data implicate GATA-2 as a regulator of genes controlling self-renewal or proliferation of early hematopoietic cells.

To address the role of GATA-2 in hematopoiesis we disrupted the gene by homologous recombination in ES cells. Both heterozygous (GATA-2⁺/⁻) and homozygous (GATA-2⁻/⁻) mutant cells were generated.Three independent approaches (analysis of homozygous mutant embryos, analysis of the contribution of GATA-2⁻/⁻ cells to various tissues and hematopoietic cells in chimeras, and in vitro differentiation of GATA-2$^{-/-}$ ES cells) were performed to examine the consequences of GATA-2 loss on hematopoiesis. GATA-2$^{-/-}$ embryos die by embryonic day 10-11 (E10-11) with severe anemia, indicative of markedly reduced primitive erythropoiesis in yolk sac. Definitive hematopoietic progenitors from E9-10 were decreased more than 50-fold. Consistent with these data, GATA-2$^{-/-}$ ES cells failed to contribute to hematopoietic progenitors and all hematopoietic compartments (spleen, bone marrow, blood) in chimeras produced by injection of GATA-2$^{-/-}$ cells into wild-type blastocysts. This defect is manifest by the fetal liver stage and persists at later developmental stages. The deficit in hematopoiesis extends to lymphopoiesis also, as GATA-2$^{-/-}$ cells contribute poorly to mature lymphoid cells following injection into blastocysts of recombinase-deficient RAG-2$^{-/-}$ blastocysts. Taken together, these findings point to a defect in the earliest hematopoietic progenitors or stem cells in the absence of GATA-2.

To characterize further the basis for hematopoietic deficits of GATA-2$^{-/-}$ cells, we performed the two-step ES *in vitro* differentiation assay. Progenitor numbers and maturation of specific lineages were assessed. GATA-2$^{-/-}$ cells failed to generate KL-dependent EryD and mast cell colonies, whereas the formation of these improved somewhat when a combination of growth factors (IL-1, IL-3, GM-CSF, and IL-11) was added. EryP and macrophage colonies, though present, were reduced in number from GATA-2$^{-/-}$ ES cells *in vitro*.

The findings summarized herein demonstate that GATA-2 is an essential nuclear regulatory factor for both primitive and definitive hematopoiesis. The results are most consistent with GATA-2 serving as a regulator of genes controlling growth factor responsiveness or the proliferative capacity of early hematopoietic cells, particularly those of definitive lineages. In addition, we postulate a potential link between the function of GATA-2 and the KL/c-kit pathway.

CONCLUSIONS

Our studies have established the critical roles of two GATA-factor regulatory proteins, GATA-1 and GATA-2, in hematopoietic development. GATA-1 is specifically required in erythroid development. As no primitive erythropoietic progenitors can be detected with current methods, it would seem that GATA-1 is essential in the earliest events of cellular commitment and differentiation. This conclusion is in accord with the pattern of embryonic GATA-1 expression in Xenopus and zebrafish. In definitive erythroid development GATA-1 likely serves a similar function, but enhanced expression of GATA-2 in the absence of GATA-1 can support maturation to the proerythroblast stage. Further maturation requires GATA-1, suggestive of dedicated functions of the protein in terminal differentiation.

The major role of GATA-2 in hematopoiesis is manifest in the earliest progenitors or the stem cell. This finding does not exclude important functions in the development of specific lineages, such as mast cells and megakaryocytes, where it is also highly expressed. Our conclusions regarding GATA-2 have immediate implications with respect to the biology of hematopoietic stem cells. First, use of GATA-2 expression as a marker may help in establishing the origin of stem cells within the embryo, a subject of considerable debate and active investigation. Second, identification of the genes that are transcriptionally activated by GATA-2 in early hematopoietic cells should define components of growth factor response systems important for proliferation or cellular survival. Finally, study of growth factors and cellular interactions which induce GATA-2 expression or modulate its functional activity in early hematopoietic cells may link environmental influences and the expansion of progenitos *in vivo*.

Our studies underscore the power of gene targeting in ES cells complemented by chimera and *in vitro* differentiation experiments to define the roles of specific proteins in various aspects of hematopoietic development and lineage commitment. As proteins often serve critical roles at multiple times in development, these additional experimental approaches permit access to particular tissues or cell types at stages beyond those compatible with viability of the whole animal.

REFERENCES

1. Moore, M.S.A., and Metcalf, D., 1970, Ontogeny of the haemopoietic system: yolk sac origin of in vivo and in vitro colony forming cells in the developing mouse embryos, Brit. J. Haemat. 18: 279-296.
2. Medvinsky, A.L., Samoylina, N.L., Muller, A.M., and Dzierzak, E.A., 1993, An early pre-liver intraembryonic source of CFU-S in the developing mouse, Nature 364: 64-66.
3. Godin, I.E., Garcia-Porrero, J.A., Coutinho, A., Dieterlen-Lievre, R., and Marcos, M.A.R., 1993, Para-aortic splanchnopleura from early mouse embryos contains B1a cell progenitors, Nature 364: 67-70.
4. Orkin, S.H., 1992, GATA-binding transcription factors in hematopoietic cells, Blood 80: 575-581.
5. Pevny, L., Simon, M.C., Robertson, E., Klein, W.H., Tsai, S-F., D'Agati, V., Orkin, S.H., and Costantini, F., 1991, Erythroid differentiation in chimeric mice blocked by a targeted mutation in the gene for transcription factor GATA-1, Nature 349: 257-260.
6. Yamamoto, M., Ko, L.J., Leonard, M.W., Beug, H., Orkin, S.H., and Engel, J.D., 1990, Activity and tissue-specific expression of the transcription factor NF-E1 multigene family, Genes Devel. 4: 1650-1662.
7. Weiss, M.J., Keller, G., and Orkin, S.H., 1994, Novel insights into erythroid development revealed through in vitro differentiation of GATA-1⁻ embryonic stem cells, Genes Dev. 8: 1184-1197.
8. Tsai, F-Y., Keller, G., Kuo, F.C., Weiss, M.J., Chen, J-Z., Rosenblatt, M., Alt, F., and Orkin, S.H., 1994, An early haematopoietic defect in mice lacking the transcription factor GATA-2, Nature 371: 221-226.
9. Simon, M.C., Pevny, L., Wiles, M.V., Keller, G., Costantini, F., and Orkin, S.H., 1992, Rescue of erythroid development in gene targeted GATA-1⁻ mouse embryonic stem cells, Nature Genetics 1: 92-98.
10. Keller, G., Kennedy, M., Papayannopoulou, T., and Wiles, M.V., 1993, Hematopoietic differentiation during embryonic stem cell differentiation in culture,Mol. Cell. Biol. 13: 471-486.
11. Zon, L.I., Youssoufian, H., Mather, C., Lodish, H.F., and Orkin, S.H., 1991, Activation of the erythropoietin receptor promoter by transcription factor GATA-1, Proc. Natl. Acad. Sci. (USA) 88: 10638-10641.
12. Aplan, P.D., Nakahara, K., Orkin, S.H., and Kirsch, I.R., 1992, The SCL gene product: a positive regulator of erythroid differentiation, EMBO J. 11: 4073-4081.
13. Crossley, M., Tsang, A.P., Bieker, J.J., and Orkin, S.H., 1994, Regulation of the erythroid Kruppel-like factor (EKLF) gene promoter by the erythroid transcription factor GATA-1, J. Biol. Chem. 269: 15440-15444.
14. Rudnicki, M.A., Schnegelsberg, P.N.J., Stead, R.H., Braun, T., Arnold, H-H., and Jaenisch, R., 1993, MyoD or Myf-5 is required for the formation of skeletal muscle, Cell 75: 1351-1359.
15. Hasty, P., Bradley, A., Morris, J.H., Edmonson, D.G., Venuti, J.M., Olson, E.N., and Kelein, W.H., 1993, Muscle deficiency and neonatal death in mice with a targeted mutation in the myogenin gene, Nature 364: 501-506.
16. Nabeshima, Y., Hanaoka, K., Hayasaka, M., Esumi, E., Li, S., Nonaka, I., and Nabeshima, Y-I., 1993, Myogenin gene disruption results in perinatal lethality because of severe muscle defect, Nature 364: 532-535.
17. Briegel, K., Lim, K-C., Plank, C., Beug, H., Engel, J., and Zenke, M., 1993, Ectopic expression of a conditional GATA-2/estrogen receptor chimera arrests erythroid differentiation in a hormone-dependent manner, Genes Dev. 7: 1097-1109.
18. Chen, J., Lansford, R., Stewart, V., Young, F., and Alt, F.W., 1993, RAG-2-deficient blastocytst complementation: an assay of gene function in lymphocyte development, Proc. Natl. Acad. Sci. (USA) 90: 4528-4532.
19. Kelley, C., Yee, K., Harlan, R.M., and Zon, L.I., 1994, Ventral expression of GATA-1 and GATA-2 in the Xenopus embryo defines induction of hematopoietic mesoderm, Dev. Biol. 165: 193-205.
20. Detrich, H.W., Kieran, M.W., Chan, F.W., Barone, L.M., Yee, K., Rundstadler, J.A., Kimmel, C., and Zon, L.I., 1994, A unifying hypothesis for hematopoietic cell migration during vertebrate embryogenesis, submitted.

DISCUSSION

R. Perlmutter

Do you believe that GATA-1, if it were expressed at the right time, could not substitute for GATA-2 in driving those very earliest progenitors?

S. Orkin

We do not know. I would think it is probably likely the other way around. A question we have posed ourselves is: "If we expressed GATA-2 or GATA-3 as if they were GATA-1, could we make a red cell?" We do not know the answer to that. You can also turn the question around and ask "Could you express GATA-1 in place of GATA-2 and regulate the early progenitors?" I would guess that you could not but we do not know the answer.

R. Perlmutter

Do you not already know the answer to the first question because you express 60 times more GATA-2 in the GATA-1 knockout and you still do not get red cells?

S. Orkin

That is right, but we do not know what 60 times means. We do not know if 60 times still approximates the level of GATA-1 in the cell as a protein. I think we are talking about RNA but we have no idea in terms of absolute protein, what it means.

D. Livingston

Do you have any evidence of a common co-activator that serves the functions of both GATA-1, GATA-2 and/or other members of the family?

S. Orkin

The question is, "Is there another co-activator?" No, we have looked for quite some time for proteins that interact with either GATA-1 or GATA-2. Initially we found very little interacted with them, in a sort of immuno precipitation or super shift assays. If one uses GST fusion proteins and looks for interaction, one can find interaction with almost any protein. The significance of that, we do not know. But there is really no evidence that there is any activator that interacts with these proteins.

D. Livingston

That would not rule it out. It could be that they exist in a massive complex and the biochemistry is unapproachable at the moment.

S. Orkin

Yes, it would not rule it out. We have certainly seen no stable complexes formed with GATA-1 *in vivo*.

D. Livingston

Is there a biological assay you could do with retro virus libraries in search of cDNAs that would overcome, for example, the GATA-1 defect, but not be GATA-1?

S. Orkin

In principle, but in practice, we have not tried it.

C. Brugnara

Did you look at the expression of band 3 in your GATA-1 minor system?

S. Orkin

We have not.

D. Baltimore

You implied that GATA-1 could repress GATA-2. Is there any difference in the binding site specificity of GATA-1 and GATA-2?

S. Orkin

We and Doug Engel, both labs, have looked carefully at various members of the family using random site selection to look at binding sites and we have basically come up with the same answer. That is, one can find very subtle differences between binding of various family members but we do not know the significance of any of that because, in fact, very few if any of the sites that distinguish the proteins *in vitro* are ever found within real genes. So, for the most part we think that the binding is identical.

J. Adams

You hinted that the GATA-2 negative cells might not be responding properly to kit ligand. Do you have direct evidence that in fact kit expression is altered?

S. Orkin

No, in fact we found the kit messages present. Certainly in embryoid bodies for example, the GATA-2 minus are expressing kit RNA so we do not think the deficit is just in expressing kit receptor. We think the block is more likely to be in some signaling pathway that is downstream of that.

J. Adams

I wonder if you would elaborate on what you think about the role of GATA-1 and GATA-2, in megakaryocytes and mast cells, particularly in view of the work of Jane Visvader, which suggests that at least in one primitive hematopoietic cell line (416B) either GATA-1 or GATA-2 can elicit megakaryocyte development (Visvader et al., EMBO J. *11*, 4557, 1992; Visvader and Adams, Blood *82*, 1493, 1993).

S. Orkin

About all we can say at this point is that one can obtain mass cells and megakaryocytes and even platelets from GATA-1 cells, so we do not think that megacaryocyte development requires GATA-1. And obviously in the cell line, I think GATA-1, forced GATA-1 expression is doing something, either replacing another GATA factor or the cell is just poised at such a position that expressing GATA-1 will trigger it in some direction. But they are not essential. We would interpret that probably mast cells made and megakaryocytes function by virtue of GATA-2 which is very abundant as well.

H. Beug

Our lab, together with that of Doug Engel, had some evidence that GATA-3 was expressed, in erythroid cells, at least, very late. That is, in the reticulocyte stage of erythroid differentiation. So I was wondering whether you have any evidence whether GATA-3 may, after all, play a role in erythropoiesis. Alternatively, you may have evidence from GATA-3 knockouts that tell you clearly that GATA-3 does not function in erythropoiesis.

S. Orkin

Yes, I have had a lot of discussion with Doug. We think that is a chicken phenomena. We have seen no expression of GATA-3 in a mammalian red cell, or MEL cells. And we certainly do not see any in any of the situations I have shown you. The GATA-3 knockout has been done and I think Doug tells me, although they die at 12 or 13 days of embryogenesis, they have red cells in their fetal liver, so I think it is not required.

S. Landolfo

Have these factors localization in the nucleus or in the cytoplasm, are they activated and translocated in the nucleus?

S. Orkin

For the most part we think they are nuclear only.

R. Dalla Favera

Are these GATA factors all clearly transcriptional activators?

S. Orkin

They in principle can act either way, as activators or repressors. There are at least two situations that have been proposed where they may act as repressors in globin expression. In one case in γ-globin or fetal gene but there is a little better evidence now that in the epsilon which is silenced, there is an embryonic gene that gets silenced in fetal life, that that silencing element may also require GATA-1.

R. Dalla Favera

So could this be one other example of the fos dual myc-max NFkB type of example where there is competitive occupance by activators and inert factors?

S. Orkin

It could be except here there are no heterodymers being formed.

J. Adams

Do you have any evidence that the GATA factors might act either as homodimers or in physical association with one another?

S. Orkin

Basically all the evidence suggests that they bind and work as monomers although we do have data that GATA-1 can form dimers with itself and we have actually mapped the dimerization region and that brings up the possibility that because the globin locus for example is littered with sites, that GATA-1 might have not only an activation function but a chromatin structure function and bring together distant elements in a loop and perhaps be a scaffold. So its major function might not be activation at all; it may be structural and that may be related to the fact that the proteins have diverged considerably. The chicken and mammalian proteins really outside of the DNA bonding domain are almost unrecognizably related and it could well be that the major function of these proteins is not so much to be strong activators but to really bind the sites, open the chromatin, and make it available for other things to come in. So the mechanism of function is not known.

C. Peschle

It may be of interest to compare your studies with our observations on primitive progenitors purified from human adult peripheral blood. As you know, we observed that GATA-2 is expressed in these cells, whereas GATA-1 is not. By transient suppression of GATA-2 mRNA with antisense oligomers, with all the appropriate controls, we could impede in part formation of all types of colonies, indicating that, as you suggested, GATA-2 plays a role in primitive hematopoietic proliferation and differentiation.

J. Azizkhan

Do you have any evidence for a role for the phosphoserine?

S. Orkin

A fellow in the laboratory has mapped all seven serine phosphorylations and in fact mutated all seven, so we have a dephospho-molecule. One of the sites, the site that is inducible during erythroid maturation, sits at the base of the DNA binding domain and the NMR structure of the finger is actually the part that wraps around into the minor groove of the DNA, so we thought by knocking out that phosphorylation we would have some major effect on either DNA binding or activation or DNA bending. But we found that in fact the dephospho-molecule binds, activates and bends DNA as well as wild type, so we have no function at all for the phosphorylation. *In vivo*, as best we can test it so far, it is normal.

H. Beug

Do you have a new GATA-protein construct with all seven phosphorylation sites mutated at the same time?

S. Orkin

Yes.

H. Beug

And that is still active, at least in the *in vitro* assays?

S. Orkin

In vitro, transactivation, DNA binding, bending, it is normal.

H. Beug

You have not had a chance to test it in cells?

S. Orkin

We tested it in cotransfection in cells and it is normal. We have also tested it with Jane Visvader; at least the serine in the DNA binding domain mutant has been tested in the 416B cell transition to megakaryocytes and it works as well as wild-type.

H. Beug

Did it function in differentiation assays?

S. Orkin

Yes, it also works. So we have no *in vivo* role for any phosphorylation.

D. Livingston

It still could be tested in ES double knockout cells and/or by breeding into knockout mice, your phosphorylation mutants

S. Orkin

Yes.

D. Baltimore

Do you have any further knowledge of what the locus control region does?

S. Orkin

No.

5

REARRANGEMENT OF THE *ALL1* GENE IN ACUTE MYELOID LEUKEMIA WITHOUT CHROMOSOMAL TRANSLOCATIONS

Steven A. Schichman and Carlo M. Croce

Jefferson Cancer Institute
Jefferson Cancer Center and
Department of Microbiology and Immunology
Jefferson Medical College of Thomas Jefferson University
Philadelphia, Pennsylvania 19107

INTRODUCTION

When the chromosomes of human leukemia cells are viewed through the microscope, aberrations are frequently seen in chromosome structure and number. These abnormalities comprise the gain, loss, or rearrangement of whole chromosomes or chromosome segments. More than 20 years ago, researchers discovered that certain types of chromosomal abnormalities were consistently associated with certain types of leukemia. These observations led to the hypothesis, now shown to be correct, that some types of chromosomal defects are causally related to the formation of leukemia.[1]

A common type of chromosomal abnormality seen in leukemia cells is the reciprocal translocation, a rearrangement that involves the exchange of genetic material between two different chromosomes. Reciprocal translocations can lead to leukemia either by the transcriptional activation of a proto-oncogene or by the fusion of parts of two different genes.[2] Gene fusion plays an important role in both chronic and acute leukemia. Reciprocal translocations giving rise to well-characterized gene fusions include t(9;22) in chronic myelogenous leukemia (CML), which fuses the *BCR* and *cABL* genes[3]; t(15;17) in acute promyelocytic leukemia, which fuses the *APL* and *RARA* genes[4]; t(8;21) in acute myelogenous leukemia (AML), associated with fusion of the *AML1* and *ETO* genes[5]; and t(1;19) in acute lymphoblastic leukemia (ALL), which fuses the *E2A* and *PBX1* genes.[6]

Rearrangements involving chromosome 11, band q23, are frequently seen in human acute leukemia. Over 15 different reciprocal translocations involving 11q23 have been reported, most commonly with chromosomes 4, 6, 9, and 19.[7,8] Using the 11q23 chromosomal breakpoint as a guide to its location, a gene called *ALL1* was identified which was subsequently shown to participate in gene fusions in most acute leukemias with 11q23 chromosomal translocations.[9-12] We were curious whether *ALL1* gene rearrangements were also present in some leukemia cases without visible chromosomal rearrangements. We found several cases of AML in adults with molecular evidence of *ALL1* gene rearrangement in the absence of cytogenetic 11q23 chromosomal translocations.[13] Characterization of the *ALL1* gene rearrangements in these cases has revealed a new type of gene defect in leukemia that we call "self-fusion".[14,15] Our studies indicate that, in adults, the *ALL1* gene may be involved in up to twice as many acute leukemia cases as was previously thought.

CLINICAL ASPECTS OF *ALL1*-ASSOCIATED LEUKEMIAS

The *ALL1* gene is involved in approximately 5 to 10% of acute leukemia cases overall in children and adults.[8] These include cases of ALL, AML, and leukemias of mixed lineage. *ALL1*-associated leukemias generally have a poor prognosis, regardless of the type of leukemia or the age of the patient. In infants less than one year of age, over 60% of ALL cases have t(4;11) involving the *ALL1* gene.[16-18] These cases are frequently characterized by hyperleukocytosis, organ infiltration, and early involvement of the central nervous system. In adults, approximately 6% of AML cases have 11q23 translocations involving the *ALL1* gene, the most common of which are t(9;11) and t(6;11).[7] These AML cases frequently have monocytic or myelomonocytic morphology (FAB-M4 or FAB-M5 according to the French-American-British classification scheme). Therapy-related leukemias following treatment with anthracyline (eg. adriamycin) or epipodophylotoxin (eg. VP-16) drugs sometimes have t(9;11) or t(11;19) involving the *ALL1* gene.[19,20]

STRUCTURE OF THE *ALL1* GENE AND ALL1 PROTEIN

The *ALL1* gene spans approximately 100 kb on chromosome band 11q23 (Figure 1, top).[11,12] *ALL1* is also called *MLL* and *HRX*.[10,12] The name *MLL* refers to mixed lineage or myeloid-lymphoid leukemia whereas the name *HRX* derives from the similarity of a portion of the ALL1-encoded protein to the trithorax protein from the fruit fly *Drosophila*. The *ALL1* gene contains at least 21 exons with transcription occuring in a centromeric-to-telomeric

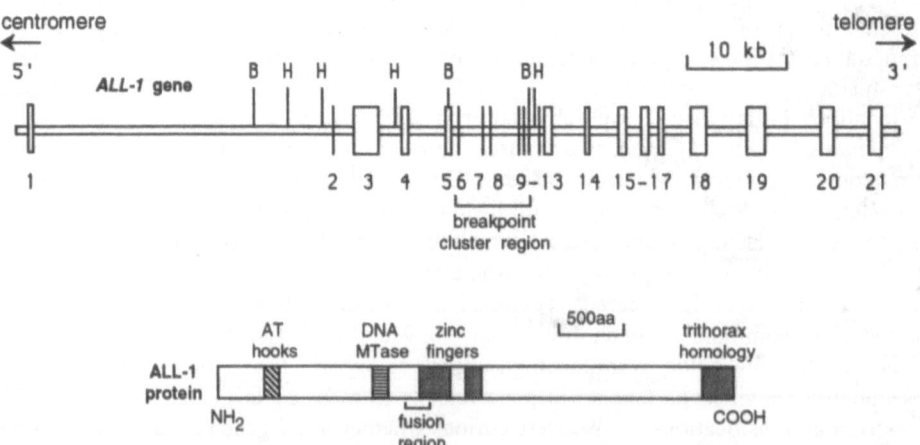

Figure 1. Physical map of the *ALL-1* gene and schematic diagram of the ALL-1 protein. *Top: ALL-1* gene map shows selected *Bam*HI (B) and *Hin*dIII (H) restriction sites. Boxes and verticle lines indicate exons. kb, kilobase. *Bottom:* Schematic representation of the ALL-1 protein. Conserved domains or motifs shared with other characterized proteins are indicated. MTase, methyltransferase.

orientation. A large intron separates the first two exons of the gene. Breakpoints in the *ALL1* gene are clustered within an approximately 8 kb region located near the mid-portion of the gene.[21]

The *ALL1* gene is capable of encoding a very large protein of about 430,000 daltons.[11,12] The functions of the ALL1 protein in normal cells are not known. The cellular location of the ALL1 protein has not yet been determined, nor is it known whether the large protein is normally processed into smaller functional pieces. The structure of the encoded ALL1 protein reveals several domains or motifs that are similar to those in previously described proteins (Figure 1, bottom). These include two different types of possible DNA binding domains - zinc fingers and "AT-hooks". Two clusters of zinc fingers, a DNA binding motif found in many transcription factors, are located near the mid-portion of the ALL1 protein.[11] AT-hook motifs, present in some high mobiltiy group proteins, are located near the amino-terminus of the encoded ALL1 protein.[12] The amino-terminal portion of the ALL1 protein also contains a domain similar to that found in the mammalian DNA methyltransferase enzyme.[22] The most notable feature of the encoded ALL1 protein is a carboxy-terminal domain with strong similarity to a portion of the *Drosophila* trithorax protein.[11,12] In *Drosophila*, trithorax plays an important role in development by positively regulating some classes of homeotic genes.

The structural features of the ALL1 protein provide some clues to the possible functions of the protein. Because the *Drosophila* trithorax protein regulates genes involved in the morphogenesis of the fruit fly, it has been suggested that the ALL1 protein may regulate genes involved in human cellular differentiation and organ development.[12] Because defects in the ALL1 gene are associated with both myeloid and lymphoid acute leukemias, it has also been postulated that the ALL1 protein may help to regulate the differentiation of hematopoietic stem cells.[7]

Although the amino acid sequence of the encoded ALL1 protein shows similarity to some known protein domains, experimental studies must be done in appropriate model systems to verify ALL1 protein functions. Recently, researchers have tested domains of the ALL1 protein for *in vitro* transcriptional activation activity.[23] These studies showed a strong activation domain located 3' to the breakpoint cluster region and a transcriptional repression domain located 5' to the breakpoint cluster region. In addition, studies showed evidence for *in vitro* binding by the AT-hook domain to cruciform DNA independent of DNA sequence.[23] It was postulated that the separation of these functional domains from one another by *ALL1* translocations could contribute to leukemogenesis.[23]

FUSION OF THE *ALL1* GENE WITH GENES ON OTHER CHROMOSOMES

Among leukemia genes, *ALL1* is unique because it participates in many different translocations. Translocations fuse the *ALL1* gene with genes on other chromosomes. The identification of the *ALL1* gene has enabled the cloning of seven different partner genes to date. These include the *AF1p* gene[24] located on chromosome band 1p32, *AF4*[11,25] on 4q21, *AF6*[26] on 6q27, *AF9*[25] on 9p22, *AF17*[27] on 17q21, *ENL*[12] on 19p13.3, and *ELL*[28] on 19p13.1. All of these genes were previously unknown. With the exception of *AF9* and *ENL*, the seven different partner genes are not homologous to one another. The cellular functions of the proteins encoded by the seven different partner genes are not yet known, nor do they generally share common structural domains or motifs. Some of the proteins have amino acid sequence features suggestive of transcription factors.

Because translocations usually create two reciprocal gene fusions, it is important to determine whether one or both potential fusion genes is involved in leukemogenesis. Molecular and cytogenetic evidence indicates that the fusion gene on the der(11) chromosome is the critical fusion in *ALL1*-associated leukemias.[29-31] The der(11) chromosome encodes a fusion protein that consists of the amino-terminal portion of ALL1 and the carboxy-terminal portion of one of the various fusion partners. Thus, the critical fusion protein contains the AT-hook and DNA methyltransferase domains derived from ALL1. The diversity of the translocation partners does not suggest a common structural domain or motif that these proteins may contribute to the carboxy-terminal portion of the various ALL1 fusions.

SELF-FUSION OF THE *ALL1* GENE

In some cases of AML lacking 11q23 chromosomal translocations, we have detected molecular rearrangement of the *ALL1* gene by Southern analysis.[13] These include, in adults, 2 of 19 cases of AML with normal karyotypes and 3 of 4 cases of AML with trisomy 11 as a sole cytogenetic abnormality (Figure 2).

We have shown by molecular characterization that *ALL1* is rearranged by tandem duplication of an internal portion of the gene.[14,15] In two of the cases with trisomy 11, the duplication involves a 16 kb region spanning exons 2-6 of the gene. In one case with a normal karyotype, the duplication consists of an approximately 20 kb region involving exons 2-8 of the gene (Figure 3). RT-PCR and DNA sequence analysis demonstrate that the partially duplicated *ALL1* gene is transcribed into RNA capable of encoding a partially duplicated protein. In the two cases of AML with trisomy 11, we detected an in-frame fusion of exon 6 with exon 2. In one case of AML with a normal karyotype, we detected a fusion of exon 8 with exon 2 that also preserves the open reading frame.

The structure of the partially duplicated *ALL1* gene resembles a gene that is fused with part of a copy of itself. We have coined the term "self-fusion" to describe this special type of gene fusion. Self-fusion of the *ALL1* gene represents a novel genetic mechanism for leukemogenesis. The self-fusion gene defect indicates that the *ALL1* gene can play a critical role in leukemia on its own without a contribution from a partner gene on a different chromosome. This observation suggests that the amino-terminal portion of the ALL1 protein is critical for leukemogenesis in all *ALL1* gene rearrangements, including those involving translocations.

Alu-mediated recombination in somatic cells may provide a molecular mechanism for self-fusion of the *ALL1* gene.[15] DNA sequence analysis in two cases of AML with trisomy 11 reveals, at the junction point of the duplication, fusion of an *Alu* sequence in intron 6 of the *ALL1* gene with unique DNA sequence in intron 1. In one case of AML with a normal karyotype, sequence analysis of the genomic junction point of the duplication shows fusion

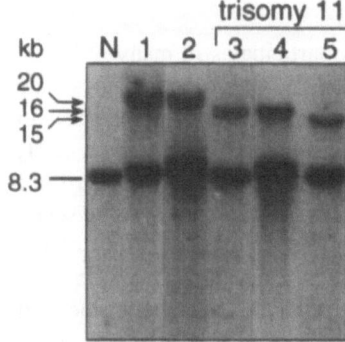

Figure 2. Southern analysis of *ALL-1* gene rearrangements in leukemias without visible chromosomal translocations. A single rearranged band is detected in *Bam*HI restriction enzyme digests of each patient sample in addition to a germline 8.3 kb band. Lanes 1 and 2, leukemia cases with normal karyotypes. Lanes 3-5, leukemia cases with trisomy 11 as a sole cytogenetic abnormality. N, normal control; kb, kilobase.

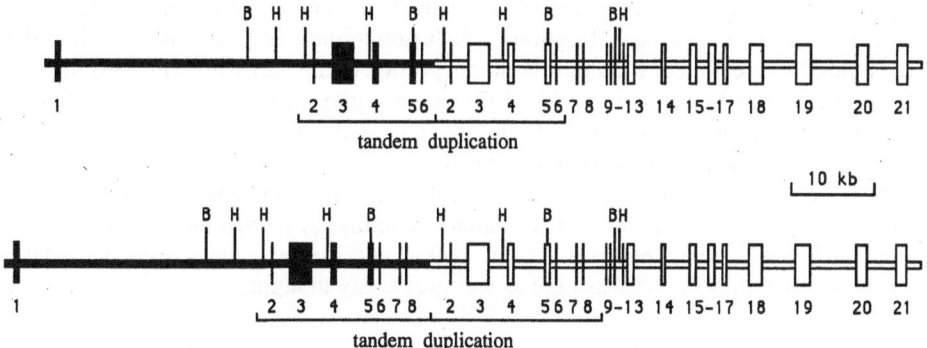

Figure 3. Physical maps of two different examples of *ALL-1* gene self-fusion. In each example, self-fusion results in a direct tandem duplication of an internal portion of the gene shown within bracketts. Genomic fusion points of the duplication are indicated by the junction of the black and white horizontal bars. Exons are indicated by boxes or verticle lines. H, *Hin*dIII; B, *Bam*HI; kb, kilobase.

of two different *Alu* elements from different introns of the *ALL1* gene. In this latter example, partial duplication of the *ALL1* gene may have resulted from homologous recombination between the two *Alu* elements (Figure 4).

The molecular detection of *ALL1* gene rearrangements in 3 of 4 AML cases with trisomy 11 suggests that trisomy 11 may be a marker for self-fusion of the *ALL1* gene.[14] As a sole abnormality, trisomy 11 is a rare recurrent finding in AML, estimated to occur at a frequency of about 1%. However, trisomy of other chromosomes (such as +8, +12, and +21) is frequently detected in leukemia cells. The mechanistic role of trisomy in leukemia is not known, but researchers have postulated a dosage effect associated with an extra copy of a chromosome. However, no specific genes - either normal or defective - have been associated previously with any trisomy in leukemia. The identification of *ALL1* self-fusion is the first example of a specific gene defect associated with trisomy in leukemia. Possibly an extra copy of the defective gene confers a growth advantage to the leukemic clone.

CONCLUSIONS

Our findings demonstrate that the *ALL1* gene can play a role in leukemogenesis on its own without a contribution from other genes. We believe that the self-fusion gene defect

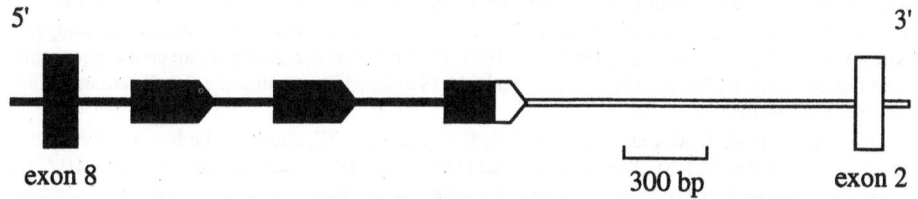

Figure 4. Physical map of the genomic junction region in a case of *ALL-1* gene self-fusion. Somatic recombination is proposed to have occurred between two different *ALL-1* gene loci (black and white). Formation of a chimeric *Alu* element (part black and part white) results from homologous recombination between two *Alu* elements from the different loci. Pentagonal symbols, *Alu* elements with the vertex denoting the 3′ end; rectangles, exons; bp, basepairs.

is an initiating event in some cases of acute leukemia. In common with fusions found on the der(11) chromosome in *ALL1* translocations, self-fusion of the *ALL1* gene may result in the "out of context" expression of the amino-terminal portion of the ALL1 protein. This implies that the amino-terminal portion of the ALL1 protein may play a critical role in the initiation of leukemia. It remains to be demonstrated whether *ALL1* gene defects act through a gain-of-function or a loss-of-function mechanism.

Our preliminary results suggest that, in adults, the self-fusion gene defect may occur in approximately 5% of AML cases. This number is similar in frequency to the combined total of all visible *ALL1* translocations in adult AML. Thus, in adults, the involvement of the *ALL1* gene in AML may be twice as high as was previously thought.

ACKNOWLEDGMENTS

The experimental studies described in this article were performed in collaboration with Matthew P. Strout, Michael A. Caligiuri, and Clara D. Bloomfield of Roswell Park Cancer Institute, Buffalo, New York. C.M.C. is supported by an Outstanding Investigator Grant (CA39860) from the National Cancer Institute and by the Falk Medical Research Trust.

REFERENCES

1. Nowell, P. C., 1994. Cytogenetic approaches to human cancer genes. *FASEB Jour.* 8: 408-413.
2. Croce, C. M., 1987. Role of chromosome translocations in human neoplasia. *Cell (Cambridge, Mass.)* 49: 155-156.
3. Shitvelman, E., Lifshitz, B., Gale, R. P., and Canaani, E., 1985. Fused transcript of abl and bcr genes in chronic myelogenous leukemia. *Nature (London)* 315: 550-552.
4. Burrow, J., Goddard, A. D., Sheer, D., and Solomon, E., 1990. Molecular analysis of acute promyelocytic leukemia breakpoint cluster region on chromosome 17. *Science* 249: 1577-1580.
5. Miyoshi, H., Shimizu, K., Kozu, T., Maseki, N., Kaneko, Y., and Ohki, M., 1991. t(8;21) breakpoints on chromosome 21 in acute myeloid leukemia are clustered within a limited region of a single gene, AML1. *Proc. Natl. Acad. Sci. U.S.A.* 88: 10431-10434.
6. Nourse, J., Mellentin, J. D., Galili, N., Wilkinson, J., Stanbridge, E., Smith, S. D., and Cleary, M. L., 1990. Chromosomal translocation t(1;19) results in synthesis of a homeobox fusion mRNA that codes for a potential chimeric transcription factor. *Cell (Cambridge, Mass.)* 60: 535-545.
7. Rowley, J. D., 1993. Rearrangements involving chromosome band 11q23 in acute leukemia. *Sem. Cancer Biol.* 4: 377-385.
8. Thirman, M. J., Gill, H. J., Burnett, R. C., Mbangkollo, D., McCabe, N. R., Kobayashi, H., Ziemin-Van Der Poel, S., Kaneko, Y., Morgan, R., Sandberg, A. A., Chaganti, R. S. K., Larson, R. A., Le Beau, M. M., Diaz, M. O., and Rowley, J. D., 1993. Rearrangement of the *MLL* gene in acute lymphoblastic and acute myeloid leukemias with 11q23 chromosomal translocations. *N. Eng. J. Med.* 329: 909-914.
9. Cimino, G., Moir, D. T., Canaani, O., Williams, K., Crist, W. M., Katzav, S., Cannizzaro, L., Lange, B., Nowell, P. C., Croce, C. M., and Canaani, E., 1991. Cloning of *ALL-1*, the locus involved in leukemias with the t(4;11)(q21;q23), t(9;11)(p22;q23), and t(11;19)(q23;p13) chromosome translocations. *Cancer Res.* 51: 6712-6714.
10. Zieman-van der Poel, S., McCabe, N. R., Gill, H. J., Espinosa, R., III, Patel, Y., Harden, A., Rubinelli, P., Smith, S. D., Le Beau, M. M., Rowley, J. D., and Diaz, M. O., 1991. Identification of a gene, *MLL*, that spans the breakpoint in 11q23 translocations associated with human leukemias. *Proc. Natl. Acad. Sci. U.S.A.* 88: 10735-10739.
11. Gu, Y., Nakamura, T., Alder, H., Prasad, R., Canaani, O., Cimino, G., Croce, C. M., and Canaani, E., 1992. The t(4;11) chromosome translocation of human acute leukemias fuses the *ALL-1* gene, related to *Drosophila trithorax*, to the *AF-4* gene. *Cell (Cambridge, Mass.)* 71: 701-708.
12. Tkachuk, D. C., Kohler, S., and Cleary, M. L., 1992. Involvement of a homolog of *Drosophila trithorax* by 11q23 chromosomal translocations in acute leukemias. *Cell (Cambridge, Mass.)* 71: 691-700.

13. Caligiuri, M. A., Schichman, S. A., Strout, M. P., Mrozek, K., Baer, M. R., Frankel, S. R., Barcos, M., Herzig, G. P., Croce, C. M., and Bloomfield, C. D., 1994. Molecular rearrangement of the *ALL-1* gene in acute myeloid leukemia without cytogenetic evidence of 11q23 chromosomal translocations. *Cancer Res.* 54: 370-373.

14. Schichman, S. A., Caligiuri, M. A., Gu, Y., Strout, M. P., Canaani, E., Bloomfield, C. D., and Croce, C. M., 1994. *ALL-1* partial duplication in acute leukemia. *Proc. Natl. Acad. Sci. U.S.A.* 91: 6236-6239.

15. Schichman, S. A., Caligiuri, M. A., Strout, M. P., Carter, S. L., Gu, Y., Canaani, E., Bloomfield, C. D., and Croce, C. M., 1994. *ALL-1* tandem duplication in acute myeloid leukemia with a normal karyotype involves homologous recombination between *Alu* elements. *Cancer Res.* 54: 4277-4280.

16. Chen, C.-S., Sorensen, P. H. B., Domer, P. H., Reaman, G. H., Korsmeyer, S. J., Heerema, N. A., Hammond, G. D., and Kersey, J. H., 1993. Molecular rearrangements on chromosome 11q23 predominate in infant acute lymphoblastic leukemia and are associated with specific biologic variables and poor outcome. *Blood* 81: 2386-2393.

17. Cimino, G., Lo Coco, F., Biondi, A., Elia, L., Luciano, A., Croce, C. M., Masera, G., Mandelli, F., and Canaani, E., 1993. *ALL-1* gene at chromosome 11q23 is consistently altered in acute leukemia of early infancy. *Blood* 82: 544-546.

18. Rubnitz, J. E., Link, M. P., Shuster, J. J., Carroll, A. J., Hakami, N., Frankel, L. S., Pullen, D. J., and Cleary, M. L., 1994. Frequency and prognostic significance of *HRX* rearrangements in infant acute lymphoblastic leukemia: a pediatric oncology group study. *Blood* 84: 570-573.

19. Hunger, S. P., Tkachuk, D. C., Amylon, M. D., Link, M. P., Carroll, A. J., Welborn, J. L., Willman, C. L., and Cleary, M. L., 1993. *HRX* involvement in de novo and secondary leukemias with diverse chromosome 11q23 abnormalities. *Blood* 81: 3197-3203.

20. Gill Super, H. J., McCabe, N. R., Thirman, M. J., Larson, R. A., Le Beau, M. M., Pedersen-Bjergaard, J., Philip, P., Diaz, M. O., and Rowley, J. D., 1993. Rearrangements of the *MLL* gene in therapy-related acute myeloid leukemia in patients previously treated with agents targeting DNA-topoisomerase II. *Blood* 82: 3705-3711.

21. Gu, Y., Alder, H., Nakamura, T., Schichman, S. A., Prasad, R., Canaani, O., Saito, H., Croce, C. M., and Canaani, E., 1994. Sequence analysis of the breakpoint cluster region in the *ALL-1* gene involved in acute leukemia. *Cancer Res.* 54: 2327-2330.

22. Ma, Q., Alder, H., Nelson, K. K., Chatterjee, D., Gu, Y., Nakamura, T., Canaani, E., Croce, C. M., Siracusa, L. D., and Buchberg, A. M., 1993. Analysis of the murine *All-1* gene reveals conserved domains with human *ALL-1* and identifies a motif shared with DNA methyltransferases. *Proc. Natl. Acad. Sci. U.S.A.* 90: 6350-6354.

23. Zeleznik-Le, N. J., Harden, A. M., and Rowley, J. D., 1994. 11q23 translocations split the "AT-hook" cruciform DNA-binding region and the transcriptional repression domain from the activation domain of the mixed-lineage leukemia (*MLL*) gene. *Proc. Natl. Acad. Sci. U.S.A.* 91: 10610-10614.

24. Bernard, O. A., Mauchauffe, M., Mecucci, C., Van Den Berghe, H., and Berger, R., 1994. A novel gene, *AF-1p*, fused to *HRX* in t(1;11)(p32;q23), is not related to *AF-4*, *AF-9* nor *ENL*. *Oncogene* 9: 1039-1045.

25. Nakamura, T., Alder, H., Gu, Y., Prasad, R., Canaani, O., Kamada, N., Gale, R. P., Lange, B., Crist, W. M., Nowell, P. C., Croce, C. M., and Canaani, E., 1993. Genes on chromosomes 4, 9, and 19 involved in 11q23 abnormalities in acute leukemia share sequence homology and/or common motifs. *Proc. Natl. Acad. Sci. U.S.A.* 90: 4631-4635.

26. Prasad, R., Gu, Y., Alder, H., Nakamura, T., Canaani, O., Saito, H., Huebner, K., Gale, R. P., Nowell, P. C., Kuriyama, K., Miyazaki, Y., Croce, C. M., and Canaani, E., 1993. Cloning of the *ALL-1* fusion partner, the AF-6 gene, involved in acute myeloid leukemias with the t(6;11) translocation. *Cancer Res.* 53: 5624-5628.

27. Prasad, R., Leshkowitz, D., Gu, Y., Alder, H., Nakamura, T., Saito, H., Huebner, K., Berger, R., Croce, C. M., and Canaani, E., 1994. Leucine-zipper dimerization motif encoded by the *AF17* gene fused to *ALL-1* (*MLL*) in acute leukemia. *Proc. Natl. Acad. Sci. U.S.A.* 91: 8107-8111.

28. Thirman, M. J., Levitan, D. A., Kobayashi, H., Simon, M. C., and Rowley, J. D., 1994. Cloning of *ELL*, a gene that fuses to *MLL* in a t(11;19)(q23;p13.1) in acute myeloid leukemia. *Proc. Natl. Acad. Sci. U.S.A.* 91: 12110-12114.

29. Rowley, J. D., 1992. The der(11) chromosome contains the critical breakpoint junction in the 4;11, 9;11, and 11;19 translocations in acute leukemia. *Genes Chrom. Cancer* 5: 264-266.

30. Kobayashi, H., Espinosa III, R., Thirman, M. J., Gill, H. J., Fernald, A. A., Diaz, M. O., Le Beau, M. M., and Rowley, J. D., 1993. Heterogeneity of breakpoints of 11q23 rearrangements in hematologic malignancies identified with fluorescence *in situ* hybridization. *Blood* 82: 547-551.

31. Downing, J. R., Head, D. R., Raimondi, S. C., Carroll, A. J., Curcio-Brint, A. M., Motroni, T. A., Hulshof, M. G., Pullen, D. J., and Domer, P. H., 1994. The der(11)-encoded *MLL/AF-4* fusion transcript is consistently detected in t(4;11)(q21;q23)-containing acute lymphoblastic leukemia. *Blood* 83: 330-335.

DISCUSSION

D. Metcalf

You did not give us any information about cell biology. What happens if you transfect a cell such as a hematopoietic cell or fibroblast with your fused cDNA? Do you transform the cells?

C. Croce

The answer is that we do not know yet. All those experiments are in progress. At the beginning we were completely sidetracked because we thought that the chimeric product would lead to malignant transformation and that you could see from the sequence of the chimeric genes. The chimeric genes were very, very large. So we had some problems in the beginning to construct the right construct. Now we think in totally different terms. We have done both retroviruses and the appropriate constructs which are in transgenic mice; but we do not know the answer yet.

P. Amati

The rate of transcription of the rearranged genes is the same as for the wildtype?

C. Croce

In general it is higher, but not in every case. In most of the cases the level of the chimeric transcript is higher than the transcript could be for the normal gene.

P. Amati

How do you explain that?

C. Croce

For the time being we cannot. We are addressing that point too.

D. Livingston

Do you know if the region on 11-Q23 is syntenic with the region of the mouse that houses all the classical homeotic genes?

C. Croce

Yes.

D. Livingston

It is? Do you have a cluster of homeotic genes?

C. Croce

Not there, not at 11-Q23.

D. Livingston

So do you know if ALL1, for example, will do a bi-thorax lethal in the fly?

C. Croce

We do not know yet. The man who cloned ALL1 is doing precisely that experiment, but we have not done that.

R. Dalla Favera

What is the normal pattern of expression of the gene in normal cells and is the normal gene expressed in the leukemia cells that carry the fusion.

C. Croce

Yes, the normal gene is expressed, as I have shown, in all the leukemias that carry the fused genes. In all of them, although the levels seem to be lower. Now the pattern of expression is very promiscuous. For example, we see expression of ALL1 in most of the cell lines we tested. In embryogenesis, there is a different pattern of expression. You find it in the lymphoid tissue in central nervous systems.

P. Pelicci

Is there a reciprocal translocation product expressed in all cases?

C. Croce

The reciprocal gene is expressed in most of the cases.

P. Pelicci

But is there a fusion?

C. Croce

Yes, there is a fusion in frame. So you can expect that in most of the leukemias, you have the expression of two aberrant products. The fact is that in a small number of acute leukemias, the derivative is lost. For example like it happened sometimes in acute lymphoblastic leukemias with the Philadelphia chromosome and so on, that the other chromosomes can be lost by the leukemia.

C. Sawyers

Do you know anything about where the protein is in a cell?

C. Croce

We expect that the ALL1 fusion product with many of these genes will be nuclear but we have no evidence yet because we had tremendous problems and many others also had tremendous problems getting antibodies against the ALL1 product. So now finally after we worked on for more than two years getting antibodies, now perhaps we have some decent antibodies; but the preliminary data indicate that there are nuclei.

C. Sawyers

Is bi-thorax nuclear?

C. Croce

Yes.

J. Adams

I wonder what you make of the impressive link to infant leukemia? Is it possible that this gene is particularly potent in the embryonic form of hematopoiesis?

C. Croce

I think that probably that is true and maybe in these acute leukemias you do not need more than one hit, or you just need very, very few hits. I think that that probably is the case. In fact, if you look at identical twins with an acute lymphoblastic leukemia, you see in both of them, the same leukemia, so there is trans-placental transmission of the leukemic cells, which indicates that this leukemia can occur in the uterus, so it can really occur very early and probably the target cell must be a very, very early hematopoietic progenitor.

T. Taniguchi

Is this area of translocation accompanied by a particular activation of oncogenes?

C. Croce

We really do not know that yet.

L. Nadler

In view of the association of ALL1 with secondary leukemias and the fact that there is a very small group of children that has a highly resistant leukemia such that when they are less than a year of age, they all die, could one postulate that this disease is secondary to an oncogenic event elicited by some chemical, for example, and that there might therefore be a predisposition to the disease?

C. Croce

It is certainly possible. The fact that secondary leukemias involve ALL1 suggests that that might be the case. That a variety of environmental agents might lead to those kinds of rearrangements or at least increase the probability of that kind of rearrangement. There is, as you know, ALL1 rearrangement involving leukemias with a very good prognosis. These leukemias do not respond very well to therapy. Now one of the interesting issues is that we know from cytogenetics that leukemias with a normal karyotype, in general have a good prognosis. Now what we have to do is to determine whether in fact the ones with the ALL1 rearrangement and the normal karyotypes have a very poor prognosis because I think we will be able to distinguish acute leukemia with a normal karyocyte which do not have ALL1 rearrangement, and those have a good prognosis, from the ones that have a ILR-1 rearrangement and they will have a poor prognosis. I strongly believe there might be environmental factors that can increase the incidence of those rearrangements.

E. Mihich

In this secondary leukemia, which is often resistant to treatment, is it known whether there is an MDR-1 dependent resistance or whether it is a topo-isomerase, iso-enzyme dependent resistance?

C. Croce

I have no idea.

A. D'Andrea

The crystal structure was recently solved for a DNA methyltransferase. Do you know where the homology is between DNA methyltransferase and ALL1?

C. Croce

We know where it is. It is in the critical region of the ALL1.

A. D'Andrea

But where does the region fall within the structure of DNA methyltransferase? Do you know if it is in the DNA binding region?

C. Croce

I do not remember.

A. D'Andrea

It might be interesting to find out if it is in the DNA binding domain. Also, you showed one of your fusions actually contained a region of myesin. Frances Collins recently presented some work that showed in the case of inversion 16, that there is a fusion of smooth muscle myesin heavy chain with CBF Beta. Do you care to speculate as to whether or not the presence of myesin in a fusion protein contributes to transforming activity?

C. Croce

No, I cannot speculate. The problem is that in the case of the inverted 14, essentially the same two genes fuse all the time. The situation is very different from this situation where ALL1 is a common denominator but the partner gene is very different. And that is what bothers me a lot in the analysis of the 11-QT3 rearrangement. These translocations are clearly different from all the others. All the others, either juxtapose an element to an oncogene or you have the fusion of two genes and in general, although there might be some variation to the common theme, these variations are very, very few. In these cases you have a gene which can fuse with at least 20 different genes and we might see only the tip of the iceberg. It might be fusing with 50 different genes and it is possible that this translocation might involve fusion of genes which are expressed during differentiation. The restriction might involve what kind of mechanism is involved in the translocation but they must make valuable protein products, a stable protein product, and because of the combination of these factors you have fusion of ALL1 with one of those many different genes and all the different fusions result essentially in a truncation in this connection of the first domain of ALL1 from the rest of the protein and that leads to a protein product that might act (we do not have formal proof that that is the case) in a dominant negative fashion leading to perhaps shutting off of normal ALL1 function.

A. D'Andrea

In the gene duplications, the finger region of the protein is retained. So this must give quite a different result from the translocation.

C. Croce

Absolutely. It is a totally different situation. In fact we think that this is really a different genetic mechanism of leukemogenesis that does not involve an activation of an oncogene but that all these different translocations result in the expression of a dominant negative of the protein products that act in a dominant negative fashion. And perhaps there are other leukemias and perhaps solid tumors which are due to similar mechanisms. And in fact I mentioned to you that now we have found ALL1 fusing with itself in gastric cancers, in fact in several gastric cancers.

P. Amati

You say that the first part of the gene is kept. In this region, are there any structures that may suggest dimerization with other proteins?

C. Croce

Yes, we think that that is the case. We think it might dimerize with other proteins.

R. Perlmutter

The part that I do not understand, I guess, is that since the endogenous locus is expressed and in some cases, the level of expression, at least at the RNA level, is not much higher for the co-dominant negative than it is for the endogenous protein and the mechanism of action of the dominant negative seemingly would require a stoichiometric effect, that is direct interaction with some part of the protein in DNA binding. I am having trouble understanding how that would work as a dominant negative.

C. Croce

You will have to judge for yourself. Essentially when we carry out northern blotting we look at the transcript. We see that the expression of the chimeric transcript is higher for the one on the derivative 11. It is higher although we do not know the mechanism. But unfortunately what we should be doing is to look at the protein product, and soon we will have that answer.

R. Perlmutter

Along those lines you said, I believe, that in the translocation cases, the fusion transcript is fused in frame.

C. Croce

Always in frame, at least in all the cases that we have sequenced, we and others have sequenced.

R. Perlmutter

So you believe that the selection is for proteins that would be relatively stable, which perhaps might explain myesin etc.

C. Croce

We think that that is the case. Now we have to prove it.

R. Perlmutter

Protein stabilization would be the selection?

C. Croce

Yes.

A. D'Andrea

Along those lines, the work that Frances Collins presented last week at Cold Spring Harbor showed that the fusion of myesin heavy chain protein with the CBF Beta very strongly stabilizes the protein. You can actually see myofibrils in the nucleus of cells expressing the fusin protein, suggesting stabilization as being important.

S. Landolfo

You mentioned before that you have analyzed twins for ALL1.

C. Croce

No, we did not do it. A group in London had a look at identical twins.

S. Landolfo

OK, my question was: Do they display the same kind of fusion proteins?

C. Croce

Yes, it is the same rearrangement, it is monoclonal. It came from one of the two twins.

J. Adams

Just a very general question: It is often said that solid tumors do not have nearly as many translocations as hematopoietic ones. Do you think that is still true or is it just a technical problem? And if it is true, do you see any rationale for it?

C. Croce

I think that that is almost true but not quite. I think that clearly the initiating event in leukemias and lymphomas, in most cases is oncogene activation, either due to transcriptional activation or chimerisation. Now in solid tumors, clearly loss of function is a dominant mechanism although during the progression of a malignancy you might have both loss of function and oncogene activation in the same malignant cell. What is fascinating is the fact that in some solid tumors, for example, the sarcomas, in most cases the chimerisation is the critical step. So it looks like in mesemchimally origined tumors, chromosomal translocations are playing a critical role, at least in the initiation of malignancies. Now, in the other tumors, let us say epithelial tumors, I know of specific translocation only in very few cases. For example, the inversion of chromosome 11 in parathyroid adenomas which involve bcl-1, prad-1, cyclin D-1. And also what is fascinating in that case is that whenever the cyclin D-1 is involved, the phenotype is a very mild phenotype. So in the case of leukemia, in the case of lymphomas, in the case of the adenomas which involve the same gene, the phenotype is the one of a very low grade malignancy. So I would say that in general, chromosomal translocation and inversion, which essentially involve the same mechanism, is the critical step in the initiation of leukemias and lymphomas. A major exception seems to be, if it is true that they work in a dominant negative fashion, is involvement of ALL1 in a variety of acute leukemias. In solid tumors, with the exception of the sarcomas, a loss of function of the tumor suppresser gene is a critical event.

J. Adams

Why should there be that difference between solid tumors and hematopoietic ones?

C. Croce

I do not know but clearly that suggests tumor death or cell death originated from mesenchima has a different kind of genetic control of proliferation and/or apoptosis. I think that the more we look into it, the more it is becoming like a team. Before we had only 1 or 2 examples of one and of the other. Now we have essentially dozens of examples.

6

NON-RECEPTOR PROTEIN TYROSINE KINASES

Pivotal Regulatory Molecules Controlling Lymphopoiesis

Roger M. Perlmutter, Steven J. Anderson, Jane A. Gross,
and Mark W. Appleby

Howard Hughes Medical Institute and
Departments of Immunology, Medicine and Biochemistry
University of Washington
Seattle, Washington 98195

INTRODUCTION

Reversible phosphorylation of proteins has proved to be the most widely-employed regulatory mechanism governing cell behavior. Seminal studies performed in the 1960s by Edwin Krebs and colleagues on the regulation of glucose homeostasis paved the way for the enumeration of literally hundreds of protein kinases acting in species as diverse as bacteria and mammals (Krebs, 1993). Within this broad group, the protein tyrosine kinases have attracted special attention as regulators of cell proliferation. In particular, the receptors for numerous polypeptide growth factors, including epidermal growth factor, platelet-derived growth factor, and insulin (among many others), are themselves transmembrane protein tyrosine kinases which stimulate replication through activation of their intrinsic kinase function (Schlessinger and Ullrich, 1992). Where detailed study has been possible, interaction with cognate ligand stimulates dimerization of the growth factor receptors in the membrane, and subsequent augmentation of phosphate transfer. Substrates for this kinase activity include the receptor molecules themselves, which when phosphorylated manifest binding sites for regulatory enzymes that stimulate membrane phospholipid catabolism, the accumulation of $p21^{ras}$ in its GTP-bound form, and the phosphorylation of transcription factors (Schlessinger and Ullrich, 1992). In this light, it is unsurprising that growth factor receptor stimulation may affect the differentiative state of the cell as well as its replicative properties. Thus, for example, stimulation of the colony stimulating factor-1 (CSF-1) receptor, a typical receptor-type protein tyrosine kinase, encourages the differentiation of myeloid progenitors into macrophages (Sherr, 1991).

Study of the growth factor receptor kinases has provoked interest in a second group of related molecules, the non-receptor protein tyrosine kinases. These kinases, though often membrane associated, lack a membrane-spanning domain. Nevertheless, it is apparent that

Normal and Malignant Hematopoiesis, Edited by Enrico Mihich and Donald Metcalf
Plenum Press, New York, 1995

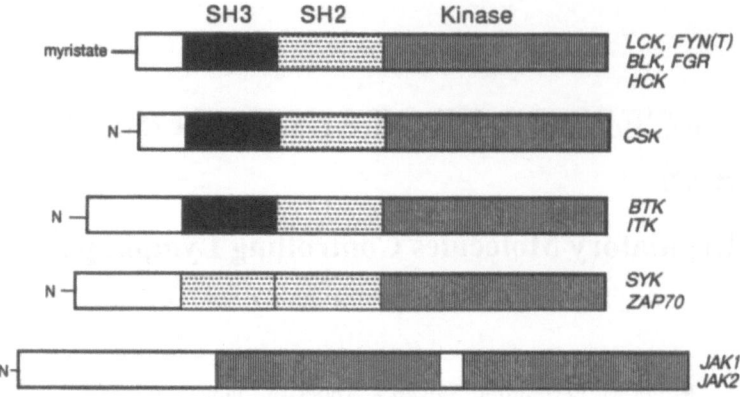

Figure 1. Familes of non-receptor protein tyrosine kinases. The kinases are represented in cartoon form with the domain structure of each defined using filled boxes. See text for details.

the non-receptor protein tyrosine kinases participate in signal transduction in most cell types. Molecular cloning studies have identified more than two dozen non-receptor protein tyrosine kinases, all of which contain a kinase domain of about 250 amino acids within which certain stereotyped sequence motifs, characteristically present in protein tyrosine kinases, are recognizable. In most cases, the protein tyrosine kinase domain is juxtaposed with one or more accessory domains involved in interactions with substrates, receptors, and regulatory proteins. These include SH2 domains, which interact with phosphotyrosine residues when these are embedded in an appropriate sequence context (Songyang et al, 1993), SH3 domains, which interact with proline-rich motifs found in some kinase-associated proteins, (Ren et al, 1993; Yu et al, 1994), and so-called "pleckstrin-homology" domains, for which a precise function has not yet been elucidated (Musacchio et al, 1993). Figure 1 presents a diagrammatic summary of the non-receptor protein tyrosine kinases which are grouped into families based on structural relationships. It is worth noting that many of the non-receptor protein tyrosine kinases are expressed primarily, or even exclusively, in hematopoietic cells (Fig. 2). This observation has focussed attention on blood cells as experimental systems in which to dissect the function of these signaling molecules. In the following pages, we summarize experiments that define roles for four different non-receptor protein tyrosine kinases in the development of T lymphocytes.

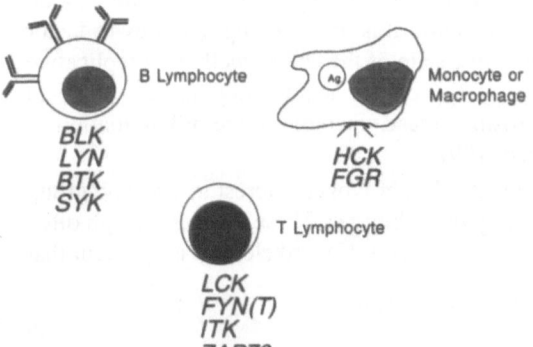

Figure 2. Expression of non-receptor protein tyrosine kinases in hematopoietic cells. Shown are examples of protein tyrosine kinase genes, the expression of which is typically restricted to the illustrated cell types.

RESULTS AND DISCUSSION

The maturation of T lineage cells proceeds according to a well-defined scheme involving the successive acquisition of cell-surface proteins that participate in antigen recognition (Petrie et al, 1990). General features of this process are depicted in Figure 3. Essentially all circulating T lymphocytes are derived from immature progenitors that mature within the thymus. During this process, both self-renewal of precursors and expansion of more mature daughter cells takes place. The least mature precursors identifiable in the adult thymus lack expression of T cell receptor components and also fail to express the CD4 and CD8 coreceptor molecules. These "triple-negative" cells give rise to cells that express both CD4 and CD8 simultaneously, a population that typically constitutes 85% of the total thymocyte pool, and about half of this population also expresses the T cell antigen receptor, a feature usually detected by assessing the surface representation of CD3 components. It is these "double-positive", CD3-bearing cells that serve as substrates for antigen-mediated repertoire selection, thereby permitting maturation of cells that express either CD4 or CD8 in a mutually exclusive fashion, cells that will ultimately exit the thymus to populate peripheral lymphoid organs. Recent studies demonstrate that thymocyte maturation is guided by exquisitely specific regulatory signals that are themselves entrained by non-receptor protein tyrosine kinases expressed primarily or exclusively in T-lineage cells. This has been documented most comprehensively in the case of p56^{lck}.

The Role of p56^{lck} in Thymocyte Development

The *lck* gene encodes a membrane-associated non-receptor protein tyrosine kinase that is expressed in lymphocytes, mainly T cells, and NK cells. Originally identified as the highly-expressed product of a transforming gene activated by retroviral promoter insertion, p56^{lck} associates physically and functionally with the CD4 and CD8 coreceptor proteins (Turner et al, 1990), and with the β subunit of the interleukin-2 receptor (Hatakeyama et al, 1991), among others. However, during thymocyte development it is apparent that p56^{lck} relays signals that arise in very immature CD4⁻8⁻ cells. Elimination of p56^{lck} function, achieved either through targeted gene disruption (Molina et al, 1992) or by deliberate overexpression of a competing catalytically inactive ("dominant negative") form of p56^{lck} (Levin et al, 1993) yields animals in which thymocyte development is arrested at the triple-negative stage. Although the very few (typically <5% of normal) cells that arise in such thymuses proved capable of catalyzing satisfactory rearrangements at the T cell receptor β locus, in other respects these thymocytes resembled those found in animals bearing crippling mutations in elements regulating antigen receptor gene rearrangement or in the antigen receptor β chain gene itself (Shinkai et al, 1992; Mombaerts et al, 1992a; 1992b). For this reason, it was hypothesized that p56^{lck} might represent a component of the signaling

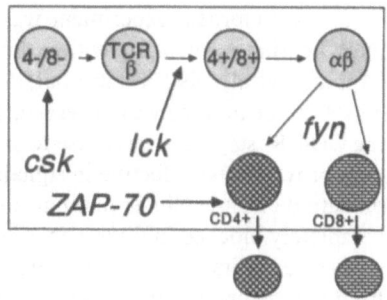

Figure 3. Diagrammatic representation of thymocyte maturation. Shown are the populations of cells represented in each thymocyte subset. Arrows denote lineage relationships. The importance of individual protein tyrosine kinases in the maturation or function of each subpopulation is noted. See text for description.

structure that ordinarily informs developing thymocytes when successful T cell receptor β locus rearrangement has occurred.

This hypothesis makes two extremely strong predictions. First, since successful β chain gene rearrangement ordinarily blocks subsequent complete rearrangements of the homologous sequences on the non-expressed homologue (a process referred to as allelic exclusion), interference with p56^{lck} function could be expected to abrogate allelic exclusion, permitting continued rearrangement of endogenous T cell receptor β chain genes even in the presence of a functional β chain transgene. This violation of allelic exclusion was demonstrated directly in mice bearing both a dominant-negative *lck* transgene and a functional T cell receptor β chain transgene (Anderson et al, 1993). The hypothesis that p56^{lck} delivers signals provoked by satisfactory expression of a functionally rearranged T cell receptor β chain gene also predicts that illegitimate augmentation of p56^{lck} activity should by itself stimulate maturation of CD4⁻8⁻ cells to the CD4⁺8⁺ stage. This prediction was confirmed using a mutant *lck* transgene that encodes a relatively active form of p56^{lck} bearing a phenylalanine-for-tyrosine substitution at position 505 (Abraham et al, 1991). The *lck*F505 transgene not only extinguished endogenous T cell receptor β chain gene rearrangement (Anderson et al, 1992), but also corrected the deficit in thymocyte cellularity observed in mutant mice incapable of supporting T cell receptor gene rearrangement (Mobaerts et al, 1994). Together these data argue persuasively that p56^{lck} serves as the gatekeeper at a checkpoint in thymocyte development, sensing the production of a satisfactory T cell receptor β chain and thereafter delivering signals that extinguish further rearrangements at the β locus, augment rearrangements at the T cell receptor α locus, and promote proliferation (Anderson et al, 1994).

Control of T Cell Receptor Signaling by p59^{fyn}

Although T lymphocytes contain numerous non-receptor protein tyrosine kinases, p56^{lck} appears unique in its ability to direct early thymocyte maturation. This is illustrated by considering the effects of analogous manipulations involving p59^{fyn}. The *fyn* gene encodes a 59 kDa non-receptor protein tyrosine kinase, closely related to p56^{lck}. Unlike the *lck* gene, however, which is expressed at high levels in immature thymocytes, *fyn* transcripts accumulate primarily in the most mature single-positive cells and in peripheral T lymphocytes (Cooke et al, 1991). Nevertheless, it seemed plausible that in light of the close structural relationship between p56^{lck} and p59^{fyn}, augmented expression of p59^{fyn} might deliver signals analogous to those seen in *lck* transgenic mice. To test this hypothesis, animals were generated in which p59^{fyn} was expressed at very high levels under the control of the *lck* proximal promoter, achieving thereby an intrathymic transgene expression pattern analogous to that normally observed with *lck* itself. This manipulation did not produce the expected block in T cell receptor β chain gene rearrangement. Instead, thymocytes expressing high levels of p59^{fyn} became inordinately sensitive to T cell receptor stimulation (Cooke et al, 1991).

A reciprocal experiment was performed by examining thymocyte development in mice bearing a mutation within the *fyn* gene that blocked expression in lymphoid cells (Appleby et al, 1992; Stein et al 1992). Once again, no defect analogous to that seen in p56^{lck}-deficient mice was observed. Thymocytes from *fyn*^{null} mice matured satisfactorily to the CD4⁺8⁺ stage and gave rise to single-positive cells in normal numbers. However these cells proved to be defective in signaling from the T cell antigen receptor. For example, mobilization of intracellular calcium stores and subsequent DNA replication were both almost completely blocked in thymocytes lacking p59^{fyn} activity (Appleby et al, 1992). These two types of experiment strongly support the view that p59^{fyn} participates in signal transduction from the mature T cell antigen receptor, impinging upon a process that proceeds primarily in single-positive thymocytes and circulating T cells (Fig. 3). Importantly, the functions of p56^{lck} and p59^{fyn}, two closely-related kinases expressed in overlapping thymocyte subpopulations, appear to be completely distinct. The former regulates maturation of early T-lineage cells, while the latter participates in signal transduction from the mature antigen receptor.

Involvement of the *csk*-Encoded Kinase in T Cell Development

The general theme that individual non-receptor protein tyrosine kinases regulate specific points in the thymocyte maturational sequence is reiterated in the case of p50^{csk}, a kinase that resembles both the *lck*- and *fyn*-encoded kinases but which lacks the amino-terminal myristoyl group and the carboxy-terminal phosphorylation site characteristic of these latter proteins. Indeed, considerable evidence supports the view that p50^{csk} selectively phosphorylates the carboxy-terminal regulatory tyrosines of p56^{lck} and p59^{fyn}. Hence it was predicted that defects in p50^{csk} function would yield animals in which the *lck*- and *fyn*-encoded kinases were confined to an activated state.

Divising an experimental test for this hypothesis proved difficult, primarily because loss of p50^{csk} expression blocks embryonic development in general at approximately fetal day 10.5 (Imamoto et al, 1993). Hence satisfactory expression of p50^{csk} protein is required for normal organogenesis, and the specific importance of p50^{csk} in T cell development could not be evaluated in such mice. To overcome this difficulty, embryonic stem cells bearing a homozygous disruption of the *csk* gene have been generated, and the differentiative potential of these cells has been tested in chimeric mice (unpublished results). Remarkably, although some T-lineage precursors derived from *csk*^{null} progenitors are detectable in the thymuses of mice analyzed at fetal day 17, these have completely disappeared by shortly after birth. Thus *csk*^{null} T lineage cells cannot compete effectively in generating more mature thymocytes following colonization of the thymic rudiment. Interestingly, this defect in thymocyte development occurs even when *csk*^{null} T lineage cells are provided with the opportunity to mature in the thymuses of animals that cannot themselves produce T lymphocytes (unpublished results).

The unique defect in T cell maturation resulting from loss of *csk* gene expression cannot easily be explained by invoking the ability of p50^{csk} to phosphorylate, and thereby suppress the activity of, p56^{lck} and p59^{fyn}. If either of these proteins were indeed the target of p50^{csk} action, an effect analogous to that which is seen after overexpression of these kinases should have been observed. Instead, it appears that p50^{csk} acts on a quite different target to regulate the very earliest steps in thymocyte maturation. Detailed dissection of this process should provide insight into the molecular events that permit hematopoietic progenitors to give rise to committed lymphoid progenitors, the earliest examples of which are detectable in thymocyte populations (Wu et al, 1991).

Later Thymocyte Maturation Controlled by ZAP-70

The functions of the non-receptor protein tyrosine kinases encoded by the *csk*, *fyn* and *lck* genes have all been tested through the generation of targeted gene disruptions in otherwise normal mice. Additional insights have recently emerged through study of patients with an immunodeficiency syndrome resulting from a mutation in a related kinase gene that encodes a protein called ZAP-70. The ZAP-70 kinase, which is expressed exclusively in T cells, is structurally distinct from the *lck*, *fyn* and *csk* kinases in that it lacks an SH3 domain and contains tandemly reiterated SH2 domains (Fig. 1). Prior studies demonstrate that in at least some human T cell lines ZAP-70 becomes physically associated with the ζ chain of the T cell antigen receptor complex following antigen-induced stimulation (Chan et al, 1992). For this reason, it is widely believed that ZAP-70 contributes to the signal transduction process leading from the antigen receptor to the cell interior. Remarkably, human patients lacking ZAP-70 expression exhibit a form of severe combined immunodeficiency in which peripheral T cells are confined to the CD4[+] subset, and are essentially unresponsive to T cell receptor-induced stimulation (Arpaia et al, 1994; Chan et al, 1994; Elder et al, 1994). Although a systematic analysis of T cell development has not yet been performed in these patients, all evidence suggests that development of thymocytes is normally through the CD4[+]8[+] stage, but that subsequent selection events, known to be mediated via the T cell antigen receptor, are compromised. Thus ZAP-70- represents a fourth non-receptor protein tyrosine kinase expressed primarily in lymphoid cells that acts to control a specific point in thymocyte development (Fig. 3). These results, placed in the context of those already described for experiments performed with the *lck*, *fyn* and *csk* genes, encourage the view that non-receptor protein tyrosine kinases will in general act to coordinate signaling events that regulate lineage commitment, especially in hematopoietic cells in which these kinases are especially abundant.

CONCLUSIONS

The non-receptor protein tyrosine kinases, like their membrane-spanning counterparts the growth factor receptors, are intimately involved in signal transduction processes that govern differentiation in many cell lineages. Recent studies performed using gene manipulation techniques permit the establishment of an ordered sequence of thymocyte developmental milestones for which a protein tyrosine kinase governs the maturation process. Although it has not yet been possible to position these kinases precisely within signal transduction pathways, gene disruption experiments (and in one case an analogous experiment of nature in the human population) document the crucial role of the non-receptor protein tyrosine kinases in orchestrating T lymphocyte differentiation. It has not yet been possible to define explicitly the nature of the substrates upon which these kinases act, or the regulatory machinery that governs their phosphorylating ability. These remain challenges for the future. Nevertheless, our studies, and those of others, encourage the general conclusion that the non-receptor protein kinases will behave in general as pivotal regulators of normal differentiation, especially in hematopoietic cells. Thus there is every reason to believe that the regulatory events governing myelopoiesis, thrombocytopoiesis, and erythropoiesis will ultimately be shown to depend upon the timely and concerted action of this diverse group of protein tyrosine kinases.

REFERENCES

Abraham, K.M., Levin, S.D., Marth, J.D., Forbush, K.A., Perlmutter, R.M., 1991b, Delayed thymocyte development induced by augmented expression of p56[lck], *J. Exp. Med* 173:1421-1432.

Anderson, S.J., Abraham, K.M., Nakayama T., Singer, A. and Perlmutter, R.M., 1992, Inhibition of T cell receptor β chain gene rearrangement by overexpression of the non-receptor protein tyrosine kinase p56[lck], *EMBO J.* 11: 4877-4886.

Anderson, S.J., Levin, S.D., and Perlmutter, R.M., 1993, Protein tyrosine kinase p56[lck] controls allelic exclusion of the T cell receptor β chain genes, *Nature* 365: 552-554.

Anderson, S.J., Levin, S.D., and Perlmutter, R.M., 1994, Involvement of the protein tyrosine kinase p56[lck] in T cell signaling and thymocyte development, *Adv. Immunol.* 56:151.

Appleby, M.W., Gross, J.A., Cooke, M.P., Levin, S.D., Qian, X. and Perlmutter, R.M., 1992 Defective T cell receptor signaling in mice lacking the thymic isoform of p59[fyn]. *Cell* 70: 751-763.

Arpaia, E., Shahar, M., Dadi, H., Cohen, A., and Roifman, C.M., 1994, Defective T cell signaling and CD8+ thymic selection in humans lacking Zap-70 kinase, *Cell* 76:947-958.

Bolen, J.B., 1993, Nonreceptor protein tyrosine kinases, *Oncogene* 8:2025-2031.

Chan, A.C., Iwashima, M., Turck, C.W. and Weiss, A., 1992, ZAP-70: a 70 kd protein tyrosine kinase that associates with the TCR zeta chain, *Cell* 71:649-662.

Chan, A.C., Kadlecek, T.A., Elder, M.E., Filipovich, A.H., Kuo, W.-L., Iwashima, M., Parslow, T.G., and Weiss, A., 1994, ZAP-70 deficiency in autosomal recessive form of severe combined immunodeficiency, *Science* 264:1599-1601.

Cooke, M.P., Abraham, K.M., Forbush, K.A., and Perlmutter, R.M.,1991, Regulation of thymocyte signal transduction by a non-receptor protein tyrosine kinase, p59$^{fyn(T)}$. *Cell* 65:281-291.

Elder, M.E., Lin, D., Clever, J., Chan, A.C., Hope, T.J., Weiss, A., and Parslow, T.G., 1994, Human severe combined immunodeficiency due to a defect in ZAP-70, a T cell tyrosine kinase, *Science* 264:1596-1598.

Hatakeyama, M., Kono, T., Kobayashi, N., Kawahara, A., Levin, S., Perlmutter, R.M. and Taniguchi, T., 1991, IL-2 receptor interacts with a *src*-family kinase, p56^{lck}; identification of novel intermolecular association, *Science* 252:1523-1528.

Imamoto, A. and Soriano, P., 1993, Disruption of the csk gene encoding a negative regulator of src-family kinases leads to neural tube defects and embryonic lethality in mice, *Cell* 73:1117-1124.

Krebs, E.G., 1993, Nobel Lecture. Protein phosphorylation and cellular regulation, *Biosci. Rep.* 13:127-142.

Levin, S.D., Anderson, S.J., Forbush, K.A., and Perlmutter, R.M., 1993, A dominant-negative transgene defines a role for p56^{lck} in thymopoiesis, *EMBO J.* 12: 1671-1680.

Molina, T.J., Kishihara, K., Siderovski, D.P., van Ewijk, W., Narendran, A., Timms, E., Wakeham, A., Paige, C.J., Hartmann, K.U., Veillette, A., Davidson,D., and Mak, T.W., 1992, Profound block in thymocyte development in mice lacking p56^{lck}, *Nature* 357:161-164.

Mombaerts, P., Iacomini, J., Johnson, R.S., Herrup, K., Tonegawa, S., and Papaioannov, V.E., 1992a, RAG-1-deficient mice have no mature T and B lymphocytes, *Cell* 68: 869-877.

Mombaerts, P., Clarke, A.R., Rudnicki, M.A., Iacomini, J., Itohara, S., Lafaille,J.J., Wang, L., Ichikawa, Y., Jaenisch, R., Hooper, M.L., and Tonegawa, S., 1992b, Mutations in T cell antigen receptor genes α and β block thymocyte development at different stages, *Nature* 360:225-231.

Musacchio, A., Gibson, T., Rice, P., Thompson, J. and Saraste, M., 1993, The PH domain: a common piece in the structural patchwork of signalling proteins, *Trends. Biomem. Sci.* 18:343-348.

Petrie, H.T., Hugo,P., Scollay, R., and Shortman, K., 1990, Linkage relationships and developmental kinetics of immature thymocytes: CD3, CD4 and CD8 acquisition in vivo and in vitro, *J. Exp. Med.* 172:1583-1593.

Perlmutter, R.M., Levin, S.D., Appleby, M.W., Anderson, S.J. and Alberola-Ila, J., 1993, Regulation of lymphocyte function by protein phosphorylation. *Ann. Rev. Immunol.* 11:451-499.

Ren, R., Mayer, B.J., Cacchetti, P., and Baltimore, D., 1993, Identification of a ten-amino acid proline-rich SH3 binding site, *Science* 259:1157-1161.

Schlessinger, J. and Ullrich, A., 1992, Growth factor signaling by receptor tyrosine kinases, *Neuron* 9:383-391.

Sherr, C.J., 1991, Mitogenic response to colony-stimulating factor 1, *Trends Genet.* 7:398-402.

Shinkai, Y., Rathbun, G., Lam, K-P., Oltz, E.M., Stewart, V., Mendelsohn, M., Charron, J., Datta, M., Young, F., Stall, A.M., and Alt, F.W., 1992, RAG-2-deficient mice lack mature lymphocytes owing to inability to initiate V(D)J rearrrangement, *Cell* 68:855-867.

Shinkai, Y., Koyasu, S., Nakayama, K., Murphy, K.M., Loh, D.Y., Reinherz, E.L., and Alt, F.W., 1993, Restoration of T cell development in RAG-2-deficient mice by functional TCR transgenes, *Science* 259:822-825.

Songyang, Z., Shoelson, S.E., Chadhuri, M., Gish, G., Pawson, T., Haser, W.G., King, F., Roberts, T., Ratnofsky, S., Lechleider, R.J., Neel, B.G., Birge, R.B., Fajardo, J.E., Chou, M.M., Hanafusa, H., Schaffhausen, B., and Cantley, L.C., 1993, SH2 domains recognize specific phosphopeptide sequences, *Cell* 72:767-778.

Stein, P.L., Lee, H.-M., Rich, S., and Soriano, P., 1992, pp59fyn mutant mice display differential signaling in thymocytes and peripheral T cells, *Cell* 70:741-750.

Turner, J.M., Brodsky, M.H., Irving, B.A., Levin, S.D., Perlmutter, R.M., and Littman, D.R., 1990, Interaction of the unique N-terminal region of the tyrosine kinase p56^{lck} with the cytoplasmic domains of CD4 and CD8 is mediated by cysteine motifs, *Cell* 60:755-765.

Wu, L., Antica, M., Johnson, G.R., Scollary, R., and SHortman, K., 1991, Developmental potential of the earliest precursor cells from the adult mouse thymus, *J. Exp. Med.* 174:1617-1627.

Yu, H., Chen, J.K., Feng, S., Dalgarno, D.C., Braver, A.W., and Schreiber, S.L., 1994, Structural basis for the binding of proline-rich peptides to SH3 domains, *Cell* 76:933-945.

DISCUSSION

D. Cantrell

Where did you put syk in your model because syk is expressed earlier in the thymocytes than ZAP-70?

R. Perlmutter

I do not know where to put syk. We have not performed any experiments with it and have not had the opportunity to look at gene disruptions for syk. I suspect that it will have a clean phenotype and that it will not overlap ZAP-70 in terms of its function.

D. Cantrell

On that point, you suggest that it is the CD3 complex in the double negative cells that activates lck but in mature T-cells, CD3 does not activate lck. Or at least that was my understanding.

R. Perlmutter

That is a controversial point. I have to argue about something for which we cannot adduce any evidence. There are many who believe that p56 lck is a crucial signal transduction molecule for the T-cell receptor. In our own experiments we have had great difficulty showing that compromising lck function in a mature T-cell does much to antigen receptor signaling. You can demonstrate that p56 lck associates with the antigen receptor complex, the CD3 complex. In Art Weiss' case, the cell line he works with does not express co-receptor and the only molecules that he believes it could associate with, that are involved in the signal transduction event, are complements of the antigen receptor complex itself. So, others make the argument that p56 actually signals directly from the T-cells receptor CD3 complex. Exactly how that is done though, at the moment, is unclear.

M. Caligiuri

Did the p56 knockouts have natural killer cells?

R. Perlmutter

We looked at those only carefully in the dominant negative case, and they did. I mention this because it is possible that the proximal lck promoter does not drive dominant negative expression at quite as high a level in the NK cells and so it is possible we would not have seen a defect in that environment. There exists no really good lck knockout. The only knockout that exists is a Exyon 12 knockout that potentially could produce a small amount of functional protein, so you cannot really address the question there either but many of us are trying to generate a more definitive lck knockout and then I think we will be able to look at that. My suspicion is that NK cells will develop satisfactorily.

P. Pelicci

I am not sure I clearly understood on which experimental basis you say that p56 lck signaling does not require MAP kinase activation and RAS activation.

R. Perlmutter

Yes, I glossed over that completely and just simply made the statement. The experiments that we have performed were first to express an asparagine 17 version of p21 RAS. We expressed a dominant negative version of p21 RAS which is a dominant negative version of H-RAS but has been shown in most other systems to compromise RAS function, presumably by interacting with nucleotide exchange factors. When you do that, you block the ability of the antigen receptor to signal in mature T cells, so if we look at a single population expressing reasonable levels of this protein, they do not proliferate in response to antigenic challenge. But the lck directed developmental step, occurs normally. Moreover, we can test that in a rather more formal way by crossing lck transgenic animals with RAS transgenic animals and ask whether the dose response curves that we have defined for the lck transgene are in any way affected by the presence of dominant negative RAS. You might expect, for example, that if dominant negative RAS was positioned in any way downstream of lck, you would need less catalytically inactive lck to achieve the same developmental block. But instead the dose response curves are perfectly skewed. That is, you see the effects of both transgenes independent of the presence of both. So that argues that those two are not on the same pathway. Now, that by itself is perhaps not strong enough evidence but in addition we have performed the same experiment using a catalytically inactive version of MAP kinase kinase. Once again, MAP kinase kinase exerts effects that are very similar to those of dominant negative p21 RAS, with some differences that are of interest. However, it does not block the ability of lck to drive development. We can show that the dominant negative proteins are made at the appropriate stage and are quite abundant at the appropriate stage. Moreover, we can show that MAP kinase activity does not increase. So MAP kinase, which of course is controlled by MAP KK, is completely flat. And yet, development proceeds. For these reasons we believe that p56 lck drives thymocyte development through a pathway that does not involve the traditional p21 RAS MAP kinase pathway.

D. Cantrell

On that point, do you need the SH3 domains or the SH2 domains of lck when you are using that activated 505 mutant. What happens if you mutate the molecule?

R. Perlmutter

We have not performed as detailed a series on this as we would like to but we have done some experiments that seem to indicate that the kinase domain is the critical feature. Let me explain that. You might argue that p56 lck exerts this effect simply because it is a tyrosine kinase and any tyrosine kinase would work. However, if you overexpress p59 fyn, it will not do this and instead it has a very different effect. It accentuates T cell activation. Similarly, if you disrupt the fyn gene, as we did, that blocks T cell activation in mature cells but does not affect development. If you overexpress hck in its activated form, with phenylalanine positioned 501, that is an extremely potent transforming gene in fibroblasts, for example, but in thymocytes, it does not really do much at all. We have made a large number of those animals and they are normal. They express large amounts of p59 hck and they seem fine. So we then performed some swap experiments in which we asked what was the part of lck that was so important for enabling it to drive development in this way. The first read on those, that has not been followed through quite as well as it should be, was that the kinase domain was what was important. That is, if we made a chimera with the amino terminal SH3 and SH2 domains of fyn and the lck kinase domain, it drove development. So, that is the one piece of information we have. Now that does not answer your question because we still have functional SH3 and SH2 domains, and they are quite similar but it simply says that, to the extent that it was analyzed, the lck kinase domain can work in the context of heterologous SH3 and SH2 domains.

J. Adams

With this type of kinase, do you think the signaling involves an external ligand and receptor dimerization as in other types of receptor kinases?

R. Perlmutter

My suspicion is that it does. In the case of the pre T cell receptor I think there is good evidence that there is a structure that comes to the surface and that its coming to the surface is required for subsequent maturation. This is true also for the receptor in immature B cells, although it has been harder to demonstrate on the surface in this case. My guess is that in the pre T cells, it comes to the surface in order to interact with the ligand and that ligand is probably a relatively stereotyped one, that is, it does not depend on clone typic variation in the beta chain. To throw out a speculation, my guess is that the very same site that super antigens identify in the external face of the beta chain, is the site that the external ligand recognizes on the pre T cell receptor and that is what it uses to drive lck. But that is completely a speculation. I think that is why super antigens work, but that is another story.

A. D'Andrea

Along those lines, I was under the impression that the T cell receptor required extensive assembly before it was translocated to the cell surface. Is there any evidence that cell lines that express dominant negative p56 lck have aberrant cell surface localization? Have you looked at that at all in established lines from these animals?

R. Perlmutter

We have not looked at it specifically although we do know that lck can affect the representation of CD4 on the surface and there is some data which comes largely from the studies done by Nakayama when he was working with Al Singer at the NIH, which are very indirect, but which suggest that p56 lck is involved in what they have termed the wake-up call, where CD3 low cells become CD3 mid and available for signaling. The experiments are indirect because they use our CD4 transgenic animals to sequester lck and one can interpret the experiments in the way that they have chosen to interpret them. It is quite clear though that the CD3 components can get to the surface even in the absence of alpha and beta chains because if you use anti CD3 in fetal thymocyte cultures from a rag null animal, you drive development to the double positive stage. So anti CD3 is seeing that surface CD3 protein even though you did not see it by flow. Now with better flow you can see it.

S. Landolfo

Do you have any evidence on whether the p56 interacts with stat factors or if there is a cross talk with JAK family proteins?

R. Perlmutter

We have no evidence of that kind. It is one of those sets of molecules which we are attempting to address as possible signaling molecules for p56 lck. There is no precedent at the moment for that but it is certainly possible.

7

THE REGULATION AND FUNCTION OF p21^{Ras} IN T CELL ACTIVATION AND GROWTH

D. A. Cantrell, M. Izquierdo, J. Nunes, N. Osman, K. Reif, and
M. Woodrow

Lymphocyte Activation Laboratory
Imperial Cancer Research Fund
Lincolns Inn Field
London WC2A 3PX
United Kingdom

Interleukin-2 (IL-2) regulates the proliferation and functional differentiation of T lymphocytes, B cells and natural killer (NK) cells [1]. IL-2 is secreted by antigen-activated T lymphocytes, and mediates its effects through interaction with a specific high-affinity receptor (IL-2R). The process of T cell activation and growth is initiated by the interaction of antigen with the T cell antigen receptor (TCR), which triggers the G_0-G_1 transition of the cell cycle, and the induction of IL-2 secretion and IL-2R expression. It is the subsequent IL-2/IL-2R interaction which ultimately drives G_1-S phase progression, T cell growth, clonal expansion and functional differentiation [2, 3].

There are signal transduction responses common to the TCR and IL2R of which the most characterised involves the activation of the low molecular weight guanine nucleotide binding proteins p21^{ras} [4, 5]. The activity of Ras proteins is regulated by a GTP binding cycle [6]. The cycle is initiated by the binding of GTP which generates the biologically active form of the protein. The bound GTP is hydrolysed to GDP. The GDP bound form of the protein is inactive and it is reactivated by exchange of bound GDP for free cytosolic GTP. The level of active p21ras -GTP complexes is determined by a balance of the rate of hydrolysis of bound GTP and the rate of exchange of bound GDP for cytosolic GTP. Triggering of the TCR or the IL2R perturbs this equilibrium resulting in a rapid and prolonged accumulation of active Ras-GTP complexes. The GTPase activity of p21^{ras} is controlled by GTPase activating proteins of which two mammalian proteins are known. p120-GAP and neurofibromin[7]. Proteins that regulate guanine nucleotide exchange on Ras have also been characterised in T cells and include the homologue of the Drosophila "son of sevenless" gene product, SOS [8, 9]. The object of the present review is to describe how the TCR and IL2R regulate the Ras guanine nucleotide cycle. The role of Ras in TCR and IL2R control of T cell activation and growth will also be discussed.

Normal and Malignant Hematopoiesis, Edited by Enrico Mihich and Donald Metcalf
Plenum Press, New York, 1995

THE REGULATION OF p21RAS BY THE TCR AND IL2R

The T cell antigen receptor (TCR) is a multichain complex comprising a disulfide-linked heterodimer of the idiotypic αβ chains noncovalently associated with a signal transduction complex composed of the invariant CD3 γ, δ, ε chains and the ζ (16 kDa) subunits [10]. The intracellular tails of the CD3 and ζ molecules contain a common motif,$EX_2YX_2L/IX_7YX_2L/I$, which is present in a single copy in each of the CD3 chains and triplicated in the ζ subunit [11, 12]. This activation motif is termed the tyrosine based activation motif (TAM) and is crucial for TCR coupling to intracellular PTKs which is absolutely required for all subsequent T cell responses including Ras activation [13]. In both B and T cells,activation of PKC with phorbol esters or diacylglycerols can mimic antigen receptor triggering and induce the accumulation of ras-GTP complexes [14]. However,this seems to reflect phorbol ester stimulation of a PKC pathway that is not controlled by the antigen receptor[13]. Thus TCR coupling to calcium/PKC signaling pathways is independent of the Ras signals that originate from the TCR.

The IL2R comprises three subunits, α, β , γ. The β and γ chains belong to the haematopoietin receptor family. IL2 regulation of Ras is dependent on cellular PTK activity and deletion analysis of the IL2R β subunit has identified a membrane distal acidic domain as crucial for IL2R coupling to src kinases and p21ras.

Experiments to explore Ras regulatory mechanisms have shown that TCR triggering inhibits the activity of ras-GAPs [15]. In contrast IL2 activation of p21ras is not associated with any detectable regulation of Ras GAP proteins and it is predicted therefore that the IL2R regulates the activity of Ras guanine nucleotide exchange proteins[5]. The mechanism that allows the TCR to inhibit ras-GAP proteins is not understood nor does TCR regulation of GAP proteins preclude TCR regulation of guanine nucleotide exchange proteins. In fibroblasts, growth factors such as Epidermal growth factor (EGF) regulate p21[ras] by stimulating guanine nucleotide exchange [16]. The Ras exchange protein in fibroblasts is SOS which is recruited to the the cell membrane by the adapter protein Grb-2/Sem5 [17-21]. Grb-2 is composed of one SH2 domain and two SH3 domains. The SH3 domains of Grb2 bind to the carboxy terminal proline rich domain of SOS and the interactions of the Grb-2 SH2 domain with tyrosine phosphorylated molecules apparently recuits SOS to the cell membrane [22].

Studies of Grb-2 in T cells have identifed three proteins that when tyrosine is phosphorylated can potentially bind to the Grb-2; Shc and a membrane located tyrosine phosphoprotein of 36kDa [9, 23]. There is also a 75kDa molecule that is constitutively associated with the SH3 domains of Grb-2 analogous to the Grb-2/SOS association [24]. This protein is a substrate for TCR activated but not IL2R activated PTKs. On the basis of its association with the Grb-2 SH3 domains this molecule is a candidate for a second Grb-2 effector molecule. Further molecular characterisation of p75 will establish whether this protein is important for Ras regulation in T cells. In TCR activated cells both Shc and p36 are tyrosine phosphorylated but only p36/Grb-2/SOS complexes can be detected. Shc can bind to the tyrosine phosphorylated TCR ζ chain and is tyrosine phosphorylated in TCR activated cells and thus could potentially recruit Grb-2 and SOS to the TCR complex[23]. However, tyrosine phosphorylation of Shc in TCR activated cells is weak. Moreover, T cell express two isoforms of Shc,p46 and p52 and we have observed that only non tyrosine phosphorylated p52 Shc molecules can bind to the ζ chain and no ζ/Shc/Grb-2/SOS complexes are formed [25]. On the basis of these data we feel that Shc is unlikely to recruit SOS to the TCR complex.In contrast, Shc may be important in IL2R coupling to Ras since in IL2 activated cells there is no tyrosine phosphorylation of p36, a high level of Shc tyrosine phosphorylation and rapid formation of Shc/Grb-2/SOS complexes [26] (Reif and Cantrell, unpublished data). The IL2R can be coprecipitated with Shc in IL2R activated T cells but it is not clear whether this association recuits Grb-2 and SOS to the IL2R complex [26]. For example, SOS coprecipitation with the IL2R has not been described. It is thus more likely that Shc association with the Il2R is transient and neccessary for Shc tyrosine phosphorylation but the tyrosine phosphorylated Shc may dissociate from the receptor and form a complex with Grb-2/SOS.

Intracellular PTKs important for TCR signal transduction include the *src* family kinases p56lck,p59^{fyn} and a 70 kDa PTK termed ZAP70 [27, 28]. The IL2R also regulates the activity of src kinases notably p56lck[29]. Shc is tyrosine phosphorylated by both the TCR and IL2R PTKs and on this basis we would predict that Shc is a substrate for the src kinases. In contrast p36 is only tyrosine phosphorylated in response to TCR triggering and in a simplistic analysis is probably a substrate for the kinase ZAP70 which is activated by the TCR and not the IL2R.

It remains to be proven by genetic studies if the regulation of SOS /Grb-2 complexes is important for TCR or IL2R regulation of p21ras. It has been described also that Vav, a protooncogene, [30] is an exchange protein for p21ras in T cells [31] even though from its structure Vav is a more likely candidate for a *rho/rac* exchange protein [32]. Vav is tyrosine phosphorylated in response to TCR triggering and this tyrosine phosphorylation appears to stimulate the *in vitro* guanine nucleotide exchange activity.of the protein [31]. Unlike SOS, there is no genetic evidence in any cell system that Vav functions as a Ras exchange protein *in vivo*. We have also observed that tyrosine phosphorylation of Vav does not correlate with p21ras activation. For example, in peripheral blood derived T cells, Vav is not phosphorylated in response to IL2 even though IL2 activates p21ras. Conversely, stimulation of T cells via CD28 with the B7-1 ligand induces a strong persistent tyrosine phosphorylation of Vav but does not activate p21ras. Instead, the failure of CD28 to regulate the Ras guanine nucleotide binding cycle correlates with the failure of CD28 to induce the tyrosine phosphorylation of p36 or Shc or any other protein capable of recruiting the Grb-2/SOS complex to the membrane [33].

Any analysis of Ras regulatory mechanisms in T cells must explain PKC regulation of p21ras since this appears to be a lymphocyte specific response [34]. It is conceivable that there is a PKC sensitive, lymphocyte specfic Ras exchange protein. Vav is an example of one potential Ras regulatory protein that is expressed only in haematopoietic cells. However, Vav cannot explain the apparent lymphoid specificity of the PKC-p21ras link because Vav is expressed in all haematopoietic cells [32] and mast cells or myeloid cells that do express Vav do not express a PKC mediated route for p21ras regulation [4, 34].Figures 1 and 2 summarise our current understanding of TCR and IL2R regulation of p21ras.

p21RAS FUNCTION IN T CELLS

The function of p21ras in T cells has been explored using transient transfection protocols that examine the consequences of expressing mutated constitutively active or dominant inhibitory ras mutants on T cell activation. The Ras oncogene p21-v-Ha- ras is mutated at codon 12 (Ser > Val) and 59 (Ala > Thr). These mutations render the Ras protein insensitive to negative regulation by ras-GAPs such that it accumulates in cells in an "active" GTP bound state. A second Ras mutant, N17 ras blocks cellular Ras function by competing for Ras guanine nucleotide exchange proteins thereby preventing endogenous Ras activation by stopping exchange of GTP onto cellular Ras.p21-v-Ha ras, when expressed in T cells, can activate transcriptional factors such as AP-1 and also synergise with a calcium signalling pathway to activate NFAT-1 and the IL-2 gene [35, 36]. The calcium signaling system that cooperates with p21ras is apparently controlled by src family kinases [37] and mediated by

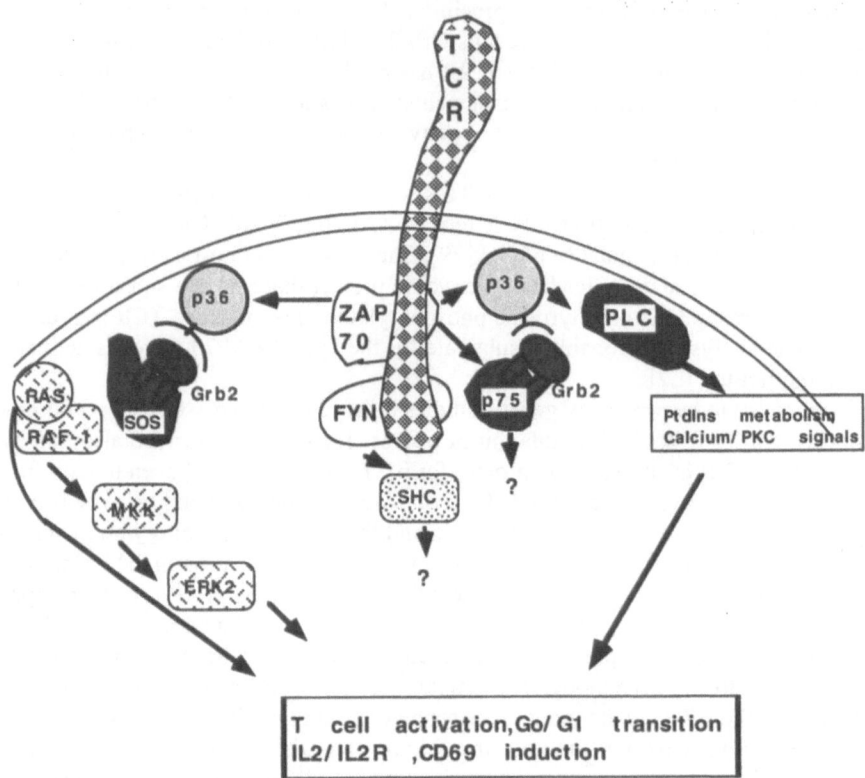

Figure 1. T cell antigen receptor signal transduction.

the calcium phosphatase calcineurin [38]. The conclusion that Ras function is important in T cells is supported further by the observations that expression of N17 ras inhibits TCR induction of the transcriptional factor NFAT-1 and the IL-2 gene [36, 39]. The regulation of lymphokine expression is one important facet of T cell activation but equally important are the signals that regulate the expression and function of lymphokine receptors and adhesion molecules. p21ras may have a role in this aspect of TCR function as demonstrated by observations that Ras regulates the expression of CD69, a T cell activation marker [40].

The transmission of signals from p21^{ras} to the nucleus is proposed to involve the regulation of the activity of Map kinases or ERKs (mitogen activated protein kinases or extracellular signal regulated kinases) [41, 42]. T lymphocytes express at least two MAP kinases ERK1 and ERK2 that are stimulated in response to TCR triggering [43]. Two intracellular pathways for ERK2 regulation co-exist in T cells: one mediated by Ras and the other by PKC [44]. The TCR stimulates both p21ras and PKC but it appears that p21ras and not PKC couples the TCR to the regulation of MAP kinases. The MAP kinases are activated by a kinase cascade involving a MAP kinase kinase (MKK) that phosphorylates and stimulates the ERK1 and 2 kinases directly [42]. The activity of the MKK is itself controlled by phosphorylation and hence a MAP kinase kinase kinase (MKKK) plays a crucial role in the the regulation of ERK1 and 2 [45]. In some cells, Raf-1 has been identified as the MAPKKK that plays a key role in coupling p21ras and hence receptors that stimulate p21ras to the MAP kinases [46, 47]. The N-terminal regulatory domain of the serine/threonine kinase Raf-1 can interact directly with "activated" GTP bound p21ras [48] and it has been shown in T cells, that stimulation of the TCR and PKC induces the formation of a complex

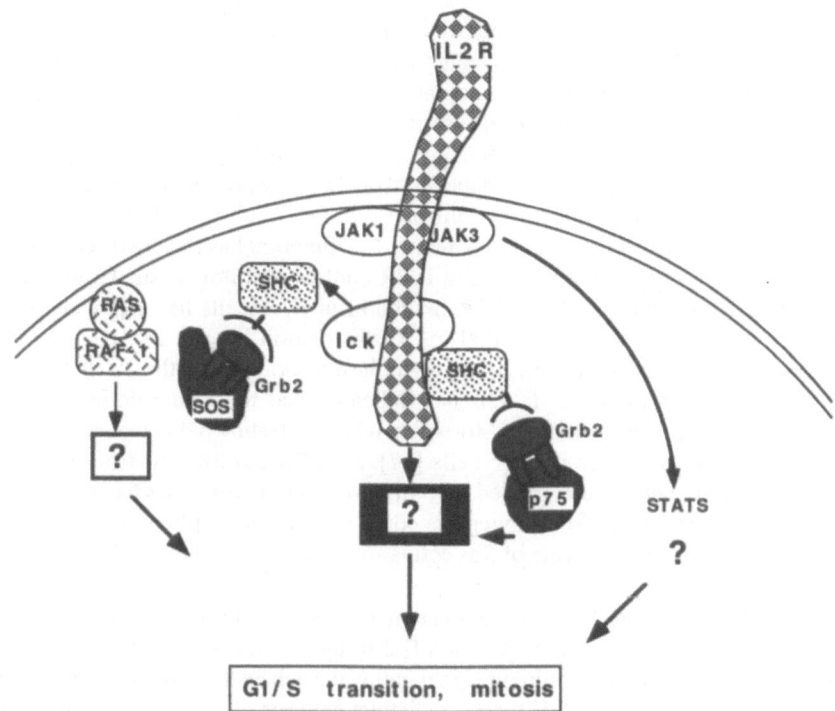

Figure 2. IL2 receptor signal transduction.

of Raf-1 with p21ras [49]. It has also been shown that constitutively active Raf-1 can mimic the effect of activated Ras and stimulate ERK2 and synergise with calcium signals to induce the IL2 gene [47, 50]. These observations collectively suggest that Raf-1 is the effector molecule for p21ras in T cells.

The MAP kinase ERK2 can phosphorylate and regulate transcriptional factors such as c-jun and Elk1[51, 52]. Accordingly, the role of Ras in coupling the TCR to ERK2 and the ability of ERK2 to translocate to the nucleus where it can directly modulate transcriptional factors could explain the role of p21ras in TCR signal transduction.It should be emphasised however that it is not proven that ERK2 is the link between Ras and the induction of lymphokine gene expression. It is known for example that the regulation of c-jun phosphorylation is complicated with both positive and negative phosphorylations that control transcriptional activity.Kinases other than ERKs1,2 can phosphorylate c-jun and hence regulate AP-1. One of these, JNK-1, is activated synergistically by the TCR and CD28. JNK-1 is probably not a Ras regulated kinase but it is a member of the ERK family and it should not be excluded that other Ras regulated ERKs will exist.

In this context it is of interest to note that IL2R triggering is associated with Ras and Raf-1 activation but there are discrepant reports as to whether IL2 stimulation of ERK2 occurs. Recently it has been shown that cell phosphatase activity may be rate limiting for regulation of ERK2 [53] . In particular, a dual specificity phosphatase ,MKP-1,that dephosphorylates ERK2 *in vivo*. thus terminating or preventing its activation has been identified [54]. A related dual specificity phosphatase, PAC-1, is expressed in T cells [55]. The expression of PAC-1 in mitogen activated i.e IL2R expressing T cells probably explains the inability of IL2 to stimulate ERK2. Whether the IL2R activates other ERKs is not known but it is becoming clear that the expression of MAP kinase phosphatases is an important factor that will determine the cellular pattern of ERK activation.

Studies of Ras function in T cells have focused on the role of Ras in T cell activation. Our studies have identified the transcritional factor NFAT-1 as a major target for Ras signals in T cells. NFAT-1 binding sites have been described in the promoters of a number of cytokine genes and Ras may therefore have a general role in controlling cytokine expression in lymphocytes Lymphocyte cell growth is dependent on cytokine production which means that Ras function would be, albeit indirectly, indispensible for T cell growth. In this context it should be emphasised that it is not known whether Ras function has a more direct impact on T cell growth by playing an essential role in IL2R control of mitogenesis. However,structure/function deletion analysis of the IL2R β subunit in BAF cells has suggested that the acidic domain of the β subunit is essential for Ras activation but not for IL2 mitogenesis [56]. However, analogous dissection of the cytoplasmic domain of the IL2R β chain in receptor chimaera experiments in T cells has indicated that the acidic domain is indeed neccessary for Il2R signal transduction. Moreover, it has been shown that that serum derived growth factors can activate Ras in BAF cells [57]. All of the IL2R mutation/ cell growth experiments in BAF cells were carried out in serum containing media and thus any requirement for Ras signaling for mitogenesis could have been supplied by serum derived growth factors. Accordingly, the role of Ras activation in haematopoietic cell growth should be re evaluated.

Finally, figures 1 and 2 summarise the data discussed within the present review and illustrate where Ras is positioned in TCR and IL2 R signal transduction.The ability of the TCR and the IL2R to regulate a common signal pathway such as the Ras pathway may seem discrepant with the observations that there are distinct patterns of gene expression regulated by the TCR and IL2R [58]. However. Ras signaling may be more complicated than depicted at present with the potential for multiple Ras effectors. It is also clear that biochemical signaling pathways specific to either the TCR or the IL2R exist. One well-documented TCR-mediated signaling pathway that is not shared by the IL2R is the activation of phosphatidylinositol hydrolysis, intracellular calcium mobilization and protein kinase C activation [59]. Similarly the IL2R but not the TCR activates a signal transduction pathway which involves Janus kinases (JAKs) and signal transducer and activator of transcription (STAT) proteins [60]. The synergistic interaction between Ras signals and calcium/PKC signals can substitute for the TCR and regulate T cell activation. It remains to be determined whether the interaction between Ras signals and the JAK/STAT pathway control T cell growth.

REFERENCES

1. Smith, K. A., 1988, Interleukin-2: Inception, impact and implications, *Science* 240:1169-1176.
2. Cantrell, D. A. and Smith K. A., 1984, The interleukin-2 T-cell system: a new cell growth model, *Science* 224:1312-1316.
3. Smith, K. A., 1992, Interleukin-2, *Curr. Opin. Immunol.* 4:271-276.
4. Satoh, T., Nafakuku, M., and Kaziro, Y., 1992, Function of Ras as a molecular switch in signal transduction, *J. Biol. Chem.* 267:24149-24152.
5. Graves, J. D., Downward, J., Izquierdo, M., Rayter, S., Warne, P. H., and Cantre ll D. A., 1992, The growth factor interleukin-2 activates p21ras proteins in normal human T lymphocytes, *J. Immunol.* 148:2417-2422.
6. Bourne, H. R., Sanders, D. A., and McCormick, F., 1991, The GTPase superfamily: conserved structure and molecular mechanism, *Nature* 349:117-127.
7. Downward, J., 1992, Regulation of p21^{ras} by GAPs and guanine nucleotide exchange proteins in normal and oncogenic cells, *Curr. Opin. Genet. Dev.* 2:13-18.

8. Bowtell, D., Fu, P., Simon, M., and Senior, P., 1992, Identification of murine homologues of the Drosophila Son of Sevenless gene:potential activators of ras, *Proc. Natl. Acad. Sci. USA,* 89:6511-6515.

9. Buday, L., Egan, S. E., Rodriguez-Viciana, P., Cantrell, D. A., and Downward, J., 1994, A complex of Grb-2 adaptor protein,SOS exchange factor and a 36kDa membrane bound tyrosine phosphoprotein is implicated in Ras activation in T cells, *J. Biol Chem.* 269;9019-9023

10. Weiss, A., 1993, T cell antigen receptor signal transduction: a tale of tails and cytoplasmic protein-tyrosine kinases, *Cell* 73:209-212.

11. Weiss, A., and Littman, D., 1994, Signal transduction by lymphocyte antigen receptors, *Cell* 76:263-274.

12. Reth, M., 1989, Antigen receptor tail clue, *Nature* 338:383-384.

13. Izquierdo, M., Downward, J., Graves J. D., and Cantrell, D. A., 1992, Role of Protein Kinase C in T-cell antigen receptor regulation of p21ras: evidence that two p21ras regulatory pathways coexist in T cells, *Mol. Cell. Biol.* 12:3305-3312.

14. Downward, J., Graves, J. D., Warne, P. H., Rayter, S., and Cantrell, D. A., 1990, Stimulation of p21ras upon T-cell activation, *Nature* 346:719-723.

15. Graves, J. D., Downward, J., Rayter, S., Warne, P., Tutt, A. L., Glennie, M., and Cantrell, D. A., 1991, CD2 antigen mediated activation of the guanine nucleotide binding proteins p21ras in human T lymphocytes, *J. Immunol.* 146:3709-3712.

16. Medema, R. H., Vries-Smits, A. M. M., van der Zon, G. C. M., Maassen, J. A., and Bos, J. L., 1993, Ras activation by insulin and epidermal growth factor through enhanced exchange of guanine nucleotides on p21ras, *Mol. Cell. Biol.* 13:155-162.

17. Lowenstein, E. J. R., Daly, J., Batzer, A. G., Li, W., Margolis, B., Lammers, R., Ullrich, A., and Schlessinger J., 1992, The SH2 and SH3 containing protein Grb-2 links receptor tyrosine kinases to ras signaling, *Cell* 70:431-442.

18. Pellici, G., Lanfrancone, L. Grignani,F., Mcglade, J., Cavallo, F., Forni, G. Nicoletti, I. Grignani, F., Pawson, T., Pellici, P. G.1992, A novel transforming protein Shc with an SH2 domain is implicated in mitogenic signal transduction, *Cell* 70:93-104.

19. Rozakis-Adcock, M., Fernley, R., Wade, S., Pawson, T., and Bowtell, D., 1993, The SH2 and SH3 domains of mammalian Grb-2 couple the EGF receptor to the the Ras activator mSOS, *Nature* 363:83-85.

20. Egan, S. E., Giddings, B. W., Brooks, M. W., Buday, L., Sizeland, A. M., and Weinberg, R. E., 1993, Association of SOS Ras exchange protein with GRB2 is implicated in tyrosine kinase signal transduction and transformation, *Nature* 363:45-51.

21. Buday, L., and Downward, J., 1993, Epidermal growth factor regulates p21ras through the formation of a complex of receptor, Grb2 adapter protein and Sos nucleotide exchange factor, *Cell* 73:611-620.

22. McCormick, F., 1994, Activators and effectors of ras p21 proteins, *Curr. Biol.* 4:71-76.

23. Ravichandran, K. S., Lee, K. K., Songyang, Z., Cantley, L. C., Burn, P., and Burakoff, S. J., 1993, Interaction of Shc with the ζ chain of the T cell receptor upon T cell activation, *Science* 262:902-905.

24. Reif, K., Buday, L., Downward, J., and Cantrell, D. A., 1994, SH3 domains of the adapter molecule Grb-2 complex with two proteins in T cells: The guanine nucleotide exchange protein SOS and a 75 kDa protein that is a substrate for T cell antigen receptor activated tyrosine kinases, *J. Biol. Chem.* 269:14081-14089.

25. Osman,.N.,Lucas,S.L.,Cantrell,D.A. 1994 The role of tyrosine phosphorylation in the interaction of cellular tyrosine kinases and adapter molecules with the T cell receptor ζchain tyrosine-based activation motif. *EMBO.J.* 1994, submitted.

26. Ravichandran, K. S., and Burakoff, S. J., 1994, The adapter protein Shc interacts with the interleukin-2 (IL-2) receptor upon IL-2 stimulation, *J. Biol. Chem.* 269:1599-1602.

27. Weiss, A., 1993, T cell antigen receptor signal transduction: a tail of tails and cytoplasmic protein-tyrosine kinases, *Cell* 73:209-212.

28. Klausner, R. D., and Samelson L. E., 1991, T cell antigen receptor activation pathways: the tyrosine kinase connection, *Cell* 64:875-878.

29. Minami, Y., Kono, T., Yamada, K., Kobayashi, N., Kawahara, A., Perlmutter, R. M., and Taniguchi, T., 1993, Association of p56^{lck} with IL-2 receptor β chain is critical for the IL-2-induced activation of p56^{lck}, *EMBO J.* 12:759-768.

30. Margolis, B., Hu, P., Katzav, S., Li, W., Oliver, J. M., Ullrich, A., Weiss, A., and Schlessinger J., 1992, Tyrosine phosphorylation of vav proto-oncogene product containing SH2 domain and transcription factor motifs, *Nature* 356:71-74.

31. Gulbins, E., Coggeshall, K. M., Baier, G., Katzav, S., Burn, P., and Altman, A., 1993, Tyrosine kinase stimulated guanine nucleotide exchange of Vav in T cell activation, *Science* 260:822-825.

32. Adams, J. M., Houston, H., Allen, J., Lints, T., and Harvey, R., 1992, The hematopoietically expressed vav proto-oncogene shares homology with the dbl GDP-GTP exchange factor, the bcr gene and a yeast gene (CDC24) involved in cytoskeletal organization, *Oncogene* 7:611-618.

33. Nunes, J., Truneh, A., Colette, Y., Olive, D., and Cantrell, D. A., 1994, CD28 triggering with antibodies but not the ligand B7-1 activated p21ras, *J. Exp. Med.* (In Press).

34. Izquierdo, M., Downward, J., Leonard, W. J., Otani, H., and Cantrell D. A., 1992, IL-2 activation of p21ras in murine myeloid cells transfected with human IL-2 receptor beta chain, *Eur. J. Immunol.* 22:817-821.

35. Baldari, C. T., Macchia, G., and Telford, J. L., 1992, Interleukin-2 promoter activation in T cells expressing activated Ha-ras, *J. Biol. Chem.* 267:4289-4291.

36. Rayter, S., Woodrow, M., Lucas, S. C., Cantrell, D., and Downward, J., 1992, p21ras mediates control of IL2 gene promoter function in T cell activation, *EMBO. J.* 11:4549-4556.

37. Baldari, C. T., Heguy, A., and Telford, J. L., 1993, Calcium-dependent cyclosporin A-sensitive activation of the interleukin-2 promoter by p56^{lck}*J.Biol.Chem.* 268:8406-8409.

38. Woodrow, M., Clipstone, N., and Cantrell, D. A., 1993, p21^{ras} and calcineurin synergise to regulate NFAT, *J. Exp. Med.* 178:1517-1522.

39. Woodrow, M., Rayter, S., Downward, J., and Cantrell, D. A., 1993, p21ras function is important for T cell antigen receptor and protein kinase C regulation of nuclear factor of activated cells, *J. Immunol.* 150:1-9.

40. D'Ambrosia, D., Cantrell, D. A., Frati, L., Santoni, A., Testi, R., . 1994, Involvement of p21ras in T cell CD69 expression, *Eur. J. Immunol.* 24:616-620.

41. Leevers, S. J., and Marshall, C. J., 1992, MAP kinase regulation-the oncogene conection, *Trends Cell. Biol.* 2:283-286.

42. Pelech, S. L., and Sanghera, J. S., 1992, MAP kinases: charting the regulatory pathways, *Science* 257:1355-1356.

43. Whitehurst, C. E., Boulton, T. G., Cobb, M. H., and Geppert, T. G., 1992, Extracellular signal-regulated kinases in T cells. Anti-CD3 and 4beta-Phorbol 12-Myristate 13-Acetate-induced phosphorylation and activation, *J. Immunol.* 148:3230-3237.

44. Izquierdo, M., Leevers, S. J., Marshall, C. J., and Cantrell, D. A., 1993, p21^{ras} couples the T cell antigen receptor to extracellular signal-regulated kinase 2 in T lymphocytes, *J. Exp. Med.* 178:1199-1208.

45. Ahn, N. G., Seger, R., and Krebs, E. G., 1992, The mitogen activated protein kinase activator, *Curr. Opin. Cell. Biol.* 4:992-999.

46. Howe, L. R., Leevers, S. J., Gomez, N., Nakielney, S., Cohen, P., and Marshall, C. J., 1992, Activation of the MAP kinase pathway by the protein kinase raf, *Cell* 71:335-342.

47. Izquierdo, M., Bowden, S., and Cantrell, D., 1994, The role of Raf-1 in the regulation of Extracellular signal-regulated kinase 2 by the T cell antigen receptor, *J. Immunol.* (in press).

48. Koide, H., Satoh, T., Nakafuka, M., and Kaziro, Y., 1993, GTP-dependent association of Raf-1 with Ha-Ras: identification of Raf as a target downstream of Ras in mammalian cells, *Proc. Natl. Acad. Sci.* 90:8683-8686.

49. Halleberg, B., Rayter, S., and Downward, J., 1994, Interaction of Ras and Raf in Intact mammalian cells upon extracellular stimulation, *J. Biol. Chem.* 269: 3913-3916

50. Owaki, H., Varma, R., Gillis, B., Bruder, J. T., Rapp, U. R., Davis, L. S., and Geppert, T. D., 1993, Raf-1 is required for T cell Il2 production, *EMBO J.* 12:4367-4373.

51. Hunter, T., and Karin, M., 1992, The regulation of transcription by phosphorylation, *Cell* 70:375-387.

52. Marais, R., Wynne, J., and Treisman, R., 1993, The SRF accesory protein Elk-1 contains a growth-factor-regulated transcriptional activation domain, *Cell* 73:381-393.

53. Samuels, M. L., Weber, M. J., Bishop, M. J., and McMahon, M., 1993, Conditional transformation of cells and rapid activation of the mitogen- activated Protein kinase cascade by an estradial dependent Human Raf-1 protein kinase, *Mol. Cell. Biol.* 13:6241-6252.

54. Sun., H., Charles, C. H., Lau, L. F., and Tonks, N. K., 1993, MKP-1 (3CH134), an immediate early gene product is a dual specificity phosphatase that dephosphorylates MAP kinase in vivo, *Cell* 75:487-493.

55. Rohan, P. J., Davis, P., Moskaluk, C. A., Kearns, M., Krutsch, H., Siebenlist, U., and Kelly, K., 1993, Pac-1: a mitogen induced nuclear protein tyrosine phosphatase, *Science* 259:1763-1766.

56. Satoh, T., Minami, Y., Kono, T., Yamada, K., Kawahara, A., Taniguchi, T., and Kaziro, Y., 1992, Interleukin 2-induced activation of ras requires two domains of interleukin 2 receptor β subunit, the essential region for growth stimulation and lck-binding domain, *J. Biol. Chem.* 267:25423-25427.

57. Sato, N., Sakamaki, K., Terada, N., Arai, K.-I., and Miyajima, A., 1993, Signal transduction by the high affinity GM-CSF receptor: two distinct cytoplasmic regions of the common β subunit responsible for different signalling, *EMBO J.* 12:4181-4189.

58. Beadling, C., and Smith, K., 1994, In search of cytokine-response genes, *Immunol. Today* 15:197-199.

59. Mills, G. B., Stewart, D. J., Mellors, A., and Gelfand, E. W. 1986, Interleukin-2 does not induce phosphatidylinositol hydrolysis in activated T cells, *J. Immunol.* 134:2431-2435.

60. Beadling, C., Guiscik, D. Schindler, C. Zieymeki, A. Kerr, I., and Cantrell, D. 1994, Interleukin 2 but not the T cell antigen receptor stimulates the JAK/STAT pathway, *EMBO J.* (Submitted).

DISCUSSION

P. Pelicci

Did you try to measure the amount of the soluble Shc Grb-2 complex? The majority of Shc proteins are phosphorylated but only a minority binds with receptors, while 80% of Shc molecules are free in the cytoplasm, phosphorylated and complexed to Grb-2. So my question is, in the IL-2 receptor system, whether you measured that complex and whether you know if it is associated with Grb-2.

D. Cantrell

I think the only thing we know about the IL-2 receptor complex is that you get a good Shc, Grb-2, SOS complex and probably, if we purify Grb-2 from a T cell, we can co-purify about 40% of the total cellular Shc in an IL-2 receptor activated cell. Whether that complex is bound to any receptor or free in the cytoplasm I have no idea. It has been reported that Shc can bind to the IL-2 receptor's cytoplasmic domain. Now certainly a fusion protein of the Shc SH2 domain can. I am not so sure at the moment how strong the evidence is that there is a good and stable complex forming between the IL-2 receptor and Shc. I should say we do all these experiments under very stringent washing conditions of high salt, so we are only maintaining very stable complexes. And we have never actually seen a Shc IL-2 receptor complex in our hands, although it is confusing because we do see Shc complexing to a 75 kDa tyrosine phosphoprotein that sort of looks a bit like the IL-2 receptor beta chain the way it runs on a gel, but it is not because it does not Western blot with IL-2 receptor beta chain antibodies.

P. Pelicci

What about the T cell receptor?

D. Cantrell

First of all, never can we find Shc binding to the T cell receptor *in vivo*. Then if we purify the Grb-2 out of a T cell, under conditions where we are purifying all the Grb-2, we really find it very difficult to detect Shc ever in a Grb-2 complex, any Grb-2 complex in a T cell receptor activated cell. It does not make a lot of sense to me why we fail to see a Shc/Grb-2 complex. It is not that Grb-2 is limiting. There is a huge amount of Grb-2 in T cells, so there should be a lot of free access sites even if p36 was binding to some of the Grb-2 SH2 domain. However, the level of Shc tyrosine phosphorylation is very, very low in a T cell receptor activated cell. I do not know what sites are phosphorylated, whether there is a potential for more than one site. I have not really started to consider that yet, but we just do not see Shc playing a role in that complex, so Shc may be playing a role in some other complex in T cell receptor triggering, but not, I think, the Grb-2 complex. Well we do not see it in our cells. This is normal human peripheral blood T cells. The only cell it has ever been seen in, I think, is Jurkat cells, some Jurkat cell lines but in peripheral blood T cells we just do not see ShcGrb-2. We have p36/Grb-2 complexes in our cells.

J. Ihle

You probably have not had a chance to do it, but I want to ask the question anyway. Have you mapped the domains on the IL-2 receptor that are required for Shc activation?

D. Cantrell

No we have not.

T. Taniguchi

I would assume that this is a so-called acidic domain which is required for the activation of lck.

D. Cantrell

We would assume so as well, but we have not actually done the experiment.

T. Taniguchi

Do you have any thoughts on how this RAS pathway is involved in IL-2? So far we have some indirect evidence that it is involved in T cell proliferation.

D. Cantrell

We do not have any evidence that it is involved. All our functional experiments so far have focused on the role of Ras in T cell activation. We have found it very difficult to introduce RAS mutants into IL-2 responsive T cells lines. We tried that to see, for example, if the dominant negative RAS mutant would block IL-2 driven proliferation and the experiments failed for technical reasons. We really could never successfully express high enough levels of n17 RAS in the T cells to feel that we could successfully block the RAS pathways. The data was negative but technically flawed, so I would say that we have no idea. It is complicated because there are a lot of discrepancies about what IL-2 receptor regulation of RAS is doing. I mean one discrepancy for example, is: Does Ras activate map kinase or not? Because as I am sure you know, in the literature there is a huge discrepancy as to whether IL-2 activates Map kinase or not.

T. Taniguchi

We have done some experiments showing this RAS activation is related to activation of MAP kinase.

D. Cantrell

I wonder whether that is cell line dependent. For example, in peripheral blood derived T cells, which are the cells we work on, if you make them IL-2 receptor positive, you can go to a situation where they express the T cell receptor and the IL-2 receptor. The level of RAS activation you get is identical whether you trigger the T cell receptor or the IL-2 receptor. RAF activation is identical. IL-2 is not really activating Map kinase under those conditions whereas the T cell receptor is. I was very impressed by Kathy Kelly's work on Map kinase phosphatases, where she shows that activated T cells can express a lot of Map kinase phosphatases and possibly they are de-sensitizing selectively the IL-2 pathway. That would be the idea we might have for that. I really hope that RAS does have a role in IL-2 receptor signaling and I think it is just going to be a case of finding it. But when it really comes down to the evidence that RAS is involved in growth factor signaling, it is actually not that good. It is involved in transformation. That is clear and it is clearly involved in cell activation but it does not necessarily have to be involved directly in mitogenesis, so I have an open mind. The other point is that IL-2 does not only regulate mitogenesis. It regulates other effector functions in T cells: cytotoxicity, and in NK cells for example, possibly the RAS regulation is involved in some aspects of the other effector functions of IL-2, not the mitogenic ones.

J. Adams

My question is probably just an extension of what you have been saying. I wondered just how solid the evidence was that RAS was a vital component of even the activation steps.

D. Cantrell

In cell lines there is very good evidence that it is very important in cell activation and Roger has very good data in transgenic mice, that RAS is essential for cytokine production of mature peripheral T cells, or mature T cells. So, I think the data's fairly solid that it is important for cytokine production but both Roger and I have only looked at cytokine production which focused on IL-2 production. There are a lot of different cytokines that T cell receptors make. We have not looked at those yet. Our study on RAS regulation of the IL-2 gene has indicated that it is really the NFAT site in the IL-2 gene that is the target for the RAS signals in the mature T cells. NFAT sites have been found in the promoters of a number of different cytokines and so possibly that is going to reflect that RAS regulates other cytokines in the hematopoietic cell system. But that has not been looked at yet.

H. Beug

Did you ever look whether the protein kinase C becomes tyrosine phosphorylated because it would probably get recruited to the plasma membrane by the interaction you looked at?

D. Cantrell

No, we are never seen tyrosine phosphorylation of protein kinase C. I do not think we have ever really looked for tyrosine phosphorylation of protein kinase C.

H. Beug

I asked this question because I saw described somewhere that PKC became tyrosine phosphorylated in response to receptor tyrosine kinases after ligand stimulation.

D. Cantrell

We have never looked at that in our system at all.

J. Griffin

Another group has implicated VAV in the activation of RAS, specifically in T cells. Do you have any thoughts there?

D. Cantrell

I guess the only thoughts I have about that are that the original model for VAV regulation of RAS exchange was that it had to be tyrosine phosphorylated to get regulation of RAS exchange. If we just take that idea, let alone get into the discussion as to whether VAV really is a RAS exchange protein, then there is absolutely no correlation in the T cell system between VAV tyrosine phosphorylation and RAS activation. IL-2 in our hands does not induce tyrosine phosphorylation of VAV, at least not to any significant extent but does activate RAS. T cell receptor does. Not very well: very low stoichiometry, very transient tyrosine phosphorylation. CD28 is probably the most powerful receptor for inducing VAV tyrosine phosphorylation. CD28 does not activate RAS at all. No correlation. There is no genetic evidence as far as I know that VAV is a RAS exchange protein in the fibroblast systems where people put oncogenic VAV and it is transforming. In contrast, there is pretty good genetic evidence in all sorts of systems that SOS is a RAS exchange protein, so I think at the moment there is no real evidence supporting that the VAV phenomena that are described *in vitro* really will mean anything *in vivo*.

D. Baltimore

Actually I was going to ask a similar question from a different point of view. Do you see VAV involved in interactions in any of these immuno precipitates?

D. Cantrell

Do we find them? VAV binds very effectively to the Grb-2 SH3 domain *in vitro*. We never see VAV complexing to Grb-2 *in vivo*.

D. Baltimore

We have seen a little bit, but have basically the same sort of impression. *In vitro* it is a very beautiful association.

D. Cantrell

It complexes very well, but so is PI-3 kinase. We actually found P75 because we were chasing PI-3 kinase association with Grb-2 which I think is an *in vitro* artifact and not really happening. Well, not in T cells, anyway. The most thing that strikes us about VAV is from some work we have done where VAV is very heavily tyrosine phosphorylated when you activate CD28. It is stably tyrosine phosphorylated; there is no RAS activation at all but clearly CD28 activates PI-3 kinase. Somehow if you put VAV, PI-3 kinase together and start thinking about interacting proteins, then I think you could begin to speculate as to some model for a VAV PI-3 kinase link.

D. Baltimore

The other thing I wanted to ask you was: The peptide that you use, the proline rich peptide. That must completely displace SOS from Grb-2.

D. Cantrell

It does, very effectively, because you could bring down all the Grb-2 out of the cells, but if you Western blot the cells, there is no SOS there. I mean we are using it at 25 micromolar concentration, so it is a pretty high concentration.

D. Baltimore

Would you have it bound to a bead?

D. Cantrell

Bound to a bead. It is one of the proline sequences in SOS that was defined as the SH3 interaction. I forget which one because there are two. There is only one of them that we have used.

D. Baltimore

The other thing I wanted to say is that it was my impression from just listening to people that nobody really has identified a receptor Shc Grb-2 complex of any kind.

D. Cantrell

That is my feeling as well. Our evidence only tells the immunologists who were saying that there was a complex, that there is not one. One is always trying to satisfy people who have different perceptions of what signaling complexes form. You would find in most cellular immunology groups, that they would draw a Shc, Grb-2, SOS complex. I just do not think it exists.

A. D'Andrea

Actually, along the lines of Dr. Baltimore's question, when you perform far Western blot with the GST Grb-2 (SH3) and you identify two proteins; M-SOS and P75, will P75 be competed by an excess of peptide?

D. Cantrell

Which peptide?

A. D'Andrea

For instance, a proline-rich peptide.

D. Cantrell

Yes, you can compete it.

A. D'Andrea

This might suggest that P75 is related to M-SOS. It is not a degradation product?

D. Cantrell

The experiment just suggests that it needs proline rich peptides to bind to the Grb-2 SH3 domain.

A. D'Andrea

Is p75 immuno reactive with an antibody to M-SOS?

D. Cantrell

No, it is not at all. I should say we have gotten other evidence that it is definitely not SOS. If you take the isolated Grb-2 SH3 domain, N terminal and the C terminal, one of them preferentially binds SOS and that does not bind P75 very well. The other one binds P75 very well, but not SOS. They are almost reciprocal. I do not think that is what is happening in the cell, but *in vitro* that is what happens. We have seen P75 in mast cells, B cells and T cells, that is all we have looked at so far. We have never seen it in fibroblast cells or in epithelial cells. So it may be a hematopoietic specific molecule and an interesting one that does something important.

R. Perlmutter

I guess the comment on that last point is that the peptide binding p75 and the peptide at the same time are incompatible. It does not even necessarily say that they work by the same mechanism because you have already shown that you have a situation where one will displace another at a distance. But my question is: p36, is this related to the p36 molecules that have been described by others that are tyrosine phosphorylated following T cell receptor activation. Some time ago one was described that was thought to be GTP binding. One can think of a mechanism whereby one might have seen that, given the associations you have described.

D. Cantrell

I really do not know about those. I guess the only one it does seem related to is the one that was identified as binding to the phospholipase C SH2 domain, I think by Ledbetters and subsequently by Art Weiss' group.

R. Perlmutter

What is the reason why you believe that one is related?

D. Cantrell

Well, I have not read Art's paper properly. First of all, I think they have done 2D gels showing that the 36kDa protein one co-purifies with phospholipase C Gamma in T cells and with Grb-2 seem to run in identical positions on 2D gel. I think they have also gotten co-precipitation data showing that they can actually pull together the whole complex of PLCγ, p36 and SOS.

R. Perlmutter

There was at one time of course a 36kDa protein that was described very early on in this activation that was tyrosine phosphorylated.

D. Cantrell

It is not that, I know that. There probably are about 4 or 5 36kDa tyrosine phospho proteins in T cell receptor activated cells. This is one of them. Which one, we do not know. We will not know until we purify it and get a sequence for it. And it is turning out to be very difficult to purify for whatever reasons. You would think with this Grb-2 fusion protein you would have a good way of pulling it out of cells but it has not been that straightforward.

P. Pelicci

I would like to comment on the receptor Shc/Grb-2 complex. Basically we did the experiment in the context of the erbB2 receptor. If you use an erbB2 receptor mutant for the Grb-2 binding site, it is very clear that there is no Grb-2 binding *in vitro*. That receptor still bind Grb-2 *in vivo*. However, you do not see any SOS. If you do use an erbB2 receptor with all 5 phosphotyrosine, you can isolate a complex Shc/Grb-2/SOS. It appears that when Shc is bound to the receptor, it does not bind SOS, while when it is associated with Grb-2 in the cytoplasm, it does.

D. Cantrell

Yes, that is interesting. We have just recently been using a peptide from track A that binds to Shc, SH2 domain, very efficiently in T cells to follow this point up further and with that peptide we can get Shc Grb- 2 but never Shc Grb- 2 SOS. Although I must say that is still in IL-2 receptor activated cells. We have great difficulty seeing Shc Grb- 2 in a T cell receptor activated cell.

D. Baltimore

Do you think there is an allosteric change there?

D. Cantrell

It has to be. I think you can get it both when you bind different things to the Grb-2 SH3 domain. I should say actually that using the fusion protein type approach to look at Grb-2/p36, the kinetics of p36 phosphorylation that you can monitor that way and then the association of p36 and Grb-2 in the cell do not actually follow, suggesting that it is not simply p36 tyrosine phosphorylation that is controlling the association.

R. Perlmutter

Could it be a matter of kinetics?

D. Cantrell

The up kinetics are the same. It is the down kinetics that are different. The complex disappears before the tyrosine phosphorylated p36 is gone. So I do not think it is that.

A. D'Andrea

You introduced your talk saying that there are some similarities and some differences between IL-2 and T cell receptor signaling. One could envision, for instance, a dominant inhibitory form of Shc which might preferentially block IL-2 receptor signaling but not T cell receptor signaling. Are there ways that one could actually dissect one pathway away from the others.

D. Cantrell

I would predict that a dominant negative form of Shc would block IL-2 signaling but not T cell receptor signaling.

A. D'Andrea

So that is a testable hypothesis?

D. Cantrell

That is a testable hypothesis and on that I believe those groups are working on.

P. Pelicci

We used as a dominant negative Shc just the SH2 domain. We have tried to express the SH2 domain and it was tested in a transient assay transfecting Jurkat cells and measuring the activation of an NFAT based promoter. In that context, the IL-2 stimulation of the CAT activity is blocked by expression of the SH2 domain of Shc and it was not blocked by, for example, the SH2 domain. So it seems to be FES specific.

D. Cantrell

I thought that was interesting because it suggested that wherever you have gotten an lck Shc may be involved, which would be where the IL-2 link came in, but with a T cell receptor. That is the difference actually. In peripheral blood T cells, the T cell receptor, if you just ligate it alone, does not activate lck. It does in some cell lines like Jerkitt but not in peripheral blood T cells.

8

IRF-1 FUNCTIONS AS A TUMOR SUPPRESSOR

Possible Involvement in Human Myelodysplasia and Leukemia

Tadatsugu Taniguchi,[1,2] Nobuyuki Tanaka,[1] Hisashi Harada,[1]
Masahiko Ishihara,[1] Takeshi Kondo,[1] Motoo Kitagawa,[1,2] Tohru Kimura,[1]
Marc S. Lamphier,[1] Tomohiko Tamura,[1] Toshifumi Matsuyama,[3]
Tak W. Mak,[3] and Hisamaru Hirai[4]

[1] Institute for Molecular and Cellular Biology, Osaka University
Suita-shi, Osaka 565, Japan
[2] Department of Immunology, The University of Tokyo Medical School
Bunkyo-ku, Tokyo 113, Japan
[3] Amgen Institute, Ontario Cancer Institute
Toronto, Ontario M4X, Canada
[4] The Third Department of Internal Medicine, The University of Tokyo
Bunkyo-ku, Tokyo 113, Japan

INTRODUCTION

Malignant cell transformation is a multistep process resulting from the progressive acquisition of structural alterations at multiple genetic loci which are involved in the regulation of cell growth. It has been well documented that gain-of-function mutations, found in dominantly-acting proto-oncogenes, are often accompanied by loss-of-function mutations in tumor suppressor genes in human malignant cells.

An interstitial deletion of the long arm of chromosome 5(del(5q) or loss of a whole chromosome 5 (-5 or monosomy 5) are among the most frequent recurrent cytogenetic abnormalities in human leukemia and the preleukemic myelodysplastic syndromes (myelodysplasia; MDS). Del(5q) or monosomy 5 is found in 30% of patients with MDS, in 50% of patients with secondary or therapy-induced acute myelogenous leukemia (AML), and in 15% and 2% of patients with de novo AML and de novo acute lymphocytic leukemia (ALL), respectively[1,2,3,4,5,6]. The del(5q) was first described as the hallmark of a unique myelodysplastic syndrome (the "5q- Syndrome") occurring predominantly in elderly females that is characterized by refractory anemia, thrombocytosis, and abnormal megakaryocytes[1]. Females with this syndrome usually have an indolent clinical course; the affected myeloid stem cell clone appears to have a slow capacity for expansion, acquires additional cytogenetic abnormalities only infrequently, and transforms to AML in only 10-20% of

Normal and Malignant Hematopoiesis, Edited by Enrico Mihich and Donald Metcalf
Plenum Press, New York, 1995

cases[2,4,7]. In contrast, patients who present with de novo or secondary AML with del(5q) usually have additional cytogenetic abnormalities at presentation and a very poor prognosis[3,8,9]. In AML, the presence of a del(5q)/-5 has also been associated with occupational exposure to carcinogens[10,11] or with previous exposure to alkylating agent chemotherapy or radiotherapy for the treatment of various malignancies[3].

A series of studies have revealed that the smallest commonly deleted segment of the del(5q), the so called "critical" region, lies in band 5q31[12,13]. Rare de novo AMLs with translocations involving 5q31 have also been described[14]. These findings suggest that the causative gene(s) lies in 5q31 and that deletion of this gene(s) may be central to the pathogenesis of leukemia and MDS.

The two structurally related transcription factors, IRF-1 and IRF-2, were originally identified as regulators of the interferon (IFN) system[15,16,17]. IRF-1 has also been identified by others in different contexts[18]. IRF-1 functions as a transcriptional activator whereas IRF-2 represses the effect of IRF-1 by competing for binding to the same DNA sequence elements (IRF-Es)[18]. IRF-Es can be found in both the *IFN-α* and *IFN-β* promoters, as well as in IFN-stimulated response elements (ISREs) found within the promoters of IFN-inducible genes[19].

One of the well-documented functions of IFNs is the inhibition of cell growth. The IRF-1 gene is constitutively expressed in a variety of cells at low levels, and it is further induced by IFN stimulation[16]. Hence, this gene may be one of the target genes critical for the antiproliferative action of IFNs. In synchronized NIH3T3 cells, constitutive IRF-1 mRNA expression is elevated when cells are arrested by serum starvation, declines sharply following serum stimulation and gradually increases thereafter[17]. Since IRF-2 mRNA levels remain constant, the IRF-1/IRF-2 expression ratio appears to oscillate throughout the cell cycle. The IFN stimulation of the cells would result in a transient increase the IRF-1/IRF-2 ratio.

In fact, alterations of the expression balance between the two genes cause profound effects on cell growth; overexpression of IRF-1 strongly inhibits cell proliferation, whereas overexpression of IRF-2 results in the oncogenic transformation of NIH3T3 cells. Furthermore concomitant overexpression of IRF-1 reverts these cells to a non-transformed phenotype, suggesting the anti-oncogenic potential of IRF-1[17]. Previously, *IRF-1* gene was found to be consistently deleted at one or both alleles in 13 cases of leukemia or MDS with 5q31 aberrations[20].

In order to assess further the function of IRF-1 as a tumor suppressor, we introduced IRF-1 cDNA in oncogene-transformed cell lines. In addition, we generated mice with a null mutation in the IRF-1 gene (IRF-1[-/-] mice) and examined the sensitivity of the embryonic fibroblasts (EFs) from these mice to transformation by an oncogene. We also examined the expression patterns of the IRF-1 mRNA in patients with MDS and MDS-derived leukemias.

RESULTS AND DISCUSSION

Reversion of Oncogene-Transformed Cells by IRF-1

The retrovirus pDGIRF1, which expresses murine IRF-1 upon infection, was constructed[18]. c-*myc*-transformed 3Y1 or *fosB*-transformed Rat-1A cells, termed RmycY1 and Rat-1A(FosB), respectively, both show anchorage-independent growth and enhanced tumorigenicity in rat or nude mice[21,22]. 5x105 of RmycY1 and Rat-1A(FosB) cells were each infected by the pDGIRF1 virus at the multiplicity of infection (MOI) of 10 and subsequently diluted to obtain single clones. We observed 1 or 2 weeks later that these cell clones had lost their transformed phenotype. In fact, the morphologies of these virally infected cells had changed to become indistinguishable from the original non-transformed cells, and had lost their tumorigenic potential in nude mice and colony formation in methylcellulose gel. In contrast, no such change was observed following similar infection by the control retrovirus pGD . Thus these observations clearly indicate that IRF-1 functions broadly as a tumor suppressor in cells transformed by at least two different oncogenes. We also observed that the levels of oncogene messenger RNAs expressed in these reverted cells remained essentially the same as the original, transformed cells (data not shown). Therefore, the observed phenotypic reversion of the cells as the result of *IRF-1* gene expression is not due to the inhibition of the promoters used to express each oncogene. We further tested the anti-oncogenic function of IRF-1 in activated *ras* or v-*abl* transformed NIH 3T3 cells, but no morphological change was observed (data not shown). These results demonstrate that the anti-oncogenic function of IRF-1 is limited with respect to the kind of oncogene. This suggests that IRF-1 may function in some specific cascades of cell growth regulation.

Inactivation of the IRF-1 Gene Sensitises Primary Mouse EFs to Transformation by Oncogenic c-Ha-ras

A powerful approach to assess the tumor suppressive function of a given gene is to create model animals by gene targeting, thus enabling the gene defect to be examined in the context of cell transformation and tumorigenesis.In this regard, mice lacking the functional *IRF-1* alleles grow to at least one year without any spontaneous development of tumors[23]. However, in view of the importance of multiple genetic events for the development of tumors *in vitro* and *in vivo*, it is conceivable that the IRF-1 deficient cells may manifest properties different from wild type cells with respect to their susceptibility to transformation by oncogenes.

In this regard, mutations in N-*ras*, c-Ki-*ras* or c-Ha-*ras* are found rather frequently in human leukemias and MDS[24,25]. Hence we asked if a defect in the *IRF-1* gene may make cells susceptible to *ras*.oncogene.

In order to examine the effect of the *IRF* null mutations on cell growth and transformation, EFs were infected with a retrovirus, pGDV12ras at multiplicity of infection (MOI) of 10. pGDV12ras directs the expression of an activated c-Ha-*ras* cDNA under the control of an viral LTR, and the *ras* protein carries an oncogenic mutation which converts Gly12 to Val. These cells showed no marked difference in their growth rates in serum-containing medium (doubling time is approximately 20 hrs, data not shown). However, three independent preparations of the *ras*-expressing IRF-1[-/-] EFs displayed anchorage-independent growth; they formed colonies in methylcellulose gel with an efficiency of about 30%, whereas virtually no colony formation was seen with the IRF-1[-/-]EFs without the *ras* gene[28]. Furthermore, when 2×10^6 of these rasEF11, 21, 31 cells were injected subcutaneously into nude mice, tumors developed within two weeks[28]. In contrast, wild type and IRF-2[-/-] EFs displayed no such properties even though *ras* gene was expressed at similar levels, an observation consistent with previous reports showing that neoplastic transformation of primary cells requires an additional onco-gene[26,27] (Figure 1). rasEF11, rasEF21, and rasEF31 cells. When these cells were infected with a retrovirus (pGDIRF1) that directs the expression of mouse IRF-1, the colony-forming potential of the cells in methylcellulose gel was markedly reduced. Thus the *ras*-induced transformation can be reverted in a significant population of these cells by the expression of IRF-1.

Our results presented here provide compelling evidence that IRF-1 functions as tumor suppressor; its functional loss predisposes the primary mouse EFs to tumorigenic transfor-

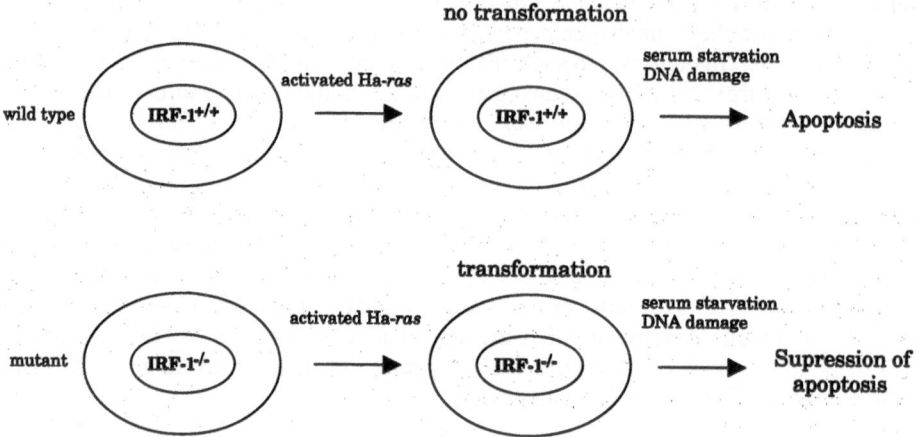

Figure 1. Cellular committment to ras-induced transformation or apoptosis is dependent on IRF-1. (See text and Ref. 28 for the details.)

mation by a *ras*-oncogene *in vitro*. The combined effects of the inactivation of the *IRF-1* alleles and activation of the c-Ha-*ras* proto-oncogene described here may provide another genetic context in which to study the biochemical interactions involved in normal growth control as well as the mechanisms underlying multistep tumorigenesis.

In addition, we found that expression of the oncogenic *ras* causes wild-type but not IRF-1-/- EFs to undergo apoptosis upon serum starvation or treatment by anticancer drugs (Adriamycin, 5-Fluorouracil) or ionizing radiation[28](Figure 1). Therefore, IRF-1 may be a critical determinant of oncogene-induced cell transformation or apoptosis.

Exon Skipping of IRF-1 mRNA in Human MDS and Leukemia

The human *IRF-1* gene consists of 10 exons, with the initiator ATG sequence located within the second exon. When we analyzed *IRF-1* mRNA expression in various cell lines by a polymerase chain reaction (PCR) assay, three bands corresponding to intact *IRF-1* mRNA, *IRF-1* mRNA lacking exon 2 (Δ2mRNA), and *IRF-1* mRNA lacking exons 2 and 3 (Δ23mRNA) were detected using the RNA from the peripheral blood mononuclear (PBM) cells (see Fig.1). This observation prompted us to examine the status of exon skipping in cells from patients with MDS or leukemia secondary to MDS. Total RNA was isolated from either bone marrow (BM) cells or PBM cells of the above patients and subjected to RT-PCR analysis.

Notably, significant number of patients gave very low levels of the intact *IRF-1* mRNA product, whereas the Δ2 and Δ23 products, particularly the Δ23 product, were still detectable. Among them, only Pt.356 showed cytogenetic aberrations in the 5q region (data not shown). Co-transfection of the Δ2 or Δ23 cDNAs with the intact *IRF-1* cDNA did not affect IRF-1-mediated gene activation, suggesting that neither IRF-1Δ2 nor IRF-1Δ23 acts in a dominant-negative manner on IRF-1-mediated transcriptional activation.

In the samples in which expression of intact *IRF-1* mRNA is low or undetectable, the Δ23 mRNA is nevertheless expressed at relatively high levels. Hence we addressed the issue of whether the IRF-1Δ23 manifests tumor suppressor activity. We then found that transfection of the cDNA for IRF-1 resulted in a profound inhibition of colony formation, whereas no such effect was seen with the cDNA for IRF-1Δ23. These results indicate that IRF-1Δ23 lacks tumor suppressor activity, presumably due to the loss of its DNA binding activity.

Inspection of the nucleotide sequence of the exons and introns of this gene (H. Harada, unpublished data) reveals some intriguing sequence elements; (i) the sequence of intron 1 is notably CG-rich near the splice donor site, a feature which may affect splicing, possibly by creating intrastrand base parings in the pre-mRNA, (ii) the polypyrimidine tract which is usually found near the splice acceptor site of intron is interrupted by G residues in intron 2. Determining how these sequences affect splicing requires further studies.

In the present study, 6 out of 20 samples showed accelerated exon skipping, the result of which should be a dramatic diminution or loss of functional IRF-1 expression. Thus far, RT-PCR/SSCP analysis of these samples did not reveal any DNA deletions or mutations within *IRF-1* exon sequences, nor did DNA sequencing reveal any alterations of intron sequences known to affect splicing, i.e. splicing donor, acceptor sites, branching sites and polypyrimidine tracts. In addition, Southern blotting analysis of the DNAs from these patients showed no sign of gross rearrangements of the *IRF-1* gene. Although we cannot strictly rule out the possibility that a mutation(s) causing exon skipping may have occurred within a region(s) which has escaped our analysis, the results presented here collectively suggest a possible alternative mechanism for tumor suppressor inactivation; dysfunction of the cell's splicing machinery which selectively affects the frequency of exon skipping in the *IRF-1* gene leading to diminution of IRF-1 function (Figure 2). In addition to the previous report, indicating the deletion and rearrangement of the IRF-1 gene in MDS/leukemia[20] (Figure 2), our present study may further point to the involvement of *IRF-1* in the development of these hematopoietic malignancies.

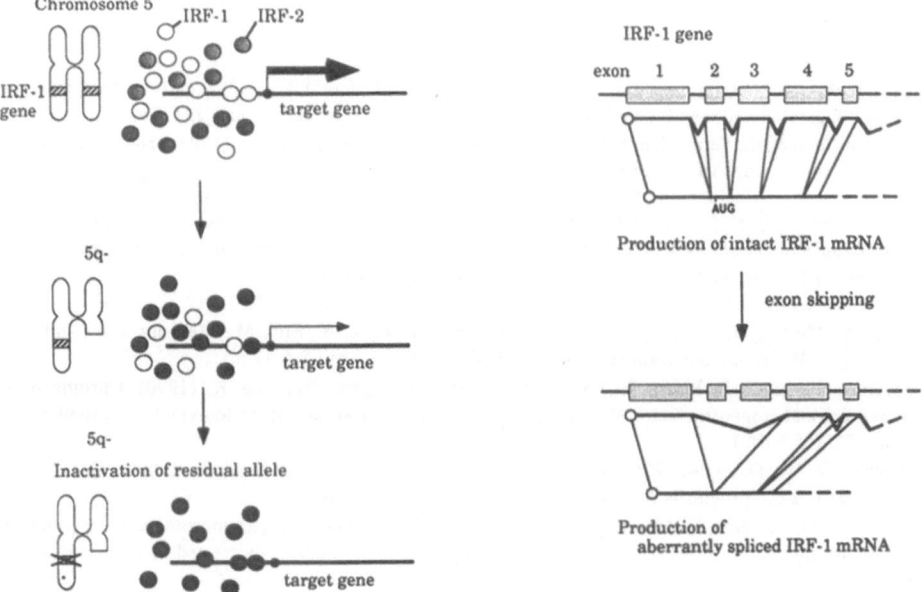

Figure 2. Two possible mechanisms for the loss of IRF-1 function. (A) Gene deletion or inactivating rearrangements would result in the loss of IRF-1 synthesis 20. (B) Accerelated exon skipping occurs at a relatively high frequency in MDS/leukemia, and the aberrantly spliced mRNAs no more encode functioned IRF-1 for tumor suppression.

CONCLUSIONS AND PERSPECTIVES

The results presented here provide compelling evidence that IRF-1 functions as a tumor suppressor, and suggest that the loss of IRF-1 expression by *IRF-1* gene deletion/re-arrangement and/or by aberrant splicing of IRF-1 pre-mRNA may be a step critical for the development of MDS and leukemia. Work is in progress to examine the effect of IRF-1 deficiency on in vivo development of leukemia by introducing various oncogenes into bone marrow cells from IRF-1$^{-/-}$ mice. Our recent study also indicates that IRF-1 is involved in mediating cell death or apoptosis. In fact, our results suggest that IRF-1 functions to cause the death of oncogene-activated cells, a function which could be important in the elimination of pre-transformed cells *in vivo* . Thus it is very interesting and important to identify the target gene(s) of IRF-1. It has been shown that IRF-1$^{-/-}$ mice are sensitive to mycobacterial infection[29]. In addition, we have recently found that IFN-induced resistance to a virus (encephalomyocarditis virus) is severely impaired in IRF-1$^{-/-}$ EFs[30]. Taken together, IRF-1 appears to be a unique transcription factor, manifesting antibacterial, antiviral and antitumor functions.

ACKNOWLEDGMENTS

We thank Ms. Y. Kasakawa for her assistance. This work was supported in part by a grant-in-aid for Special Project Research, Cancer Bioscience, from Ministry of Education, Science and Culture of Japan, and Takeda Science Foundation.

REFERENCES

1. Van den Berghe, H., Cassiman, J. J., David, G., Fryns, J. P., Michaux, J. L., and Sokal, G. (1974). Distinct haematological disorder with deletion of long arm of no. 5 chromosome. Nature *251*, 437.
2. Van den Berghe, H., Vermaelen, K., Mecucci, C., Barbieri, D., and Tricot, G. (1985). The 5q- anomaly. Cancer Genet. Cytogenet. *17*, 189-255.
3. Le Beau, M. M., Albain, K. S., Larson, R. A., Vardiman, J. W., Davis, E. M., Blough, R. R., Golomb, H. M., and Rowley, J. D. (1986b). Clinical and cytogenetic correlations in 63 patients with therapy-related myelodysplastic syndromes and acute onlymphocytic leukemia: Further evidence for characteristic abnormalities of chromosomes no. 5 and 7. J. Clin. Oncol. *4*, 325-345.
4. Nimer, S. D., and Golde, D. W. (1987). The 5q- Abnormality. Blood *70*, 1705-1712.
5. Kerim S., Mecucci. C., Cuneo, A., Vandenberghe, E., Louwagie, A., Stul, M., Michaux, J. L., and Van den Berghe, H. (1990). 5q- anomaly in lymphoid disorders. Leukemia *4*, 12-15.
6. Pedersen-Bjergaard, J., Philip, P., Larsen, S. O., Jensen, G., and Byrsting, K. (1990). Chromosome aberrations and prognostic factors in therapy-related myelodysplasia and acute nonlymphocytic leukemia. Blood *76*, 1083-1091.
7. Dewald, G. W., Davis, M. P., Pierre, R. V., O'Fallon, J. R., and Hoagland, H. C. (1985). Clinical characteristics and prognosis of 50 patients with a myeloproliferative
8. Rowley, J. D., Golomb, H. M., and Vardiman, J. W. (1981). Nonrandom chromosome abnormalities in acute leukemia and dysmyelopoietic syndromes in patients with previously treated malignant disease. Blood *58*, 759-767.
9. Samuels, B. L., Larson, R. A., Le Beau, M. M., Daly, K. M., Bitter, M. A., Vardiman, J. abnormalities in acute nonlymphocytic leukemia correlate with drug susceptibility in vivo. Leukemia *2*, 79-83.
10. Mitelman, F., Brandt, L., and Nilsson, P. G. (1978). Relation among occupational exposure to potential mutagenic/carcinogenic agents, clinical findings, and bone marrow chromosomes in acute nonlympho-cytic leukemia.
11. Golomb, H. M., Alimena, G., Rowley, J. D., Vardiman, J. W., Testa, J. R., and Sovik, C. (1982). Correlation of occupation and karyotype in adults with acute nonlymphocytic leukemia. Blood *60*, 404-411.

12. Le Beau, M. M., Lemons, R. S., Espinosa III, R., Larson, R. A., Arai, N., and Rowley, J. D. (1989). Interleukin-4 and interleukin-5 map to human chromosome 5 in a regionencoding growth factors and receptors and are deleted in myeloid leukemias with a del(5q). Blood 73, 647-650.

13. Pederson, B., and Jensen, I. M. (1991). Clinical and prognostic implications of chromosome 5q deletions: 96 high resolution studied patients. Leukemia 5, 566-573.

14. Fourth International Workshop on Chromosomes in Leukemia (1982).

15. Miyamoto, M., Fujita, T., Kimura, Y., Maruyama, M., Harada, H., Sudo, Y., Miyata, T., and Taniguchi, T. (1988). Regulated expression of a gene encoding a nuclear factor, IRF-1, that specifically binds to IFN-β gene regulatory elements. Cell 54, 903-913.

16. Harada, H., Fujita, T., Miyamoto, M., Kimura, Y., Maruyama, M., Furia, A., Miyata, T., and Taniguchi, T. (1989). Structurally similar but functionally distinct factors, IRF-1 and IRF-2, bind to the same regulatory elements of IFN and IFN-inducible genes. Cell 58, 729-739.

17. Tanaka, N., and Taniguchi, T. (1992). Cytokine gene regulation: Regulatory cis-elements and DNA binding factors involved in the interferon system. Adv. Immunol.52, 263-281.

18. Harada, H., Kitagawa, M., Tanaka, N., Yamamoto, H., Harada, K., Ishihara, M., and Taniguchi, T. (1993). Anti-oncogenic and oncogenic potentials of interferonregulatory factors-1 and -2. Science 259, 971-974.

19. Tanaka, N., Kawakami, T., and Taniguchi, T. (1993). Recognition DNA sequences of interferon regulatory factor 1 (IRF-1) and IRF-2, regulators of cell growth and the interferon system. Mol. Cell. Biol. 13, 4531-4538.

20. Willman, C. L., Sever, C. E., Pallavicini, M. G., Harada, H., Tanaka, N., Slovak, M. L., Yamamoto, H., Harada, K., Meeker, T. C., List, A. F., and Taniguchi, T. (1993). Deletion of IRF-1, mapping to chromosome 5q31.1, in human leukemia and preleukemic myelodysplasia. Science 259, 968-971.

21. Shiroki, K., Sera, K., Koita, Y. and Shibuya, M. (1986) Neoplastic transformation of rat 3Y1 cells by a transcriptionally activated human c-myc gene and stabilization of p53cellular tumor antigen in the transformed cells. Mol. Cell. Biol., 6, 4379-4386.

22. Nakabeppu, Y., Oda, S. and Sekiguchi, M. (1993) Proliferative activation of quiescent Rat-1A cells by ΔFosB. Mol. Cell. Biol., 13, 4157-4166

23. Matsuyama, T., Kimura, T., Kitagawa, M., Pfeffer, K., Kawakami, T., Watanabe, N.,Kündig, T. M., Amakawa, R., Kishihara, K., Wakeham, A., Potter, J., Furlonger, C.L., Narendran, A., Suzuki, H., Ohashi, P. S., Paige, C. J., Taniguchi, T., and Mak, T. W. (1993). Targeted disruption of IRF-1 or IRF-2 results in abnormal type I IFN gene induction and aberrant lymphocyte development. Cell 75, 83-97.

24. Bos, J. L. (1989). ras oncogenes in human cancer: A review. Cancer Res. 49, 4682-4689.

25. Carter, G., Rigde, S., and Padua, R. A. (1992). Genetic lesions in preleukemia. Crit. Rev. Oncog. 3, 339-364.

26. Land, H., Parada, L. F., and Weinberg, R. A. (1983). Tumorigenic conversion of primary embryo fibroblasts requires at least two cooperating oncogenes. Nature 304, 596-602.

27. Ruley, H. E. (1990). Transforming collaborations between ras and nuclear oncogenes. Cancer Cells 2, 258-268.

28. Tanaka, N., Ishihara, M., Kitagawa, M., Harada, H., Kimura, T., Matsuyama, T., Lamphier, M. S., Aizawa, S., Mak, T. W., and Taniguchi, T. (1994). Cellular commitment to oncogene-induced transformation or apoptosis is dependent on the transcription factor IRF-1. Cell, in press.

29. Kamijo, R., Harada, H., Matsuyama, T., Bosland, M., Gerecitano, J., Shapiro, D., Le, J., Koh, S. I., Kimura, T., Green, S. J., Mak, T. W., Taniguchi, T., and Vilcek, J. (1994). Requirement for transcription factor IRF-1 in NO synthase induction in macrophages. Science 263, 1612-1615.

30. Kimura, T., Nakayama, K., Penninger, J., Kitagawa, M., Harada, H., Matsuyama, T., Tanaka, N., Kamijo, R., Vilcek, J., Mak, T. W., and Taniguchi, T., (1994). Involvement of the IRF-1 transcription factor in antiviral responses to interferons. Science, in press.

DISCUSSION

J. Ihle

The IRF-1 and IRF-2 genes are structurally very related to p48 which is involved in the IFN-α response complex in which they participate with STAT1 and STAT2 to form the transcriptional activation complex. My question is: Have you looked at the complexes in which IRF-1 and IRF-2 participate for comparable proteins?

T. Taniguchi

No, we cannot see any complex formation with IRF-1 and IRF-2 with other factors as far as we have examined.

J. Ihle

A second unrelated question. In the IRF-1 knockout mice, have you taken bone marrow cells, cultured in IL-3, to look at apoptosis. The prediction would be that they would not undergo apoptosis at the same rate as normal cells.

T. Taniguchi

Those are still ongoing experiments.

H. Beug

Did you observe any increased genetic instability or a more easy formation of immortalized cell lines in the IRF knockout mice?

T. Taniguchi

We have not done very systematically these kinds of relevant experiments. I cannot answer your question.

I. Weissman

In the IRF-1 knockout mice, if you look at thymocytes, for example, and ask the question: Do other treatments that induce apoptosis like glucocordicoids or x-ray fail to induce apoptosis in these cells?

T. Taniguchi

Yes, we have examined the induction of apoptosis of thymocytes. Unlike p53 knockout cells, the apoptosis occurs perfectly there. By the way, I have to mention that there are some abnormalities in T cell developments and the CD8+ T cells are depleted, or not depleted completely but they are very low compared to viral types, for whatever reason.

I. Weissman

In that depletion, the partial depletion of CD8+, single positive cells, are there remaining cells that are CD8+, CD3-, are the early progenitor that remain?

T. Taniguchi

They remain normal. Double positive cells are also normal.

I. Weissman

Do you know if they have the transitional intermediate between the double positive and CD8 only?

T. Taniguchi

I do not know.

D. Baltimore

Have you tried making double transgenic mice with RAS, activated RAS oncogenes or myc or any other oncogenes?

T. Taniguchi

We are doing this now. The mouse is born but it is only about one month old and so it takes some time to see any abnormalities. It is a very interesting experiment.

E. Mihich

In these knockout mice, Dr. Weissman asked about the cortisone and x-ray, but did you try TNF by any chance, because TNF also can induce apoptosis in the thymus through mechanisms that are not necessarily the same as those of hydrocortisone.

T. Taniguchi

Very interesting experiment. We have not tried but I must say that TNF is a very good inducer of IRF-1 genes. So it is an interesting possibility. We do not know the mechanism by which TNF induces IRF-1 genes, but it can be tested.

A. D'Andrea

For the IRF-1 knockout cells, when you treat them with ionizing radiation, do you get induction of p53?

T. Taniguchi

p53 expression is normal.

A. D'Andrea

Is the kinetics of induction normal as well?

T. Taniguchi

As far as we can tell, there is no difference.

A. D'Andrea

This phenotype is very similar to ataxia telengiectasia. When you treat AT cells with ionizing radiation, the cells do not arrest at G-1; they continue to progress. Perhaps there is a delay of p53 induction.

T. Taniguchi

We are focusing on this, as you know, and we do not have complete answers to those questions.

A. D'Andrea

In the BF3 cells that overexpress IRF-2, there is a delay in apoptosis; in fact, they are transformed.

T. Taniguchi

Yes, it is like bcl-2 overexpression.

A. D'Andrea

Do they also overexpress bcl-2?

T. Taniguchi

In those cells no, but if we overexpress bcl-2 in these cells, apoptosis of course is prevented.

A. D'Andrea

Did you look at bcl-2 expression in the IRF-2 transformed cells?

T. Taniguchi

We are currently examining the bcl-2 expression. It is an interesting point.

I. Weissman

You were saying that bcl-2 overexpression prevents the apoptosis that you were just talking about. What happens in the glucocorticoid induced thymic lymphocyte apoptosis which you said still occurs in the knockout in the absence of IRF-1. The only pathway that we have found not inhibitable with bcl-2 is target of cytotoxic lymphocyte killing by apoptosis. That was not bcl-2 inhibitable. I wonder if your IRF-1 knockout cells are resistant or sensitive to apoptosis induced by killer T lymphocytes.

T. Taniguchi

We have not examined this, it is an interesting experiment to do. Thank You.

R. Perlmutter

Can IRF-1 and IRF-2 heterodymerize?

T. Taniguchi

We have no evidence for this.

R. Perlmutter

The site for IRF-1 is potentially a dymer binding site, is it not? So in addition to direct competition by IRF-1 and IRF-2, in principle, those two proteins could interact.

T. Taniguchi

One cannot exclude that but we have never found it.

R. Perlmutter

So you believe that the relationship between these two proteins is regulated largely by the transcription of the two genes and there is nothing beyond that?

T. Taniguchi

We cannot exclude the possibility of other factors affecting the IRF's function at present.

R. Perlmutter

Right, but in terms of regulating IRF-1 and IRF-2 activity, you just control how much IRF-1 and IRF-2 protein is made by controlling transcription. So what is the half life of those two proteins.

T. Taniguchi

That is an interesting point. IRF-2 has a long half life of about 6 hours whereas IRF-1 has a very short half-life of about 30 minutes in virus infected cells. We have not made that in normally growing cells but presumably the situation is very similar.

R. Perlmutter

But in the basal state, the bias favors IRF-2, promoting growth.

T. Taniguchi

Presumably, in certain stages of the cell cycle, IRF-1 activates some fatal genes which negatively regulate cell growth. But the question of how this functions to displace IRF-2 is not clear at present.

D. Livingston

Is there a specific point in G-1 at which IRF-1 interrupts the cell cycle? And is that true in cycling cells as well as cells that are driven out of G-0?

T. Taniguchi

We are still working on this. I cannot tell you exactly at which point but perhaps it is at G-1.

S. Landolfo

Do you think that IRF-1 and ISGF-3 are mutually exclusive and if your answer is yes, what is the signal that decides which one is activated after interferon binding to the receptor?

T. Taniguchi

It is possible that initial stages of transcription are sustained or activated by IS GF-3 and they are also sustained by IRF-1 for example. That cannot be excluded. In fact the data with GBP gene induction argues for this possibility, that both proteins are required.

S. Landolfo

But for the l'-5'A synthetase there is evidence that other factors beside ISGF-3 are implied. The work of I. Tam is saying that C1, C6 and so on are responsible for 2'-5'A synthetase activation.

T. Taniguchi

Which transcription factor binds nearby the ISRE might determine the dependency of the transcription factor, either IRF-1 or IS GF-3, or possibly other factors. It is a little complicated but an interesting issue to be addressed.

I. Weissman

In the apoptosis pathway, the ced-3 homologue is supposed to be interleukin 1 Beta convertase? Is there an IRF-1 or IRF-2 site in front of Interleukin 1 Beta convertase.

T. Taniguchi

We are testing this now.

J. Ihle

Of course interferon is a very strong inducer of IRF-1 expression and as you are well aware there is a very mixed literature that says that interferon suppresses normal hematopoiesis and apparently it suppresses proliferation in very specific windows, so the question is: Have you looked at the ability of interferon to suppress bone marrow stem cell development from the IRF-1 knockout mice. Presumably one would not see suppression of growth in those windows.

T. Taniguchi

When we treat the ras expressing fibroblasts by interferon gamma, which would result in the augmentation of IRF-1, apoptosis does not occur, so the situation is a little bit more complicated. I am not saying that IRF-1 has no role in interferon mediated cell growth regulation but the mechanism is perhaps more complex and IRF-1 might be a part of the mechanism by which interferons inhibit cell proliferation, but there might be some other mechanism which we do not know yet.

C. Schneider

Have you looked at the level of BAX expression in conjunction with Bcl-2 analysis?

T. Taniguchi

We are now looking at it.

9

A FAMILY OF HIGH MOLECULAR WEIGHT PROTEINS ACTIVE IN DIFFERENTIATION AND GROWTH CONTROL

Richard Eckner, Zoltan Arany, and David M. Livingston

Dana-Farber Cancer Institute and
Harvard Medical School
Boston, Massachusetts 02115

INTRODUCTION

When adenoviruses infect an epithelial cell arrested in G0/G1, they must induce host cell DNA synthesis and cell division or risk failure of propagation. The E1A gene participates actively in the replication induction process, and its two products have been well studied in this regard (Dyson and Harlow, 1992).

Three sequence units in these proteins contribute to the induction process, and, in all cases, they operate, at least in part, by serving as specific binding sites for a limited number of cellular proteins. Genetic analyses of both the protein binding sites and the phenotypical consequences of producing mutated E1A versions in infected cells have shown that certain interactions result in specific biological effects (reviewed in Dyson and Harlow, 1992; Nevins, 1992; Moran, 1993).

In particular, E1A interacts with a nuclear protein of high molecular weight, termed p300. The genetics indicate that this interaction is key to release an infected cell from G0 and for subsequent cell division (Howe et al., 1990; Stein et al., 1990). In addition, E1A represses certain transcriptional enhancers (Borelli et al., 1984; Velcich and Ziff, 1985), including several involved in specific pathways of differentiation (Hen et al., 1985; Webster et al., 1988; Stein et al., 1990). Genetic studies indicate that the E1A-p300 interaction is fundamental to these differentiation perturbing effects (Mymrik et al., 1992; Boulukos and Ziff, 1993). Thus, it can be argued that p300 is a protein involved in the maintenance of cell cycle arrest and in the promotion of certain pathways of differentiation, most probably via a transcriptional control route. How p300 operates in these two processes is now under investigation, facilitated by the recent cloning of its cDNA (Eckner et al., 1994).

Normal and Malignant Hematopoiesis, Edited by Enrico Mihich and Donald Metcalf
Plenum Press, New York, 1995

RESULTS

Structure of Human p300

The protein is the product of a cDNA which was cloned from human cDNA libraries (Eckner et al., 1994). It is composed of 2414 residues. Several discrete and striking sequence motifs were apparent from a comparative analysis of its primary structure (Figure 1A). First, there are three, well-separated cysteine/histidine (C/H)-rich regions. The most N-terminal of them can be drawn to fit the consensus sequence of a Zn finger structure (Harrison, 1992). The other two are more complex. Just N-terminal to the second of these units is a perfect bromodomain (Figure 1B). These structural units have now been identified in at least 20 different proteins, all of which share an ability to influence transcription, without the requirement of specific recognition of a cognate DNA sequence (Haynes et al., 1992; Tamkun et al., 1992; Winston and Carlson, 1992). That p300 also contains such a motif suggests that it, too, has such a property, i.e. can serve a transcriptional coactivator function. In keeping with this notion, p300 contains an additional short region of homology to another transcriptional adaptor protein, the yeast ADA2 protein (Berger et al., 1992). This 30 amino acid motif is located within the third C/H rich region (Figure 1A and 1C).

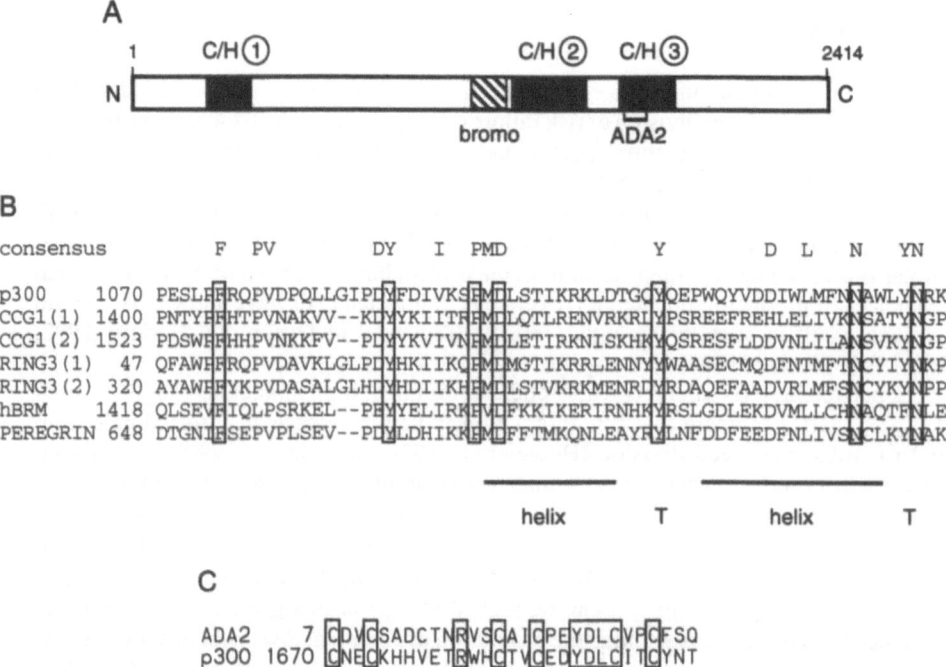

Figure 1. Domain structure of p300 and sequence comparison of the bromodomain and ada2 motifs. A) The relative locations of the three cysteine/histidine rich (C/H) regions is indicated. Embedded in the third C/H rich region is a segment of homology to the yeast ADA2 protein. B) Sequence comparison of bromodomain motifs contained in human proteins. The residues that are strictly conserved in all seven bromodomains are boxed; those that are present in at least six of the seven proteins are included in the consensus sequence (top line). In their C-terminal halves, bromodomains are likely to exhibit two putative α-helical regions (H) followed by reverse turns (T) (Haynes et al., 1992), as indicated at the bottom. C) Comparison of the region of similarity between human p300 and yeast ADA2. (Reprinted, with permission, from Eckner et al., 1994, copyright by Cold Spring Harbor Laboratory Press).

Aside from these structural features, there were no homologies to other members of the E1A binding protein family, i.e. cyclins, cdk species, or pocket proteins.

E1a Binding by Cloned p300

When the cloned product was transfected into suitable cell lines, it was shown to be capable of binding E1A, and the genetics of its interaction with the viral protein resembled those of the endogenous polypeptide (Eckner et al., 1994). Moreover, when various truncated derivatives of p300 were analyzed for E1A binding, it was possible to show that the third C/H rich region serves as the minimal E1A binding sequence of the protein. It, alone, is capable of binding E1A specifically.

p300 Can Sustain the Action of the SV40 Enhancer

With the foreknowledge that E1A binding to a protein(s) through its p300 binding domain led to inactivation of the SV40 enhancer, there was a strong indication for testing the hypothesis that p300 supports the action of this element. In this regard, we introduced a reporter carrying the SV40 enhancer/promoter linked to a luciferase gene, along with an E1A-encoding expression plasmid. This combination led to marked repression of the reporter (Figure 2A). However, when either of two mutants of p300, each lacking the ability to bind E1A due to a deletion in the third C/H rich domain, were co-transfected with the SV40-luciferase reporter and E1A, the inhibitory effect of E1A was significantly blunted (Figure 2B). Both p300 mutants (p300del-30 and del-33) were as stable as wildtype p300 and were localized to the nucleus (Eckner et al., 1994).

These results strongly suggest that p300 can sustain action of the SV40 enhancer and is a major target of E1A function. Furthermore, the prediction, from the primary sequence analyses, that p300 is a coactivator was strongly reinforced by these results.

p300 Is a Member of a Family of Structurally Related Proteins

When the data base of protein sequences was consulted, it became clear that, in several discrete areas, human p300 is closely related in primary sequence to a protein termed CBP (Figure 3 shows a schematic representation). CBP is a coactivator for CREB, a transcription factor known to deliver cAMP-mediated nuclear signals (reviewed in Lalli and Sassone-Corsi, 1994). Recently, CBP has been shown to form complexes in vitro with CREB on a traditional CRE (Chrivia et al., 1993), and these complexes are transcritpionally active (Kwok et al., 1994). Indeed, CBP has intrinsic activation function, and its homology to p300 is concentrated in each of the four structurally interesting areas of p300 noted above. These include the three C/H rich domains and the bromodomain. Within each of these regions, p300 (human) and CBP (murine) are >85% identical. The spacing of these segments is also identical in the two proteins. A partial cDNA clone of murine p300 has recently been isolated in our laboratory (Arany et al., 1994), and sequence analysis of it indicates that murine p300 and murine CBP are products of different genes.

In addition, our data base searches revealed a close homologue of p300 and CBP in the invertebrate, C. elegans (Figure 3). The C. elegans protein sequence was identified during the sequencing of chromosome III of this organism (Wilson et al., 1994). Again, there was marked preservation of p300/CBP sequence within the C/H rich units which were present in the C. elegans clone, and, from the limited information available, the spacing of these units appears to be virtually identical to that noted in CBP and p300.

These results indicate that there is a well-conserved family of related eukaryotic proteins of apparently high molecular weight, which includes CBP, p300, and the C. elegans

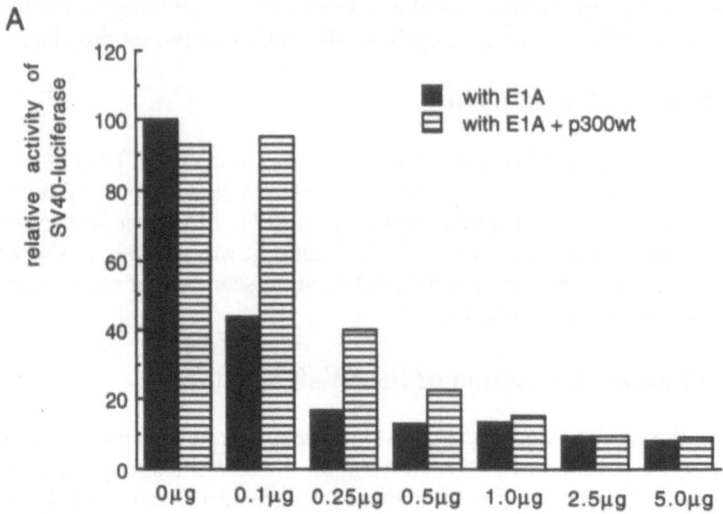

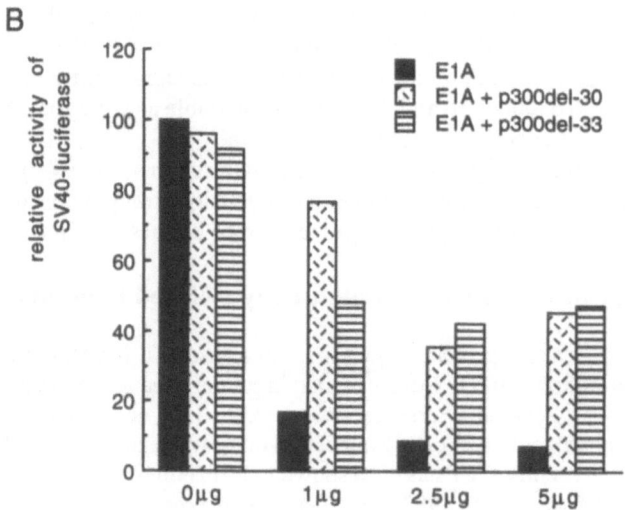

Figure 2. p300 can bypass E1A-dependent repression of the SV40 enhancer. A) Wildtype p300 can only overcome the repressive effect of low concentrations of E1A. B) p300 mutants bearing a deletion in the E1A binding region, thereby abolishing their ability to bind E1A (p300del-30 or p300del-33), can overcome E1A-mediated repression of the SV40 enhancer, even at high concentrations of an E1A-encoding expression plasmid. (Reprinted, with permission, from Eckner et al., 1994, copyright by Cold Spring Harbor Laboratory Press).

homologue. The size of the family is presently unclear. Other major questions, presently unanswered, are whether CBP is an E1A binding protein, and, if so, whether E1A binding affects its ability to carry out CREB-directed transcription signalling. Similarly, it is important to know whether p300 is a CBEB/ATF binding protein. In both cases, the underlying question is whether E1A interrupts certain cAMP-initiated signals. Most impor-

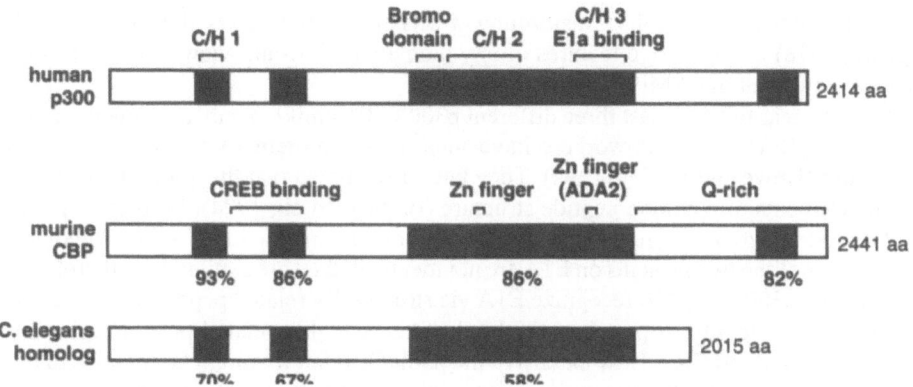

Figure 3. An evolutionarily conserved family of p300-related proteins. Schematic representation of human p300, mouse CBP and a C. elegans homolog. The percentage of identical amino acid residues relative to human p300 is indicated below each region of high homology (shaded boxes). The designation of the individual domains of p300 and CBP is given above the shaded boxes and follows the names given in the respective publications describing the cloning of the two proteins (Eckner et al., 1994; Chrivia et al., 1993). Abbreviations: C/H, region rich in cysteine and histidine residues; Zn finger, zinc finger; Q-rich, glutamine-rich domain. (Reprinted, with permission, from Arany et al., 1994, copyright by Cell Press).

tant in this regard are signals which lead to growth arrest, promotion of differentiation, and suppression of transformation-maintenance by activated ras. Clearly, it will be important to know whether cAMP signalling is a major E1A target, and, if so, how powerfully it contributes to normal growth control in cells in which E1A has a significant transforming effect.

DISCUSSION

That E1A can bring about profound changes in resting cells and efficiently block various pathways of differentiation can now be traced, at least in part, to interactions with one or more members of the p300/CBP family. The latter appear to be transcription factors, but not of the variety which specifically recognize one or another canonical DNA sequence. Rather, the available evidence implies that they depend upon specific interactions with other proteins, through which they are recruited to selected promoters. At least one member of a specific family of such factors (the CREB family) forms specific complexes with CBP to deliver its transcription modulating signal(s) (Chrivia et al., 1993). In keeping with the notion that p300 is a coactivator, as opposed to a specific DNA binding protein with activating properties (Rikitake and Moran, 1992), we were unable to simplify a wholly random oligonucleotide into discernible specific progeny after exposure to cloned p300 protein followed by PCR amplification of the bound sequences (our unpublished results). Hence, at least two members of this family may not be specific DNA binding proteins, just as has been argued for another set of transcription coactivators, the TAFs (TATA-binding protein associated factors) (Tijan and Maniatis, 1994).

Does p300 also contribute to delivery of cAMP signals, like CBP? The question is now under investigation, as is the possibility that CBP binds E1A and that E1A has a significant functional effect on CBP function. Positive results in either case would elucidate the mechanism by which E1A interferes with certain aspects of cAMP signalling (Kalvakolanu et al., 1992). That cAMP is known to contribute to the growth controlling systems of cultured cells and to neutralize the transforming effects of an activated ras allele (Pastan and Willingham, 1978) are attractive features of any model which would view a transforming protein as a perturbant of cAMP signalling.

There appear to be at least three different p300/CBP family members in mammalian cells. Recently, Bayley and co-workers have identified a protein of ~400 Kd in E1A co-precipitates (Howe and Bayley, 1992). They have also shown that this protein and p300 share some elements of common peptide structure (Barbeau et al., 1994). Moreover, p300 and p400 bind to overlapping regions in the N-terminal 80 amino acids of E1A. Although the binding sites of the two proteins on E1A are not identical, they are similar. It is, therefore, conceivable that p300 and p400 recognize E1A via structurally related peptide motifs, as is the case for the Rb family of proteins which all utilize a highly related pocket region in binding to E1A CR1 and CR2. How extensive the p300/CBP family might be remains to be determined, e.g. from in situ chromosomal hybridization analyses. Moreover, given the growth regulating and differentiation promoting properties of p300, one wonders whether any member(s) of this family are the products of tumor suppressor genes.

Finally, there is evidence suggesting that cAMP can affect the growth of human hematopoietic cells (Johnson et al., 1988; Vairo et al., 1990). Hence, it is reasonable to suspect that the p300/CBP family plays a role in the normal development of certain hematopoietic lineages and/or in the functioning of their progeny. Questions like these may well be settled by analyzing the behavior of mice which lack function of one or more members of the p300/CBP family. Experiments of this type are now beginning. Similarly, from an analysis of the genomic locations of each family member could come insights into the potential for specific disease production by altered or lost function of one or more of these proteins.

REFERENCES

Arany, Z., Sellers, W.R., Livingston, D.M., and Eckner, R, 1994, E1A-associated p300 and CREB-associated CBP belong to a conserved family of coactivators, *Cell* 55:799.

Barbeau, D., Charbonneau, R., Whalen, S.G., Bayley, S.T., and Branton, P.E., 1994, Functional interactions within adenovirus E1A protein complexes, *Oncogene* 9:359.

Berger, S.L., Pina, B., Silverman, N., Marcus, G.A., Agapite, J., Regier, J.L., Triezenberg, S.J., and Guarente, L., 1992, Genetic Isolation of ADA2: a potential transcriptional adaptor required for function of certain acidic activation domains, *Cell* 70:251.

Borelli, E., Hen, R., and Chambon, P., 1984, Adenovirus-2 E1A products repress enhancer-induced stimulation of transcription, *Nature* 312:608.

Boulukos, K.E., and Ziff, E.B., 1993, Adenovirus 5 E1A proteins disrupt the neuronal phenotype and growth factor responsiveness of PC12 cells by a conserved region 1-dependent mechanism, *Oncogene* 8:237.

Chrivia, J.C., Kwok, R.P.S., Lamb, N., Hagiwara, M., Montminy, M.R., and Goodman, R.H., 1993, Phospo-rylated CREB binds specifically to the nuclear protein CBP, *Nature* 365:855.

Dyson, N., and Harlow, E., 1992, Adenovirus E1A targets key regulators of cell proliferation, in: Cancer Surveys Volume 12:pp. 161-195. Tumour Suppressor Genes, the Cell Cycle and Cancer, A. Levine, ed. Cold Spring Harbor Laboratory Press, Cold Spring Harbor, New York.

Eckner, R., Ewen, M.E., Newsome, D., Gerdes, M., DeCaprio, J.A., Lawrence, J.B., and Livingston, D.M, 1994, Molecular cloning and functional analysis of the adenovirus E1A-associated 300-kD protein (p300) reveals a protein with properties of a transcriptional adaptor, *Genes & Dev.* 8:869.

Harrison, S.C, 1991, A structural taxonomy of DNA-binding domains. *Nature* 353:715.

Haynes, S.R., Dollard, C., Winston, F., Beck, S., Trowsdale, J., and Dawid, I.B., 1992, The bromodomain: a conserved sequence found in human, Drosophila and yeast proteins, *Nucl. Acids Res.* 20:2603.

Hen, R., Borelli, E., and Chambon, P., 1985, Repression of the immunoglobulin heavy chain enhancer by the adenovirus-2 E1A products, *Science* 230:1391.

Howe, J.A., Mymryk, J.S., Egan, C., Branton, P.E., and Bayley, S.T., 1990, Retinoblastoma growth suppressor and a 300kDa protein appear to regulate cellular DNA synthesis, *Proc. Natl. Acad. Sci. USA* 87:5883.

Howe, J.A., and Bayley, S.T., 1992, Effects of Ad5 E1A mutant viruses on the cell cycle in relation to the binding of cellular proteins including the retinoblastoma protein and cyclin A, *J. Virol.* 186:15.

Johnson, K.W., Davis, B.H., and Smith, K.A., 1988, cAMP antagonizes interleukin 2-promoted T-cell cycle progression at a discrete point in early G1, *Proc. Natl. Acad. Sci. USA* 85:6072.

Kwok, R.P.S., Lundblad, J.R., Chrivia, J.C., Richards, J.P., Bächinger, H.P., Brennan, R.G., Roberts, S.G.E., Green, M.R., and Goodman, R.H., 1994, Nuclear protein CBP is a coactivator for the transcription factor CREB, *Nature* 370:223.

Kalvakolanu, D.V.R., Liu, J., Hanson, R.W., Harter, M.L., and Sen, G.C, 1992, Adenovirus E1A represses the cyclic AMP-induced transcription of the gene for phoshoenolpyruvate carboxykinase (GTP) in hepatoma cells, *J. Biol. Chem.* 267:2530.

Lalli, E., and Sassone-Corsi, P., 1994, Signal transduction and gene regulation: the nuclear response to cAMP, *J. Biol. Chem.* 269:17359.

Moran, E., 1993, DNA tumor virus transforming proteins and the cell cycle, *Curr. Opin. Genet. Dev.* 3:63.

Mymrik, J.S., Lee, R.W.H., and Bayley, S.T., 1992, Ability of adenovirus 5 E1A proteins to suppress differentiation of BC3H1 myoblasts correlates with their binding to a 300-kilodalton cellular protein, *Mol. Biol. Cell* 3:1107.

Nevins, J., 1992, A link between the Rb tumor suppressor protein and viral oncoproteins, *Science* 258:424.

Pastan, I., and Willingham, M., 1978, Cellular transformation and the 'morphological phenotype' of trans-formed cells, *Nature* 274:645.

Rikitake, Y., and Moran, E., 1992, DNA-binding properties of the E1A-associated 300-kilodalton protein, *Mol. Cell Biol.* 12:2826.

Stein, R.W., Corrigan, M., Yaciuk, P., Whelan, J., and Moran, E., 1990, Analysis of E1A-mediated growth regulation functions: binding of the 300-kilodalton cellular product correlates with E1A enhancer repression function and DNA synthesis-inducing activity. *J. Virol.* 64:4421.

Tamkun, J.W., Deuring, R., Scott, M.P., Kissinger, M., Pattatucci, A.M., Kaufman, T.C., and Kennison, J.A., 1992, brahma: a regulator of Drosophila homeotic genes structurally related to the yeast transcrip-tional activator SNF2/SWI2, *Cell* 68:561.

Tijan, R., and Maniatis, T., 1994, Transcriptional activation: a complex puzzle with few easy pieces, *Cell* 77:5.

Vairo, G., Argyriou, S., Bordun, A.-M., Whitty, G., and Hamilton, J.A., 1990, Inhibition of the signaling pathways for macrophage proliferation by cAMP, *J. Biol. Chem.* 265:2692.

Velcich, A. and Ziff, E., 1985, Adenovirus E1a proteins repress transcription from the SV40 early promoter. *Cell* 40:705.

Webster, K.A., Muscat, G.E.O., and Kedes, L., 1988, Adenovirus E1A products suppress myogenic differen-tiation and inhibit transcription from muscle-specific promoters, *Nature* 332:553.

Wilson, R., Ainscough, R., Anderson, K., Baynes, M. Berks, M., Bonfield, J., Burton, J., Connell, M., Copsey, T., Cooper, J., 1994, 2.2 Mb of contigous nucleotide sequence from chromosome III of C. elegans, *Nature* 368:32.

Winston, F., and Carlson, M., 1992, Yeast SNF/SWI transcriptional activators and the SPT/SIN chromatin connection, *Trends Genet.* 8:387.

DISCUSSION

D. Baltimore

Your talk brings to mind a variety of small questions and some big ones. I will just ask small ones. Does p300 bind to TBP and is it ever found in association in the TAF complex?

D. Livingston

There is one experiment from Elisabeth Moran that says the following: She has an antibody to p300 that seems to be able to co-precipitate TBP and associated proteins. Thus, there is a possibility of an association.

D. Baltimore

There is something close to a true zinc finger, which suggests that the protein could bind to DNA. Have you looked at the association of that with DNA and do you need the zinc finger for activity?

D. Livingston

We have done an oligonucleotide simplification experiment, and it does not simplify. So, thus far, we have no evidence for specific DNA binding by this polypeptide.

D. Baltimore

So you think it is also a protein interaction?

D. Livingston

Yes.

D. Baltimore

You said you can prevent the reporter from functioning in myoblasts with a bromo domain deleted p300. Does that prevent also the fusion of myoblasts?

D. Livingston

That is the big question. Those experiments are still in progress. Hence, I do not want to say much about them at the moment.

R. Perlmutter

That blocking experiment with the bromo deleted p300 is a very interesting one. And the question is: You now have a system in which, in principle, you could generate a set of mutants that have those characteristics and which would be informative with respect to the mechanism whereby p300 acts: so, is that domain deletion unique in terms of its ability to antagonize the action? Will that domain deletion antagonize the transcriptional activation effects of the DNA binding domains fused p300 protein in a co-transfection experiment? That is, will it antagonize directly a p300 effect?

D. Livingston

A very good question. The experiment you suggest is on the books but has not yet been done. Naturally, we are interested in searching for a protein(s) with which the bromo domain interacts in the performance of its function(s).

H. Beug

Is there any evidence that this p300 or its mutants may interact with, interfere with, or enhance responses to members of the steroid thyroid hormone receptor superfamily? At least the retinoic acid receptor was shown to interact directly with the transcription machinery via an E1A-like activity.

D. Livingston

That is a good thought, Hartmut, and Richard Eckner should think about asking such a question(s).

D. Metcalf

Some of the functions that you had on your list for p300 were odd but they reminded me of the actions of agents like leukemia inhibitory factor and interleukin 6, namely the action on adipocyte maturation, neuronal differentiation and so on. I was wondering whether activation of gP130 by such regulators might influence p300. Do you know anything about what induces the transcription or alters the stability of p300?

D. Livingston

What we know, so far, is that p300 seems to be produced in every cell we have looked at in tissue culture. We have not done a whole mouse blot, for example, to ask whether it is everywhere, but my suspicion is that most cells will have it and that it is present throughout the cell cycle. There are some mobility shifts in M, when the protein seems to migrate a bit more slowly. That was first shown by Betty Moran. The protein is heavily phosphorylated. More than that, we really know very little about its metabolism.

J. Adams

Can you remind us whether there is evidence, for example, from E1A mutants, as to whether p300 is necessary for the oncogenic action of E1A. In other words, is p300 likely to be a real oncogene or is it just a molecule that drives cells into proliferation?

D. Livingston

It is a partner for E1A and E1A-p300 complex formation is necessary for E1A transformation in the presence of activated ras.

J. Adams

In view of the apparent importance of RB and its relatives for control of G1 progression, how do you reconcile that model with the RB knockout results. The RB minus embryos survive for over 12 days, and, as I understand it, there is a p107 minus mouse that is actually alive and healthy. Thus, the RB and p107 knockout mice suggest that these genes might not be vital for general cellular growth.

D. Livingston

Well, as you know, the RB knockout mouse does not survive gestation. An RB-1-embryo does not pass beyond 13 and 1/2 days. Certain cells of these embryos die by apoptosis in the brain and probably in the bone marrow as well. There is gross hematopoietic and CNS dys-differentiation. So, you can ask: But they get to 13 and 1/2 days, so is RB a necessary growth control element? It may not be or, if it is, there may be other proteins which provide overlapping functions which compensate for the loss of its function. In the case of p107, the animals are healthy. Interestingly, the tissue culture cells are not normal in that they grow faster than wild type cells.

J. Adams

The animals are healthy, but tissue culture reveals a problem, so you conclude that tissue culture is more real than the animal.

D. Livingston

The same is true for cells of p53 -/- mice. It may be that tissue culture growth conditions evoke growth differences which result from the absence of a growth regulator like p53 or p107. Clearly an effect such as this is masked in the animal.

J. Adams

If I could just add one point about that. Even though the RB minus mice die at an early stage, as I understand it, if one makes chimeric animals, in fact, the RB minus cells can contribute to essentially all tissues in an adult mouse.

D. Livingston

Yes, that is right.

J. Adams

So that seems to argue that RB cannot have a necessary role in most tissues. It might have a subtle role. It might have a fine tuning effect, or it might have a role, for example, in an apoptotic function that only is needed in certain circumstances, but the observation does seem to me to argue that RB cannot be vital for the routine cell cycle.

D. Livingston

Presumably, it is not. Interestingly, since Levine's laboratory and our own have observed that RB has an anti-apoptotic effect in cultured cells. This is consistent with the observation that RB -/- embryos reveal profound apoptosis in the developing CNS.

R. Perlmutter

The formal interpretation of the constitution chimeras is that there is no cell autonomous defect.

D. Livingston

Correct. There may well be some form of cell-cell communication which is RB dependent.

R. Perlmutter

RB dependent communication?

D. Livingston

It is also a professional tumor suppressor, which has been highly conserved through evolution.

D. Baltimore

There is or is not a tumor suppressor?

D. Livingston

The RB $\frac{+/+}{-/-}$ chimeric mice of both Tyler Jacks and Anton Burns developed pituitary tumors, and the tumor cells are exclusively RB –/–. What I am referring to is that Burns and Jacks have taken blastocysts and injected them with RB –/– ES cells. The resulting chimeric animals contain roughly equal number of +/+ and –/– cells in their tissues. After several months, a substantial number of those animals develop pituitary tumors composed of RB –/– cells.

J. Griffin

Have you looked at senescent cells or terminally differentiated cells for either expression or phosphorylation? Are there any major changes?

D. Livingston

Not of p300.

E. Mihich

Have you had the possibility to locate the action of the p300 in the progression through the cell cycle in relation to the decision points where the p53 acts, for instance, towards apoptosis or towards S? Is that an early event, the p300 action in this progression, or is it a late event? And the other question is: You alluded to the TGF Beta. In some cells such as promyelocytic leukemia cells for instance, both TGF Beta and TNF can induce apoptosis or can induce differentiation, depending on the co-factors present.

D. Livingston

In the case of p300, the indirect evidence, the circumstantial evidence suggests a possible role for p300 very early in the G-0 exit parade if not in the maintenance of G-0, itself, given its potential role in promoting terminal differentiation. And with respect to p300 playing other kinds of roles, it is too early to tell. The protein has only recently been cloned, and useful reagents are just becoming available.

E. Metcalf

This is a very naive question. In your progression slide from Go into S phase I did not see cyclin D_1. What happened to that? Has it been deleted or what?

D. Livingston

It was part of much detailed information which was omitted to maximize the simplicity of the slide. D_1 is an important participant in Go $\rightarrow$ S progression but, so far, there is no evidence pointing to a specific connection between D_1 and p300 function.

10

HEMATOPOIETIC SIGNALING BY THE CYTOKINE RECEPTOR SUPERFAMILY

J. N. Ihle

St. Jude Children's Research Hospital
Department of Biochemistry
332 North Lauderdale
Memphis, Tennessee 38101

ABSTRACT

Hematopoiesis is regulated through the interaction of a variety of cytokines with receptors of the cytokine receptors superfamily. Although lacking catalytic domains, signaling by this receptor family has been shown to require tyrosine phosphorylation. Recent studies have demonstrated that this is mediated through a novel family of cytoplasmic protein tyrosine kinases termed the *Janus* kinases (Jaks) which currently consists of four members (Jak1, Jak2, Jak3, Tyk2). The Jaks associate with one or more of the receptor subunits and are tyrosine phosphorylated and catalytically activated following ligand binding. Once activated, the Jaks phosphorylate one or more of the receptor subunits creating docking sites for SH2 containing proteins including SHC, the p85 subunit of phosphotidylinositol 3' kinase (PI3-Kinase) and hematopoietic cell phosphatase (HCP). The Jaks also phosphorylate cellular substrates including one or more of a family of proteins termed signal transducers and activators of transcription (Stats). When phosphorylated, Stat proteins dimerize, migrate to the nucleus and bind DNA. The Jak/Stat signaling pathway represents a novel pathway that is primarily utilized by the cytokine receptor superfamily.

INTRODUCTION

Hematopoiesis is regulated, in part, through the interaction of hematopoietic growth factors with receptors of the cytokine receptor superfamily (Clark and Kamen, 1987; Metcalf, 1989). This receptor family consists of the type I subgroup which is characterized by four positionally conserved cysteines and a WSXWS motif in the extracellular domain and the type II subgroup which is characterized by cysteine pairs at both the amino and carboxyl-termini (Bazan, 1991). The type I receptors subgroup includes receptors for most of the hematopoietic growth factors, while the type II subgroup consists of the receptors for the interferons.

Normal and Malignant Hematopoiesis, Edited by Enrico Mihich and Donald Metcalf
Plenum Press, New York, 1995

The cytoplasmic domains of members of the cytokine receptor superfamily contain no catalytic domains and only limited sequence similarity which is found in the membrane proximal domain of one or more receptor subunits. This region is frequently referred to as the box 1 and box 2 motifs (Murakami, Narazaki *et al.* 1991). Although lacking kinase catalytic domains, a variety of studies support the concept that the membrane proximal domains are required to couple ligand binding to the induction of protein tyrosine phosphorylation. Moreover, one or more of the receptor subunits is a consistent substrate for tyrosine phosphorylation. During the past two years a number of studies have demonstrated that the members of the *Janus* family of protein tyrosine kinases (Jaks) are essential for coupling ligand binding to protein tyrosine phosphorylation within the cytokine receptor superfamily.

FUNDAMENTAL CONSTRAINTS

Janus Kinases: A Novel Family of Cytoplasmic Protein Tyrosine Kinases

The *Janus* family of protein tyrosine kinases were identified during studies to identify unique kinases. The first member to be cloned was Tyk2 (tyrosine kinase 2) which was obtained by low-stringency screening of a T cell library with a c-DNA fragment, containing the catalytic domain, of c-fms (Firmbach-Kraft, Byers *et al.* 1990). Subsequently two structurally related protein tyrosine kinases were characterized and termed *Janus* kinase 1 and 2 (Wilks, Harpur *et al.* 1991). Fragments of Jak1 and Jak2 had been initially identified in PCR approaches to identify novel kinases (Wilks, 1989; Wilks, 1991). The term *Janus* refers to the ancient, two-faced, Roman god of gates and doorways. Alternatively, Jak is used as an acronym for just another kinase to reflect their identification during an era when a variety of new protein tyrosine kinases were cloned. Jak1, Jak2 and Tyk2 are ubiquitously expressed.

More recently a fourth member of the Jak family has been characterized. As with Jak1 and Jak2, fragments of Jak3 had been identified by PCR amplications to identify novel kinases (Cance, Craven *et al.* 1993; Sanchez, Tapley *et al.* 1994; Takahashi and Shirasawa, 1994; Kawamura, McVicar *et al.* 1994). Full length cDNAs for murine Jak3 (Witthuhn, Silvennoinen *et al.* 1994) rat Jak3 (Takahashi and Shirasawa, 1994) and human Jak3 (Kawamura, McVicar *et al.* 1994) have been recently obtained and sequenced. Unlike the other Jaks, Jak3 is restricted in its expression to myeloid cells (Witthuhn, Silvennoinen *et al.* 1994) and lymphoid cells (Kawamura, McVicar *et al.* 1994). In resting T cells and monocytes, Jak3 expression is increased dramatically following activation (Kawamura, McVicar *et al.* 1994).

A *Drosophila* Jak kinase has also recently been identified as the gene associated with the *hopscotch* locus (Binari and Perrimon, 1994). The maternal product of the *hopscotch* locus is required for the proper levels of expression of particular stripes of the pair-rule genes. A dominant, gain of function, mutation of the *hopscotch* locus has also been identified termed the *tumorous lethal gene*. This mutation causes the abnormal proliferation and differentiation of the larval hematopoietic system, regulating a lethal form of leukemia in flies.

The *Janus* family of kinases are relatively large proteins of 120-135 kDa and are uniquely characterized by the presence of two kinase domains in the carboxyl-half of the molecule. The most carboxyl domain contains the characteristic motifs, while the more amino terminal domain has the characteristic subdomains of kinases but lacks a number of residues that are critical for catalytic activity. The function of this kinase-like domain is unknown. Unlike most cytoplasmic protein tyrosine kinases there are no *src* homology 2 or 3 domains (SH2,SH3) although the amino terminal region of the Jaks contain blocks of homology that have been referred to as the Jak homology domains. The significance of these domains are unknown but are being examined for their ability to associate with the membrane proximal domain of the receptors.

Cytokine Receptors Associate with and Activate Jaks

A central role for the Jaks in cytokine signaling has been indicated by 1) their rapid tyrosine phosphorylation following ligand binding; 2) activation of their *in vitro* catalytic activity following ligand binding; and 3) their ability to physically associate with the critical membrane proximal cytoplasmic domains. In addition, genetic evidence has provided compelling support for a critical role for Jaks in signaling by the IFNs as detailed below.

The receptors for erythropoietin (Epo), growth hormone (GH) and prolactin (PH) consists of single chains. In each case, ligand binding results in the specific tyrosine phosphorylation and activation of Jak2 (Witthuhn, Quelle *et al.* 1993; Artgetsinger, Campbell *et al.* 1993; Campbell, Argetsinger *et al.* 1994; Rui, Kirken *et al.* 1994; Dusanter Fourt, Muller *et al.* 1994). Jak2 has also been shown to either constitutively associate with the receptor (PH) or to associate with the receptor following ligand binding (Epo, GH). In the case of the Epo receptor, Jak2 activation and association requires the membrane proximal, box 1/2, region of the receptor.

The receptor for granulocyte colony stimulating factor (G-CSF) also consists of a single chain. It has been reported that G-CSF induces the tyrosine phosphorylation and activation of Jak1 (Nicholson, Oates *et al.* 1994). In our experience (B. Witthuhn, S. Nagata, J.N. Ihle, in preparation), G-CSF strongly induces the tyrosine phosphorylation of Jak2 as well as, to a lesser extent, Jak1. Activation of Jak2, requires the membrane proximal region of the cytoplasmic domain.

The receptor for interleukin 3 (IL-3) consists of a ligand binding, α, subunit which associates with a signaling β subunit following ligand binding (Miyajima, Mui *et al.* 1993). The β chain is shared with the receptors for IL-5 and granulocyte-macrophage colony stimulating factor (GM-CSF). IL-3 induces the tyrosine phosphorylation and activation of Jak2 and to a lesser extent Jak1 (Silvennoinen, Witthuhn *et al.* 1993). Jak2 specifically associates with the common β signaling subunit of the receptor and the membrane proximal, box1/2, region of the β chain cytoplasmic domain is required for Jak2 activation.

The cytokines IL-6, oncostatin M (OSM), leukemia inhibitory factor (LIF), ciliary neurotrophic factor (CNTF) and IL-11 utilize receptors that may contain ligand specific α chains, but which associate with either gp130 or a gp130-related (LIF receptor β) signaling receptor subunit (Stahl and Yancopoulos, 1993) which contains the characteristic box1/2 motif. Recent studies have demonstrated that these cytokines induce the tyrosine phospho-rylation and activation of Jak1, Jak2 and to some extent Tyk2 (Stahl, Boulton *et al.* 1994; Narazaki, Witthuhn *et al.* 1994). Association occurs between gp130 or the gp130 like β, signaling chains and requires the box1/2 motif containing region. Unlike other receptor subunits, all three Jak family members associate with this region. Curiously, the extent of individual Jak activations is cell type specific suggesting the possibility that in some cell types, one or more of the Jaks is limiting for this group of cytokines.

The receptors for the interleukins, IL-2, IL-4 and IL-7 contain both ligand specific subunits and a shared γ chain that was first identified as a component of the IL-2 receptor (Taniguchi and Minami, 1993). In addition the receptor for IL-2 contains α and β chains; while the receptors for IL-4 and IL-7 contain an additional subunit with structural similarity to the IL-2 receptor β chain and thus is termed a β subunit. No subunits homologous to the IL-2 receptor α chain have been identified. Among the receptor subunits, the IL-2 receptor α chain does not contain any of the characteristic extra- or intracellular motifs; while the β and γ chains are clearly members of the cytokine receptor superfamily. The β chains contain the characteristic membrane proximal box1/2 motif while the shared γ chain has an SH2-like domain. Recent studies have shown that this family of cytokines induce the tyrosine phosphorylation and activation of Jak3 and to a lesser extent and in a cell type specific manner, Jak1 (Witthuhn, Silvennoinen *et al.* 1994; Johnston, Kawamura *et al.* 1994). Current studies are attempting to determine which subunits are required for Jak association with the receptor complex, although it has been shown that the membrane proximal, box1/2, region of the IL-2 β chain is required for Jak3 activation.

Perhaps the most intriguing results concerns the role of Jaks in signaling by the IFNs. The IFN receptors consist of at least two chains which are members of the type II cytokine receptor superfamily (Soh, Donnelly *et al.* 1994; Novick, Cohen *et al.* 1994; Uze, Lutfalla *et al.* 1990; Hemmi, Bohni *et al.* 1994; Aguet, Dembic *et al.* 1988). The first indication that the IFNs might utilize Jaks in signaling came from genetic studies with mutants that were defective in signaling (McKendry, John *et al.* 1991; John, McKendry *et al.* 1991; Pellegrini, John *et al.* 1989). Using expression cloning, the Tyk2 gene was identified by its ability to restore the IFN response of one complementation group of mutants (Velazquez, Fellous *et al.* 1992). Subsequent studies found that two addition complementation groups were defective in Jak2 (Watling, Guschin *et al.* 1993) or Jak1 (Muller, Briscoe *et al.* 1993). Together the studies demonstrated that the response to IFN γ required Jak1 and Jak2 while the response to IFN α/β requires Jak1 and Tyk2.

The basis for the interdependence of two Jaks is not known; however the current data suggest that activation of a receptor complex requires Jak heterodimerization. In particular, tyrosine phosphorylation and activation of any of the Jaks, requires that both be catalytically active. Thus a kinase cascade does not appear to be involved. Since the receptors consist of at least two subunits, it can be envisioned that individual receptor chains associate with specific Jaks. IFN binding and receptor complex formation thus brings two different Jaks into proximity.

Receptor Dimerization/Oligomerization and Jak Activation

Studies with a number of the cytokine receptors support the general concept that ligand induced dimerization/oligomerization is critical for activation of associated Jaks. In the case of the Epo receptor, dimerization/oligomerization of the receptor is induced by ligand (Watowich, Hilton *et al.* 1994; Miura and Ihle, 1993) a viral glycoprotein that binds the receptor (Li, D'Andrea *et al.* 1990; Showers, Moreau *et al.* 1992) or mutations that introduce cysteines in the extracellular domain and make the receptor ligand independent for activation (Watowich, Yoshimura *et al.* 1992). In each case, receptor dimerization/oligomerization results in the association and activation of Jak2.

The role of dimerization of the cytoplasmic domains of the signaling subunits of cytokine receptors has been well demonstrated with chimeric receptor constructs. In general these have involved making fusion proteins with the extracellular domains from receptors such as the epidermal growth factor (EGF), c-*kit*, the receptor for stem cell factor or extracellular domains that can be dimerized with antibodies. The intracellular domains of the fusion proteins are derived from a cytokine receptor. EGF/Epo receptor chimerics allow EGF induced Jak2 activation and support both the proliferation and differentiation functions of the Epo receptor (Ohashi, Maruyama *et al.* 1994; Maruyamam, Miyata *et al.* 1994). Similar constructs containing the cytoplasmic domain of the IL-3 receptor β chain function comparably. Lastly, chimeric receptors containing the cytoplasmic domains of the IL-2 β and γ chains have been shown to mediate signaling (Nelson, Lord *et al.* 1994; Watowich, Hilton *et al.* 1994). Interestingly, in T cells, the cytoplasmic domains of both the β and γ chains must be present while in myeloid cells, proliferation can occur when only the cytoplasmic domain of the β chain is present.

Together the results support a common mechanism for cytokine receptor function in which ligand induces the dimerization/oligomerization of the receptor chains. As a consequence there is a increased affinity of Jaks and the Jaks are juxtapositioned due the receptor aggregation. This may involve homodimerization of Jaks (i.e. Epo receptor) or heterodimerization (i.e. IFNs, IL-6) depending upon the ability of individual receptor subunits to associate with specific Jaks. As a consequence of Jak aggregation there is an opportunity for phosphorylation at the "autophosphorylation" site resulting in an activation of catalytic activity.

Tyrosine Phosphorylation of Cytokine Receptors and Signaling

One of the consistent substrates of ligand induced tyrosine phosphorylation is one or more of the receptor subunits. Both the Epo receptor and the IL-3 receptor β chain have been shown to be direct substrates of Jak2 and it is hypothesized that, in general, activated Jaks are directly responsible for receptor phosphorylation. Recent studies have demonstrated that tyrosine phosphorylation of the receptors create "docking" sites for signaling proteins containing SH2 domains.

In the responses to IL-3 and Epo, phosphorylation of the receptor is associated with the activation of the *ras* signaling pathway through the creation of binding sites for SHC. Following binding, SHC is tyrosine phosphorylated which, in turn, creates a binding site for GRB2. Although not directly demonstrated, it is hypothesized that SHC phosphorylation is mediated by a Jak. In addition to SHC phosphorylation and its association with GRB2, ligand binding results in increases in *ras* bound GTP, activation of raf-1, tyrosine phosphorylation of mitogen activated protein (MAP) kinases and induction of immediate early genes (Carroll, Spivak *et al.* 1991; Satoh, Nakafuku *et al.* 1991; Sakamaki, Miyajima *et al.* 1992; Sato, Sakamaki *et al.* 1993; Cutler, Liu *et al.* 1994; Damen, Liu *et al.* 1993; Carroll, Clark-Lewis *et al.* 1990).

The relevance of activation of the *ras* pathway has been indicated in studies with truncated receptors. In particular, appropriate truncations of the Epo receptor remove the sites of tyrosine phosphorylation but retain the ability to induce proliferation and differentiation (Miura, Miura *et al.* 1994; Maruyamam, Miyata *et al.* 1994). These truncations also uncouple ligand binding from the activation of all the changes associated with activation of the *ras* pathway. Comparable truncations of the IL-3 receptor β chain have the same effect (Quelle, Sato *et al.* 1994). Thus in these systems, the membrane distal region of the receptor signaling chain is required for *ras* activation, presumably through the phosphorylations that occur in this region. However, activation of the *ras* pathway is not required for the receptor functions that have been examined.

In addition, to providing docking sites for SHC, receptor phosphorylations also provide docking sites for the amino terminal SH2 domains of hematopoietic cell phosphatase (HCP) (Yi, Quelle *et al.* 1993; Yi, Mui *et al.* 1993). The consequences of recruiting a protein tyrosine phosphatase to the activated receptor complex might be anticipated to negatively affect receptor signaling. Several observations support such a function. First, the membrane distal region of the Epo receptor negatively affects the response to Epo in certain cell lines (D'Andrea, Yoshimura *et al.* 1991). Secondly, *in vivo*, a mutation of the Epo receptor which results in a 70 amino acid carboxyl-truncation, has been identified in genetically acquired erythrocytosis. In both cases it is hypothesized that the carboxyl-truncations prevent the recruitment of HCP to the activated receptor complex.

The third observation derives from the phenotype of mice which are genetically deficient in HCP, a mutation termed *methadon* (Shultz, Schweitzer *et al.* 1993; Tsui, Siminovitch *et al.* 1993). Homozygous *methadon* mice die within approximately three weeks after birth due to overproliferation of macrophages, particularly in the lungs. However, a number of hematopoietic lineages are affected including the erythroid lineage which becomes relatively independent of Epo for proliferation and differentiation. There is also excessive proliferation of lymphoid lineages leading to the appearance of autoimmune like pathologies. This phenotype is consistent with the hypothesis that HCP is required to negatively regulate the proliferation of most myeloid and lymphoid lineages.

Src Kinases and Cytokine Signaling

A number of reports have implicated *src* kinase gene family members in cytokine signaling by ligand induced increases in catalytic activity of 2-3 fold. Among these the most studied has been the role of src kinase in IL-2 signaling. In particular, IL-2 activates lck kinase activity in T cells (Taniguchi and Minami, 1993) and fyn and lyn in myeloid cells (Kobayashi, Kono *et al.* 1993). Association of the src family kinases is constitutive and requires the amino-terminal half of the kinase domain of lck and an acidic domain of the IL-2 receptor β chain (Hatakeyama, Kono *et al.* 1991). The acidic domain is membrane distal from the box1/2 motif which is referred to as the serine-rich domain of the IL-2 receptor β chain.

The role of lck activation and Jak3 activation in IL-2 receptor β chain functioning has been examined through the use of mutant receptors lacking either the membrane proximal serine rich domain or the more distal acidic domain (Witthuhn, Silvennoinen *et al.* 1994). Deletion of the acidic domain disrupts association of lck with the receptor and its activation following ligand binding but does not alter the ability of the receptor to couple ligand binding to activation of Jak3. Conversely, deletion of the serine rich region does not disrupt lck association with the β chain but does inactivate the receptor for activation of Jak3. Importantly, deletion of the acidic domain does not alter the mitogenic response while deletion of the serine rich domain inactivates the receptor.

Stats; Down Stream Signaling Molecules for Jaks

The signal transducer and activators or transcription (Stat) family of proteins were identified in studies concerned with IFN induced gene transcription (Darnell, Jr., Kerr *et al.* 1994). Stat1 and Stat2 were initially identified as proteins in IFN inducible, DNA binding complexes. The genes encoding these proteins were obtained from the sequence of affinity purified proteins. The predicted amino acid sequences of Stat1 and Stat2 indicated that they were structurally related. Both contain an SH2 domain in the carboxyl-terminus and an SH3-like domain immediately amino terminal to the SH2 domain. In addition blocks of homology existed throughout the amino terminal half of the proteins. Stat1 encodes a 91 kDa protein with an alternatively spliced form of 84 kDa while Stat2 encodes a 113 kDa protein, the difference in size being due to additional sequences at the carboxyl terminus of Stat2.

In the IFN responses Stat2 and/or Stat1 are rapidly tyrosine phosphorylated at a single site, translocate to the nucleus and bind DNA directly (Stat1) or participate in DNA binding complexes (Stat2). IFN γ induces the activation of Stat1 while IFN α/β activates both Stat1 and Stat2. Mutants defective in either Stat1 or Stat2 demonstrate their essential role in the responses to IFNs (Darnell, Jr., Kerr *et al.* 1994).

The requirement for both Jaks and Stats in the IFN responses led to the hypothesis that the Stats are substrates for the activated Jaks. This hypothesis has been supported by recent studies with baculovirus produced Stat1 and Jak1, Jak2 or Tyk2 (F. W. Quelle and J.N. Ihle, in preparation). When co-expressed in insect cells, Jak1, Jak2 or Tyk2 phosphorylate Stat1 at a single tyrosine residue (aa 701) and activate their DNA binding activity. Importantly each of the Jaks is equally capable of phosphorylating Stat1, indicating that the specificity of Stat1 phosphorylation seen *in vivo* (see below) is due to other determinants such as the ability of a receptor complex to recruit Stat1 to the activated Jaks.

Stats and More Stats; A Common Theme in Cytokine Signaling

The existence of a Jak/Stat pathway for gene regulation in IFN signaling suggested the possibility that this pathway might be utilized by a number of cytokines. Initially is was found that while most cytokines did not induce tyrosine phosphorylation of Stat1 or Stat2, they did induce the appearance of complexes capable of binding γ activated sequences (GAS) and/or DNA binding complexes which were tyrosine phosphorylated and serologically related to Stat1. A number of recent studies have focused on the cloning and characterization of these cytokine induced Stat-like proteins.

IL-6 induces the tyrosine phosphorylation of Stat1 and a Stat-like DNA binding protein initially termed acute phase reactive protein (APRF). Importantly, APRF has been shown to co-immunoprecipitate with the IL-6 receptor signaling subunit, gp130 and Jak1 (Zhong, Wen *et al.* 1994). The gene encoding APRF activity was obtained by protein sequence information (Akira, Nishio *et al.* 1994) and by low stringency screening of a cDNA library with a Stat1 probe (Zhong, Wen *et al.* 1994). The gene was found to encode a protein with 59% identity with Stat1 and contained the characteristic SH2 and SH3-like domains as well as other blocks of homology. Because of the similarity with Stat1, the term Stat3 has been proposed for this gene.

Stat4 was independently cloned by either PCR approaches (Yamamoto, Quelle *et al.* 1994) or low stringency hybridization with a Stat1 probed (Zhong, Wen *et al.* 1994). Stat4 is 52% identical to Stat1. Unlike other Stats however, no ligands have been identified that can induce the tyrosine phosphorylation or activation of Stat4 DNA binding activity. However, Stat4, when co-expressed with Jak1 or Jak2 in COS cells, becomes tyrosine phosphorylated and acquires DNA binding activity for a GAS element from the IRF-1 gene (Yamamoto, Quelle *et al.* 1994), demonstrating that it is functionally, as well as structurally, a Stat family member.

Stat4, unlike the other Stat genes examined is highly restricted in it's expression. Stat4 is expressed in the myeloid lineages in which expression is limited to early progenitors. Within the erythroid and granulocytic lineages, expression is lost concomitant with lineage commitment and initiation of terminal differentiation by Epo or G-CSF (Yamamoto, Quelle *et al.* 1994). Stat4 is also expressed in developing spermatogonia. This is of considerable interest since the *Stat4* gene genetically co-localizes with the *juvenile spermatogonial depletion* (jsd) gene. Whether *Stat4* is the gene involved remains to be determined.

Stat5 was recently cloned as the gene associated with a prolactin inducible GAS binding activity termed mammary gland factor (MGF). The gene was cloned based on sequence data for the protein. The deduced amino acid sequence for the entire gene, clearly indicated that it was a member of the Stat family of proteins although more distantly related than Stat1-4. Nevertheless the SH2 and SH3-like domains are conserved as well as several other blocks of homology.

The genes for a number of additional, cytokine induced Stat-like proteins remain to be cloned. These include the Stat-like activities or proteins induced by Epo (Larner, David *et al.* 1993), IL-3 (Larner, David *et al.* 1993), IL-4 (Kotanides and Reich, 1993; Schindler, Kashleva *et al.* 1994) and IL-2 (C. Beadling, D. Guschin, C. Schneidler, A. Ziemeccki, I.M. Kerr and D. Cantrell, submitted).

The existence of multiple Stat proteins may help to explain the pleiotropic responses commonly associated with cytokines. In particular, it can be hypothesized that the response to a particularly cytokine in a specific cell type, may be due to the Stat proteins that are present and available for phosphorylation. The expression pattern of Stat4 provides a striking example. In this context it should also be noted that there exist more than ten variants of GAS-like sequences. Different Stat homo- and heterodimers show very different affinities with different subsets of GAS elements.

CONCLUSIONS

During the last two years much has been learned concerning the mechanisms by which the cytokine receptor superfamily members initiate cellular responses. Two distinct pathways have been identified; namely the Jak/Stat pathway and the *ras* pathway. Activation of the *ras* pathway correlates with SHC tyrosine phosphorylation which is, in turn, dependent upon receptor tyrosine phosphorylation. The role for the activation of this pathway is unclear from the studies to date, since activation of the pathway is not required for the types of responses, including a mitogenic response, that have been examined.

The ability of the cytokine receptor superfamily to couple ligand binding to induction of tyrosine phosphorylation is mediated by the association of one or more Jaks with one or more of the receptor subunits. In all cases examined, association occurs through the membrane proximal region containing the box1/2 motif. However, the domains of the Jaks required for association have yet to be defined. Central to receptor function is the concept that ligand binding induces receptor/Jak aggregation resulting in the activation of Jaks by tyrosine phosphorylation of the "autophosphorylation"/activation site.

Although less well established, the concept that an activated receptor complex in hematopoietic cells can recruit the tyrosine phosphatase, HCP has considerable importance. The dramatic effects that the genetic deficiency of HCP has on multiple myeloid and lymphoid lineages supports a very general role in negatively influencing the response in hematopoietic cells. It can be hypothesized that HCP is essential to insure that a critical number of receptor complexes must form to initiate a response thus ensuring a ligand dependence.

Cytokines induce the tyrosine phosphorylation and activation of members of the Stat family of transcription factors. Within the hematopoietic systems we have examined, the Jak/Stat pathway is uniquely activated by the cytokine receptor superfamily members and specifically is not activated by protein tyrosine kinase receptors including c-kit, c-fms or c-met. Nor is the Jak/Stat pathway activated in the responses to IL-1, TGFs, IL-8 or TNF. Whether the Jak/Stat pathway is utilized by other receptor families in other lineages needs to be examined although EGF induces Stat1 activation.

Essential to understanding the function of cytokines, will be the characterization of additional Stat proteins. Current information indicates that a number of Stat-like activities are activated by cytokines and that these exhibit both lineage and stage of differentiation specificity. The identification of these new signaling pathways will provide important targets for drug development.

REFERENCES

Aguet, M., Dembic, Z., and Merlin, G., 1988, Molecular cloning and expression of the human interferon-gamma receptor. *Cell* 55:273-280.

Akira, S., Nishio, Y., Inoue, M., Wang, X., Wei, S., Matsusaka, T., Yoshida, K., Sudo, T., Naruto, M., and Kishimoto, T., 1994, Molecular cloning of APRF, a novel ISGF3 p91-related transcription factor involved in the gp130-mediated signaling pathway. *Cell* 77:63-71. @REFAUSTY = Artgetsinger, L.S., Campbell, G.S., Yang, X., Witthuhn, B.A., Silvennoinen, O., Ihle, J.N., and Carter-Su, C., 1993, Identification of JAK2 as a growth hormone receptor-associated tyrosine kinase. *Cell* 74:237-244.

Bazan, J.F. 1991, Haemopoietic receptors and helical cytokines. *Immunol. Today* 10:350-354.

Binari, R. and Perrimon, N., 1994, Stripe-specific regulation of pair-rule genes by *hopscotch*, a putative Jak family tyrosine kinase in *Drosophila. Genes & Development* 8:300-312.

Campbell, G.S., Argetsinger, L.S., Ihle, J.N., Kelly, P.A., Rillema, J.A., and Carter-Su, C., 1994, Activation of JAK2 tyrosine kinase by prolactin receptors in Nb$_2$ cells and mouse mammary gland explants. *Proc. Natl. Acad. Sci. USA* 91:5232-5236.

Cance, W.G., Craven, R.J., Weiner, T.M., and Liu, E.T., 1993, Novel protein kinases expressed in human breast cancer. *Int.J.Cancer* 54:571-577.

Carroll, M.P., Clark-Lewis, I., Rapp, U.R., and May, W.S., 1990, Interleukin-3 and granulocyte-macrophage colony-stimulating factor mediate rapid phosphorylation and activation of cytosolic c-raf. *J.Biol.Chem.* 265:19812-19817.

Carroll, M.P., Spivak, J.L., McMahon, M., Weich, N., Rapp, U.R., and May, W.S., 1991, Erythropoietin induces raf-1 activation and raf-1 is required for erythropoietin-mediated proliferation. *J.Biol.Chem.* 266:14964-14969.

Clark, S.C. and Kamen, R., 1987, The human hematopoietic colony-stimulating factors. *Science* 236:1229-1237.

Cutler, R.L., Liu, L., Damen, J.E., and Krystal, G., 1994, Multiple cytokines induce the tryosine phosphorylation of Shc and its association with Grb2 in hemopoietic cells. *J.Biol.Chem.* 268:21463-21465.

D'Andrea, A.D., Yoshimura, A., Youssoufian, H., Zon, L.I., Koo, J.W., and Lodish, H.F., 1991, The cytoplasmic region of the erythropoietin receptor contains nonoverlapping positive and negative growth-regulatory domains. *Mol. Cell Biol.* 11:1980-1987.

Damen, J.E., Liu, L., Cutler, R.L., and Krystal, G., 1993, Erythropoietin stimulates the tyrosine phosphorylation of shc and its association with grb2 and a 145-Kd tyrosine phosphorylated protein. *Blood* 82:2296-2303.

Darnell, J.E., Jr., Kerr, I.M., and Stark, G.R., 1994, Jak-STAT pathways and transcriptional activation in response to IFNs and other extracellular signaling proteins. *Science* 264:1415-1421.

Dusanter Fourt, I., Muller, O., Ziemiecki, A., Mayeux, P., Drucker, B., Djiane, J., Wilks, A., Harpur, A.G., Fischer, S., and Gisselbrecht, S., 1994, Functional analysis of prolactin receptor and prolactin-erythropoietin receptor chimera expressed in lymphoid cells: identification of jak kinases as signaling molecules for prolactin. *EMBO J.* in press.

Firmbach-Kraft, I., Byers, M., Shows, T., Dalla-Favera, R., and Krolewski, J.J., 1990, Tyk2, prototype of a novel class of non-receptor tyrosine kinase genes. *Oncogene* 5:1329-1336.

Hatakeyama, M., Kono, T., Kobayashi, N., Kawahara, A., Levin, S.D., Perlmutter, R.M., and Taniguchi, T., 1991, Interaction of the IL-2 receptor with the src-family kinase p56^{lck}: Identification of novel intermolecular association. *Science* 252:1523-1528.

Hemmi, S., Bohni, R., Stark, G., DiMarco, F., and Aguet, M., 1994, A novel member of the interferon receptor family complements functionality of the murine interferon gamma receptor in human cells. *Cell* 76:803-810.

John, J., McKendry, R., Pellegrini, S., Flavell, D., Kerr, I.M., and Stark, G.R., 1991, Isolation and characterization of a new mutant human cell line unresponsive to alpha and beta interferons. *Mol.Cell.Biol.* 11:4189-4195.

Johnston, J.A., Kawamura, M., Kirken, R., Chen, Y., Blake, T.B., Shibuya, K., Ortaldo, J.R., McVicar, D.W., and O'Shea, J.J., 1994, Phosphorylation and activation of the JAK3 Janus kinase in response to IL-2. *Nature* 370:151-153.

Kawamura, M., McVicar, D.W., Johnston, J.A., Blake, T.B., Chen, Y., Lal, B.K., Lloyd, A.R., Kelvin, D.I., Staples, J.E., Ortaldo, J.R., and O'Shea, J.J., 1994, Molecular cloning of L-JAK, a Janus family protein-tyrosine kinase expressed in natural killer cells and activated leukocytes. *Proc. Natl. Acad. Sci. USA* in press.

Kobayashi, N., Kono, T., Hatakeyama, M., Minami, Y., Miyazaki, T., Perlmutter, R.M., and Taniguchi, T., 1993, Functional coupling of the *src*-family protein tyrosine kinases p59^{fyn} and p53/56^{lyn} with the interleukin 2 receptor: Implications for redundancy and pleiotropism in cytokine signal transduction. *Proc. Natl. Acad. Sci. USA* 90:4201-4205.

Kotanides, H. and Reich, N.C., 1993, Requirement of tyrosine phosphorylation for rapid activation of a DNA binding factor by IL-4. *Science* 262:1265-1267.

Larner, A.C., David, M., Feldman, G.M., Igarashi, K., Hackett, R.H., Webb, D.S.A., Sweitzer, S.M., Petricoin, E.F.I., and Finbloom, D.S., 1993, Tyrosine phosphorylation of DNA binding proteins by multiple cytokines. *Science* 261:1730-1733.

Li, J.P., D'Andrea, A.D., Lodish, H.F., and Baltimore, D., 1990, Activation of cell growth by binding of Friend spleen focus-forming virus gp55 glycoprotein to the erythropoietin receptor. *Nature* 343:762-764.

Maruyamam, K., Miyata, K., and Yoshimura, A., 1994, Proliferation and erythroid differentiation through the cytoplasmic comain of the erythropoietin receptor. *J.Biol.Chem.* 269:5976-5980.

McKendry, R., John, J., Flavell, D., Muller, M., Kerr, I.M., and Stark, G.R., 1991, High-frequency mutagenesis of human cells and characterization of a mutant unresponsive to both alpha and gamma interferons. *Proc. Natl. Acad. Sci. USA* 88:11455-11459.

Metcalf, D. 1989, The molecular control of cell division, differentiation commitment and maturation in haemopoietic cells. *Nature* 339:27-30.

Miura, O. and Ihle, J.N., 1993, Dimer- and oligomerization of erythropoietin receptor by disulfide bond formation and significance of WSXWS motif on intracellular transport. *Arch. Biochem. Biophys.* 306:200-208.

Miura, Y., Miura, O., Ihle, J.N., and Aoki, N., 1994, Activation of the mitogen-activated protein kinase pathway by the erythropoietin receptor. *J.Biol.Chem.* submitted.

Miyajima, A., Mui, A.L.-F., Ogorochi, T., and Sakamaki, K., 1993, Receptors for granulocyte-marcophage colony-stimulating factor, interleukin-3, and interleukin-5. *Blood* 82:1960-1974.

Muller, M., Briscoc, J., Laxton, C., Guschin, D., Ziemiecki, A., Silvennoinen, O., Harpur, A.G., Barbieri, G., Witthuhn, B.A., Schindler, C., Pellegrini, S., Wilks, A.F., Ihle, J.N., Stark, G.R., and Kerr, I.M., 1993, The protein tyrosine kinase JAK1 complements defects in interferon-α/β and - signal transduction. *Nature* 366:129-135.

Murakami, M., Narazaki, M., Hibi, M., Yawata, H., Yasukawa, K., Hamaguchi, M., Taga, T., and Kishimoto, T., 1991, Critical cytoplasmic region of the interleukin 6 signal transducer gp130 is conserved in the cytokine receptor family. *Proc. Natl. Acad. Sci. U. S. A.* 88:11349-11353.

Narazaki, M., Witthuhn, B.A., Yoshida, K., Silvennoinen, O., Yasukawa, K., Ihle, J.N., Kishimoto, T., and Taga, T., 1994, Activation of JAK2 kinase mediated by the IL-6 signal transducer, gp130. *Proc. Natl. Acad. Sci. USA* 91:2285-2289.

Nelson, B.H., Lord, J.D., and Greenberg, P.D., 1994, Cytoplasmic domains of the interleukin-2 receptor beta and gamma chains mediate the signal for T-cell proliferation. *Nature* 369:333-336.

Nicholson, S.E., Oates, A.C., Harpur, A.G., Ziemiecki, A., Wilks, A.F., and Layton, J.E., 1994, Tyrosine kinase JAK1 is associated with the granulocyte-colony-stimulating factor receptor and both become tyrosine-phosphorylated after receptor activation. *Proc. Natl. Acad. Sci. USA* 91:2985-2988.

Novick, D., Cohen, B., and Rubinstein, M., 1994, The human interferon α/β receptor: characterization and molecular cloning. *Cell* 77:391-400.

Ohashi, H., Maruyama, K., Liu, Y., and Yoshimura, A., 1994, Ligand-induced activation of chimeric receptors between the erythropoietin receptor and receptor tyrosine kinases. *Proc. Natl. Acad. Sci. USA* 91:158-162.

Pellegrini, S., John, J., Shearer, M., Kerr, I.M., and Stark, G.R., 1989, Use of a selectable marker regulated by alpha interferon to obtain mutations in the signaling pathway. *Mol.Cell.Biol.* 9:4605-4612.

Quelle, F.W., Sato, N., Witthuhn, B.A., Inhorn, R., Ernst, T.J., Miyajima, A., Griffin, J.D., and Ihle, J.N., 1994, JAK2 associates with the β_c chain of the receptor for GM-CSF and its activation requires the membrane proximal region. *Mol.Cell.Biol.* 14:4335-4341.

Rui, H., Kirken, R.A., and Farrar, W.L., 1994, Activation of receptor-associated tyroisne kinase JAK2 by prolactin. *J.Biol.Chem.* 269:5364-5368.

Sakamaki, K., Miyajima, I., Kitamura, T., and Miyajima, A., 1992, Critical cytoplasmic domains of the common β subunit of the human GM-CSF, IL-3 and IL-5 receptors for growth signal transduction and tyrosine phosphorylation. *EMBO J.* 11:3541-3549.

Sanchez, M.P., Tapley, P., Saini, S.S., Pulido, D., and Barbacid, M., 1994, Multiple tyrosine protein kinases in rat hippocampal neurons: isolation of Ptk-3, a receptor expressed in proliferative zones of the developing brain. *Proc. Natl. Acad. Sci. USA* 91:1819-1823.

Sato, N., Sakamaki, K., Terada, N., Arai, K.I., and Miyajima, A., 1993, Signal transduction by the high-affinity GM-CSF receptor: two distinct cytoplasmic regions of the common β subunit responsible for different signaling. *EMBO J.* 12:4181-4189.

Satoh, T., Nakafuku, M., Miyajima, A., and Kaziro, Y., 1991, Involvement of ras p21 protein in signal-transduction pathways from interleukin 2, interleukin 3, and granulocyte/macrophage colony-stimulating factor, but not from interleukin 4. *Proc. Natl. Acad. Sci. USA* 88:3314-3318.

Schindler, C., Kashleva, H., Pernis, A., Pine, R., and Rothman, P., 1994, STF-IL-4: a novel IL-4-induced signal transducing factor. *EMBO J.* 13:1350-1356.

Showers, M.O., Moreau, J., Linnekin, D., Druker, B., and D'Andrea, A.D., 1992, Activation of the erythropoietin receptor by the Friend spleen focus-forming virus gp55 glycoprotein induces constitutive protein tyrosine phosphorylation. *Blood* 12:3070-3078.

Shultz, L.D., Schweitzer, P.A., Rajan, T.V., Yi, T., Ihle, J.N., Matthews, R.J., Thomas, M.L., and Beier, D.R., 1993, Mutations at the murine motheaten locus are within the hematopoietic cell protein tyrosine phosphatase (Hcph) gene. *Cell* 73:1445-1454.

Silvennoinen, O., Witthuhn, B., Quelle, F.W., Cleveland, J.L., Yi, T., and Ihle, J.N., 1993, Structure of the JAK2 protein tyrosine kinase and its role in IL-3 signal transduction. *Proc. Natl. Acad. Sci. USA* 90:8429-8433.

Soh, J., Donnelly, R.J., Kotenko, S., Mariano, T.M., Cook, J.R., Wang, N., Emmanuel, S., Schwartz, B., Miki, T., and Pestka, S., 1994, Identification and sequence of an accessory factor required for activation of the human interferon gamma receptor. *Cell* 76:793-802.

Stahl, N., Boulton, T.G., Farruggella, T., Ip, N.Y., Davis, S., Witthuhn, B.A., Quelle, F.W., Silvennoinen, O., Barbieri, G., Pellegrini, S., Ihle, J.N., and Yancopoulos, G.D., 1994, Association and activation of Jak-Tyk kinases by CNTF-LIF-OSM-IL-6 beta receptor components. *Science* 263:92-95.

Stahl, N. and Yancopoulos, G.D., 1993, The alphas, betas and kinases of cytokine receptor complexes. *Cell* 74:587-590.

Takahashi, T. and Shirasawa, T., 1994, Molecular cloning of rat JAK3, a novel member of the JAK family of protein tyrosine kinases. *FEBS. Lett.* 342:124-128.

Taniguchi, T. and Minami, Y., 1993, The IL-2/IL-2 receptor system: A current overview. *Cell* 73:5-8.

Tsui, H.W., Siminovitch, K.A., de Souza, L., and Tsui, F.W.L., 1993, Motheaten and viable motheaten mice have mutations in the haematopoietic cell phosphatase gene. *Nat. Gen.* 4:124-129.

Uze, G., Lutfalla, G., and Gresser, I., 1990, Genetic transfer of a functional human interferon α receptor into mouse cells: cloning and expression of its cDNA. *Cell* 60:225-234.

Velazquez, L., Fellous, M., Stark, G.R., and Pellegrini, S., 1992, A protein tyrosine kinase in the interferon alpha/beta signaling pathway. *Cell* 70:313-322.

Watling, D., Guschin, D., Muller, M., Silvennoinen, O., Witthuhn, B.A., Quelle, F.W., Rogers, N.C., Schindler, C., Stark, G.R., Ihle, J.N., and Kerr, I.M., 1993, Complementation by the protein tyrosine kinase JAK2 of a mutant cell line defective in the interferon-τ signal transduction pathway. *Nature* 366:166-170.

Watowich, S.S., Hilton, D.J., and Lodish, H.F., 1994, Activation and inhibition of erythropoietin receptor function: role of receptor dimerization. *Mol.Cell.Biol.* 14:3535-3549.

Watowich, S.S., Yoshimura, A., Longmore, G.D., Hilton, D.J., Yoshimura, Y., and Lodish, H.F., 1992, Homodimerization and constitutive activation of the erythropoietin receptor. *Proc. Natl. Acad. Sci. U. S. A.* 89:2140-2144.

Wilks, A.F. 1989, Two putative protein-tyrosine kinases identified by application of the polymerase chain reaction. *Proc. Natl. Acad. Sci. USA* 86:1603-1607.

Wilks, A.F. 1991, Cloning members of protein-tyrosine kinase family using polymerase chain reaction. *Methods Enzymol.* 200:533-546.

Wilks, A.F., Harpur, A.G., Kurban, R.R., Ralph, S.J., Zurcher, G., and Ziemiecki, A., 1991, Two novel protein-tyrosine kinases, each with a second phosphotransferase-related catalytic domain, define a new class of protein kinase. *Mol.Cell.Biol.* 11:2057-2065.

Witthuhn, B., Quelle, F.W., Silvennoinen, O., Yi, T., Tang, B., Miura, O., and Ihle, J.N., 1993, JAK2 associates with the erythropoietin receptor and is tryosine phosphorylated and activated following EPO stimulation. *Cell* 74:227-236.

Witthuhn, B.A., Silvennoinen, O., Miura, O., Lai, K.S., Cwik, C., Liu, E.T., and Ihle, J.N., 1994, Involvement of the JAK3 Janus kinase in IL-2 and IL-4 signalling in lymphoid and myeloid cells. *Nature* 370:153-157.

Yamamoto, K., Quelle, F.W., Thierfelder, W.E., Kreider, B.L., Gilbert, D.J., Jenkins, N.A., Copeland, N.G., Silvennoinen, O., and Ihle, J.N., 1994, Stat4: A novel GAS binding protein expressed in early myeloid differentiation. *Mol.Cell.Biol.* in press.

Yi, T., Mui, A.L.-F., Krystal, G., and Ihle, J.N., 1993, Hematopoietic cell phosphatase associates with the interleukin-3 (IL-3) receptor β chain and down-regulates IL-3-induced tyrosine phosphorylation and mitogenesis. *Mol.Cell.Biol.* 13:7577-7586.

Yi, T., Quelle, F.W., and Ihle, J.N., 1993, HCP associates with the tyrosine phosphorylated negative regulatory region of the Epo receptor. In preparation.

Zhong, Z., Wen, Z., and Darnell, J.E.J., 1994, Stat3 and Stat4: Members of the family of signal transducers and activators of transcription. *Proc. Natl. Acad. Sci. USA* 91:4806-4810.

DISCUSSION

D. Metcalf

I am always puzzled about the equilibrium situation. In many of your diagrams you have an early induction of expression or induction of phosphorylation and then the change passes off. What is the equilibrium state with respect to some of these changes? More specifically, suppose you transform a FDCP-1 cell to an autonomous cell and you examine such cells. What do you actually find phosphorylated? Is there any general features that distinguish such a transformed cell from a factor-dependent normal or immortalized cell.

J. Ihle

First of all with regards to the equilibrium question. I did not really show the kinetics which have been published. Nevertheless, in these experiments the cells are starved of growth factors and induced with high levels of growth factors to obtain the highest levels of induction of tyrosine phosphorylation. Following this initial high level of induction, after about 30 minutes, the levels reach a much lower but detectable steady state level phosphorylations. These levels may be sufficient to support cell cycle progression which in the case of IL-3, like other growth factors such as CSF-1, requires approximately 10 hours in G-1. That gets the cell through all its checkpoints and now allows it to go through the cell cycle. Whether, and how, the consistent levels of tyrosine phosphorylation that are seen during this period contribute to cell cycle progression are not known.

D. Metcalf

And the neoplastic state?

J. Ihle

Some of these aspects are currently being examined but we have identified certain kinds of transforming mutant of the EPO receptor isolated by Harvey Lodish's group. When this mutant is introduced into cells, the cells are growth factor independent and constitutive activation of Jak-2 is seen. Whether other types of transforming events that abrogate growth requirements, such as transformation by bcr-abl, would result in constitutive activation of Stats or Jaks is not known although I think that there is a good chance, in some cases, they might not. In particular, myeloid cells can be grown in stem cell factor. Under these conditions you do not see activation of Jaks or Stats. Thus, mitogenesis is activated through a non Jak/Stat pathway. So, I would assume transformation may affect other pathways as well. In this regard, one question is whether these other pathways, or the pathways utilized by stem cell factor, require ras signaling for mitogenesis.

D. Metcalf

Would it be an oversimplification to say that an autonomous cell cannot differentiate because it is exclusively using the Stat4 pathway in contrast to a normal cell which has a second pathway able to induce differentiation? I noticed you did not find Stat4 in M-1 leukemic cells but what happens if you induce them to differentiate? Does it then appear?

J. Ihle

You still do not see Stat4. And I should mention that you also see Stat4 expression lost during G-CSF induced 32Dcl differentiation which, unlike the M-1 cells which differentiate to macrophages, differentiate to granulocytes.

D. Cantrell

I have two questions. In your model for the involvement of Jak and shc, do you think Jaks are actually phosphorylating shc? Second question: In the CTLL cells co-expressing EPO receptors and IL-2 receptors where EPO is not mytogenic, have you looked at the pattern of Stat activation by EPO in those cells? Do you get any, or is that the reason why it is not mitogenic?

J. Ihle

To tell the truth, we have not done the experiments although they are quite simple studies to do. However, I can say that the Stat protein that is activated in CTLL cells in response to IL-2 is different from the EPO induced Stat protein by gel shift assays.

D. Cantrell

In different cell types?

J. Ihle

Yes, in different cell types.

T. Taniguchi

You said you have a dominant negative form of Jak 2 kinase. If you overexpress this kinase, do you see any effect on EPO signaling?

J. Ihle

Those experiments are just ongoing. We do not have any definitive answers yet.

T. Taniguchi

Another question is: You said that the GAS binding activity is detected after IL-2 stimulation. Do you see any GAS sequence in c-myc or other genes involved in cell growth?

J. Ihle

I did not mean to imply a functional relationship. At this point all there is, is a correlation such that everything we do to the receptor that inactivates the Jak-Stat pathway correlates with the loss, or retention, of the mitogenic signaling. It is quite possible that, in fact, there are additional pathways being activated. Now, to the second part of your question. The myc promoter does not contain any obvious GAS elements. There is a GAS-like element that is upstream which, interestingly, was originally identified as being important for the suppression of myc expression during B cell differentiation. We have specifically taken that element to see if we could see an EPO induced binding activity against the element. We have not been able to. However, I think an important part of the question is whether or not there are Stat proteins that are functionally of the Stat family but which recognize sequences that are quite different from GAS. Obviously, everything we have been looking at right now is a very biased view because we only take the known GAS elements to look at. So, I think it will be particularly interesting to clone additional family members, based on homologies in other regions that might not be involved in DNA binding activity, to see if there are not other family members that recognize quite different sequences and then go back to look at genes like c-myc for their potential direct involvement.

R. Perlmutter

One of the interesting features about the IL-2 receptor Beta chain, which actually Tada Taniguchi elucidated, was that if you get rid of the serine rich domain the lck activation via the acidic domain no longer takes place. Do you have any insight into the way by which that process occurs?

J. Ihle

Obviously one of the interesting possibilities is that Jak-2 is somehow involved in the activation of the src kinases and we really do not have any answers to that. However, I think it is a very important question that need to be addressed.

R. Perlmutter

Have you been able to generalize those kinds of observations to any other members of the cytokine receptor family?

J. Ihle

In terms of the association of src kinases with cytokine receptors? I must say that in myeloid cells we looked very closely for the association of fyn or lyn with the IL-2 receptor β chain, particularly when we saw the dramatic phenotypic effects that the acidic domain has on the cells, without success. Of course, lck is not expressed in the cells we used. However, as you are aware, the conditions for looking at associations are quite critical. So, I do not think that we have developed the conditions that we would feel comfortable in excluding the association of src kinases with other cytokine receptors.

I. Weissman

Have you looked at any of the radiation induced or non-oncogene containing leukemia virus induced T lymphomas for Jak-3 expression or any other Jak expression? I mean phosphorylation.

J. Ihle

No, but certainly we have the cell lines which we spent a number of years developing. I will give one example, however, that I failed to mention with regard to the question that Don Metcalf asked regarding oncogenes. In particular, in the past we had looked at various oncogenes for their ability to abrogate IL-3 dependence. One that we found could abrogate was the activated c-trk gene, which was somewhat unique in that, unlike v-abl, it did not cause a massive increase in tyrosine phosphorylation in the cells. In fact, the pattern was very similar to that seen in response to IL-3. Quite interesting, no activation of Jaks, or Stats, is seen in these cells. However, the pathway is intact since both Jak2 and Stats are activated following stimulation with IL-3. So, in this case, transformation is utilizing quite a different pathway. I think that everything we have seen, emphasizes the fact that protein tyrosine kinases and the cytokine receptor superfamily have evolved to use unique pathways. In this regard it is possible that the different pathways account for the synergy that is often seen between ligands that use these two classes of receptors.

H. Beug

You said you cannot induce activation of any of these known Stats by the EPO receptor. I suppose none of these experiments have been done in erythroid cells?

J. Ihle

We have looked in erythroid cells. The only Stat we cannot exclude at this point is Stat5 because we do not have an appropriate antibody.

H. Beug

Could you tell me in which type of erythroid cells?

J. Ihle

The best example that we have used are the 32Dcl (Epo1) cells which respond to EPO with expression of the globin gene.

J. Adams

Because the membrane-proximal region of the receptor seems to be all that is required for the proliferative response and because this is the region that causes association of the Jak kinase, it would seem to follow that if one simply had dimerization of the Jak kinases that one might get a full proliferative signal. Is that your expectation and have you tried to do anything to explore that?

J. Ihle

Yes, we are trying to ask exactly that question which is a very important point. I should also mention that there is a wealth of experiments that confirm the role of aggregation of the cytoplasmic domains of one or more of the receptor subunits. These have been done primarily with chimeric receptors. For example, chimerics containing the cytoplasmic domain of the IL-2 receptor β chain and the extracellular and transmembrane domains of c-kit will allow stem cell factor induced growth in myeloid cells by bringing together the β chains. Jim Griffin's group has similar studies with the IL-3 receptor β chain. The same has been done with the EPO receptor in which case the cytoplasmic domains were linked to the EGF extracellular domain and EGF was used to drive dimerization. So there is a wealth of information that indicates that bringing the domains together starts the process. And I think there is going to be some very interesting experiments to explore the possibility that pharmacological approaches might be able to bring about dimerization and receptor activation.

J. Adams

A second, rather more general, question is: Since this type of signal transduction leads all the way to the nucleus and gene activation, does this model somewhat downplay the role of some of the chemical messengers, such as the phosphoinositols, in receptor action?

J. Ihle

I think that there are some questions as to if, and how, PLC-γ, and PI-3 kinase and so forth, are involved in mitogenesis, particularly because the regions of receptors associated with their activation can often be deleted without affecting mitogenesis. But, of course, a cell response is complex and a number of functions such as membrane biogenesis are required and for which these activities may be necessary. The key question that we need to address is whether the Stat proteins have anything to do in mitogenesis. I should also note that since this pathway has only recently been identified, it is very likely that other, more important pathways, remain to be discovered..

A. D'Andrea

Now that you know that cytokine receptors assemble combinatorially with Jaks, are there consensus sequence in the box-1, box-2 region which confer specific Jak binding?

J. Ihle

No, it is not obvious from the linear sequences and I think it is going to take some structural work to really see what is going on there.

A. D'Andrea

Also, I am puzzled as to why you cannot identify a growth factor that activates Stat-4 and I am just wondering whether or not it is for some trivial reason. Is it possible that the antibody you are using cannot recognize the activated form of Stat-4? They are telling you functionally that you are getting activation of Stat-4 in the COS cells and you are getting a new gel shift. But, is it not possible that the antibody that you are using does not recognize the tyrosine phosphorylated form of Stat-4.

J. Ihle

It is unlikely. In particular, as I mentioned, Stat4 can be co-expressed in COS cells with Jak1 or Jak2 and becomes tyrosine phosphorylated and acquires DNA binding activity. The antibody we have can detect both the tyrosine phosphorylated and non-phosphorylated forms and can be used in supershift type experiments. Thus the reagents we have can see the activated form. The current model for Stat activation is that phosphorylation occurs on a single tyrosine and that this is required for the dimerization. Specifically, homo or hetero, dimerization of Stats occurs through the interaction of the SH-2 domain on one partner and the phosphorylated tyrosine on the other partner. If correct, one could obtain a heterodimer in which only one partner was phosphorylated, perhaps with a weaker association. Thus the question arises as to whether unphosphorylated Stat5 contributes to heterodimers with another phosphorylated protein. To assess this we have looked at the ability of our antibody to supershift complexes from a various cytokine induced GAS binding activities. None of these have been affected, thus I think it is unlikely that Stat4 contributes in this manner either.

E. Mihich

The work of Heinz Bauman is concerned with the effect of cytokines on the expression of acute phase proteins in liver cells. And that seems to be a nice model to compare with your work on the Jak/Stat in relation to a non mitogenic effect of certain cytokines.

J. Ihle

There are two parts to the answer. First, the acute phase response factor that you refer to is Stat-3, and thus his system is another example of cytokines using the Stat pathways. Secondly, you raise a very important point which I can emphasize. For example, like the model you mention, interferons are also not associated with mitogenesis and, in some cases, are associated with a negative growth signal. However, whether one can deduce from this that, in general, the Stats are not involved in growth is unclear. There are a number of Stats, which can affect genes in different manners, the differential ability to induce IRF-1 expression is one clear example. Thus I think, at this point, we need to design experiments that can directly assess the importance of individual Stats in the mitogenic responses with which they are currently correlated.

D. Metcalf

Did you say that gp-130 activates Stat-3 to mediate all cell functions, or are you still postulating a difference between mitogenic pathways and functional pathways.

J. Ihle

In the cases of Stat-3, specifically in the cells in which it has been examined, its activation is often not associated with mitogenesis. So I would say that with Stat-3, just as with Stat-1, there is no association with mitogenesis. However, whether IL-6 is activating other Stats in cells in which it is mitogenic is another question. In these cases, a particular Stat may be essential for mitogenesis.

11

POSITIVE AND NEGATIVE GROWTH EFFECTS OF Abl GENES

Charles L. Sawyers[1],* Andrei Goga[2], Daniel Afar[2], Jami McLaughlin[2], and Owen Witte[2,3]

Department of Medicine, Hematology-Oncology
Microbiology and Molecular Genetics
Howard Hughes Medical Institute
University of California, Los Angeles, California

INTRODUCTION

The Abl gene is well known for its role in the pathogenesis of Abelson virus induced murine leukemia and Philadelphia chromosome positive human leukemias. In both cases structural alterations in the normal c-Abl gene occur due to retroviral transduction or chromosome translocation. Gag-abl (v-Abl) induces lymphoid leukemias in mice, and Bcr-Abl is associated with chronic myelogenous leukemia and acute lymphocytic leukemia in humans. The fusion of either Bcr or Gag sequences to Abl converts c-Abl from a low specific activity, nuclear tyrosine kinase to a highly active cytoplasmic tyrosine kinase (reviewed in Kurzrock et al, 1988; Sawyers, 1992). Our work has focused on understanding the Bcr-Abl signal transduction pathway that leads to leukemia. We have used genetic approaches to define molecules involved in transmission of the Bcr-Abl tyrosine kinase signal from the cytoplasm to the nucleus and have identified Ras and Myc as essential proteins in the Bcr-Abl transformation pathway. We have also found that the normal c-Abl protein suppresses growth and that mutant proteins such as v-Abl and Bcr-Abl convert c-Abl from a growth suppressive to a transforming protein.

RESULTS AND DISCUSSION

Bcr-Abl Signals through Ras

The Ras family of G-proteins play a central role in relaying growth stimulatory signals from outside the cell to the nucleus. Mutations in Ras disrupt this normal signaling

*Address correspondence to: Charles L. Sawyers, 11-934 Factor Building, Hematology-Oncology, University of California Los Angeles, 10833 LeConte Avenue, Los Angeles, California, 90024. Phone: 310-206-5585; Fax: 310-825-6192.

Normal and Malignant Hematopoiesis, Edited by Enrico Mihich and Donald Metcalf
Plenum Press, New York, 1995

pathway and can lead to cancer by allowing deregulated cell growth. Ras mutations are a common molecular characteristic of several myeloid leukemic diseases such as acute myelogenous leukemia and certain myelodysplastic syndromes (Janssen et al, 1987). Curiously, Ras is rarely mutated in Philadelphia chromosome positive CML (Cogswell et al, 1989; Urbano-Ispizua et al, 1992). One explanation might be that the Ras protein in CML cells, although normal, is regulated in an abnormal fashion by the Bcr-Abl fusion protein.

Ras is activated by stimulatory GDP/GTP exchange molecules which convert Ras from its inactive GDP-bound state to its active GTP-bound state. Inhibitory GTPase activating molecules such as RasGAP and neurofibromin (NF-1) convert Ras back to its inactive state. The adaptor molecule Grb2 provides a potential biochemical connection between Bcr-Abl and Ras because Grb-2 can bind Bcr at tyrosine residue 177 (Pendergast et al, 1993a; Puil et al, 1994). By analogy with growth factor receptor tyrosine kinases, this interaction would lead to Ras activation through the Sos protein (reviewed in Schlessinger, 1993).

We have used a genetic approach to evaluate if Ras is essential for this transformation signal. When Ras function was interrupted in Rat-1 fibroblasts by overexpression of the catalytic domain of RasGap (Gap C terminus), the transforming activity of Bcr-Abl and v-Abl was blocked (Sawyers et al, in press). The ability to suppress transformation was not a result of toxicity because expression of the Gap C terminus did not slow the growth of Rat-1 cells and did not block transformation by v-Mos, which functions independent of Ras. Transformation of mouse bone marrow by Bcr-Abl was also blocked by co-expression of either Gap C terminus or a dominant negative mutant of Ras (Ras Asn 17). These results show that Ras is required for the oncogenic activity of Bcr-Abl and v-Abl. Grb-2 may be one molecule that connects Bcr-Abl to Ras.

The Bcr-Abl SH2 Domain Mediates a Signal to Myc

V-Abl and Bcr-Abl both activate transcription of the c-Myc gene (Cleveland et al, 1989) and cooperate with Myc to transform fibroblasts (Lugo et al, 1989). These findings raise the issue of whether Myc is required for the Bcr-Abl transformation signal. Similar to our studies with Ras, we have used dominant negative forms of Myc to interfere with its function in cells expressing Bcr-Abl. We have found that transformation of fibroblasts as well as bone marrow cells by either Bcr-Abl or v-Abl is dependent on threshold levels of Myc (Sawyers et al, 1992). This finding genetically defines a connection between Bcr-Abl and Myc, but the mechanism for this link is unclear.

To explore this link we have used complementation analysis to identify a domain within Bcr-Abl responsible for a signal to Myc. Point mutations were generated in domains of Bcr-Abl that are expected to participate in connecting its tyrosine kinase signal to downstream molecules. If the mutation impairs the transforming activity of Bcr-Abl by uncoupling a critical signaling pathway, then overexpression of a downstream molecule would be expected to restore transforming activity.

Point mutations were generated in the major tyrosine autophosphorylation site of the kinase domain (Pendergast et al, 1993b), the SH2 domain (Afar et al, 1994) and the Grb-2 binding site within Bcr (Pendergast et al, 1993a). Each of these mutants lost the ability to transform Rat-1 fibroblasts. However, overexpression of c-Myc specifically restored transforming activity to cells expressing the SH2 domain point mutant (figure 1). Biochemical analysis of cells expressing the SH2 domain mutant showed both a reduction in the overall content of tyrosine phosphorylated molecules and the absence of several prominent tyrosine phosphorylated bands such as p62. These changes in tyrosine phosphorylation were not due to the lack of transforming activity of the SH2 domain mutant because they persisted even

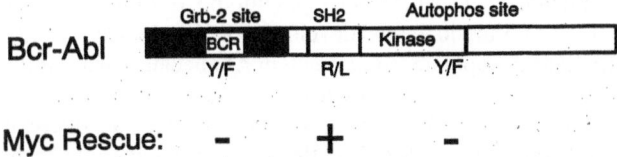

Myc Rescue:

Figure 1. Transformation by a Bcr-Abl molecule with a mutant SH2 domain is specifically rescued by c-Myc. A cartoon of the Bcr-Abl protein containing single amino acid changes in the Grb-2 binding site, phosphotyrosine binding site of the SH2 domain, or major tyrosine autophosphorylation site is pictured. The results of complementation experiments with c-Myc are shown below.

in Rat-1 cells transformed by addition of c-Myc. These observations provide evidence that Bcr-Abl transforms cells by at least two signaling pathways: an SH2 signal that functions upstream of Myc and a Myc independent signal, perhaps transmitted by the Grb-2 binding site or major tyrosine autophosphorylation site (figure 2).

Overexpression of c-Abl Induces Growth Arrest

Despite this understanding of the oncogenic Abl fusion proteins, very little is known about the function of normal c-Abl. C-Abl is located predominantly in the nucleus (Van Etten et al, 1989), and the tyrosine kinase activity of c-Abl is tightly regulated. Despite *in vitro* autokinase activity, detectable levels of phosphotyrosine on c-Abl are not observed *in vivo* unless it is expressed at very high levels (Pendergast et al, 1991). C-Abl is differentially phosphorylated during the cell cycle at consensus sites for the serine/threonine kinase cdc2 (Kipreos and Wang, 1990), suggesting a potential role in cell cycle progression. C-Abl contains a DNA binding domain (Kipreos and Wang, 1992) and is reported to bind a specific regulatory sequence in the hepatitis B enhancer known as EP (Dikstein et al, 1992). These findings indicate a possible role for c-Abl in transcription or DNA replication.

Gene disruption experiments have failed to clarify the normal function of c-Abl. Abl deficient mice proceed normally through fetal development. However, animals are born runted, and most die within a few weeks from a poorly understood malnutrition syndrome (Schwartzberg et al, 1991; Tybulewicz et al, 1991). Genetic studies of the drosophila Abl homologue suggest a role in the nervous system. Abl deficient flies have subtle abnormalities of neural development (Henkemeyer et al, 1987). When crossed with a recessive mutation

Figure 2. Multiple signaling pathways are required for the transforming activity of Bcr-Abl. Experiments using dominant negative mutants have shown that Ras and Myc are required for transformation by Bcr-Abl. Complementation experiments show that c-Myc specifically restores transforming activity to an SH2 domain Bcr-Abl mutant, defining a pathway between Bcr-Abl and Myc. Tyrosine phosphorylated (PY) proteins are suspected to mediate this signal based on different patterns of tyrosine phosphorylation in cells expressing wild-type versus the SH2 mutant Bcr-Abl. Grb2 is postulated to be a connection between Bcr-Abl and Ras. The major tyrosine autophosphorylation site may connect Bcr-Abl to an additional signaling pathway, but the proteins participating in this pathway have not been identified.

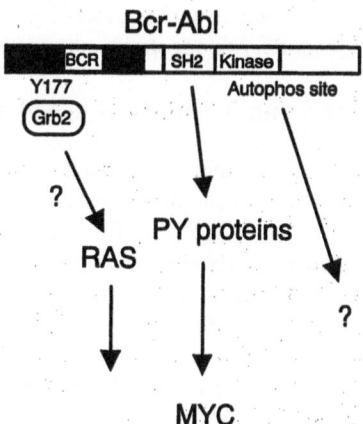

in a second locus known as Disabled, this phenotype is dramatically converted to an embryonic lethal due to a failure in axonal development (Gertler et al, 1989). The drosophila results may be difficult to extrapolate to mammalian Abl because drosophila Abl is cytoplasmic and its C-terminus differs considerably from the mammalian form.

To further investigate the function of mammalian Abl, we overexpressed c-Abl or various mutants in fibroblasts. NIH3T3 cells were infected with high titer retrovirus expressing wild-type c-Abl or a kinase defective c-Abl mutant and grown in antibiotic selection media to derive clones which overexpressed the respective proteins. Despite efficient gene transfer and expression of wild-type c-Abl two days after infection of NIH3T3 cells, very few cells infected with wild-type c-Abl expanded to form colonies after antibiotic selection when compared to the kinase defective c-Abl mutant. Furthermore, high level expression of wild-type c-Abl in the survivors was quite low (Sawyers et al, 1994). These results suggested that c-Abl either inhibits growth or is toxic to cells. To distinguish these possibilities we generated a cell line in NIH3T3 cells (3T3/MMTV c-Abl cells) in which c-Abl expression can be induced by addition of dexamethasone. Induction of wild-type c-Abl, but not the kinase defective mutant, led to growth arrest. To determine the point of cell cycle arrest, the DNA content of growth arrested 3T3/MMTV c-Abl cells was analyzed by propidium iodide staining. After 48-72 hours in dexamethasone, 90% of cells were arrested in G1 (Sawyers et al, 1994).

These findings demonstrate a new biological activity for c-Abl, growth arrest, that contrasts with that of the fusion oncogenes Bcr-Abl and v-Abl. The biologic activity of c-Abl resembles that of tumor suppressor genes such as p53 and RB (reviewed in Levine, 1993). If c-Abl functions to control growth in a negative way, events which block or disrupt Abl function should lead to growth deregulation. To test this hypothesis we examined the growth characteristics of fibroblasts expressing a dominant negative (kinase inactive) c-Abl protein. Although dominant negative c-Abl did not significantly affect the growth rate of Rat-1 fibroblasts, it did render them more susceptible to transformation by secondary oncogenes such as Myc, v-Abl, Ras and v-Fms (Sawyers et al, 1994). These findings demonstrate that loss of c-Abl function can facilitate cell transformation, consistent with a role in negative regulation of cell growth.

Based on these findings one might also expect that Abl deficient mice would be predisposed to form tumors. However, this does not appear to be the case. Potential explanations may be the failure of these mice to live long enough to develop tumors due to a malnutrition syndrome or redundant function among Abl related genes such as Arg (Kruh et al, 1990). Alternatively, multiple genetic events may be required to generate tumors in addition to loss of c-Abl. The latter hypothesis may explain the lack of tumor formation in mice deficient in the IRF-1 tumor suppressor gene. Similar to the dominant negative c-Abl cells, fibroblasts lacking IRF-1 grow normally but are more susceptible to transformation by secondary oncogenes such as v-Ras (Tanaka et al, 1994).

Last Exon Mutants of c-Abl Transform Cells

To further understand the biological activity of c-Abl, we have used a random mutagenesis scheme to isolate mutant c-Abl molecules which lose growth suppression activity and gain transforming activity (Goga et al, 1993). A c-Abl molecule which transforms NIH3T3 fibroblasts was recovered in this screen and contains an in-frame deletion within the C-terminal exon (c-Abl ΔLX). This mutant differs from previously described transforming Abl mutants such as the v-Abl and Bcr-Abl fusion proteins or a c-Abl molecule lacking the SH3 domain (Jackson and Baltimore, 1989; Franz et al, 1989). First it defines a novel role for the Abl C terminus, a domain which has not been previously implicated in activating the transforming activity of c-Abl. Second, c-Abl ΔLX localizes

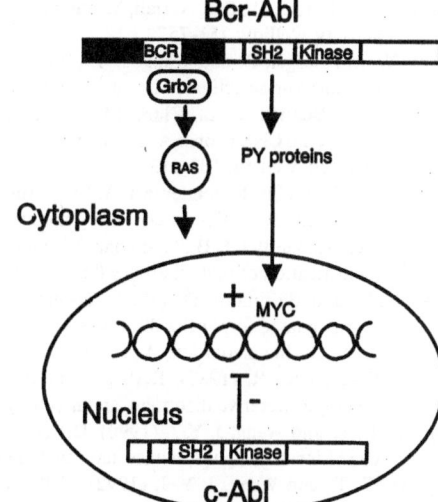

Figure 3. Growth stimulatory versus growth suppressive effects of Bcr-Abl and c-Abl. Bcr-Abl is a cytoplasmic tyrosine kinase which leads to cellular transformation by activation of Ras and Myc pathways. c-Abl is a nuclear tyrosine kinase which suppresses cell growth by an unknown pathway.

to both the nucleus and cytoplasm whereas all other transforming Abl proteins are exclusively cytoplasmic. This finding suggests that c-Abl ΔLX may transform cells by a distinct mechanism. In fact, deletion of the nuclear localization signal of c-Abl ΔLX creates a protein which is exclusively cytoplasmic and no longer transforming (A. Goga and O. Witte, unpublished observations). In light of the growth suppressive activity of the normal c-Abl protein in the nucleus, it will be of interest to define the mechanism of transformation by c-Abl ΔLX.

CONCLUSION

We have used genetic approaches to define proteins in the Bcr-Abl signaling pathway required for its transforming activity. Ras and Myc have been identified as critical signaling intermediates in the cytoplasm and nucleus. Bcr-Abl communicates with Ras, at least in part, through the Grb-2 adaptor protein. The SH2 domain of Bcr-Abl mediates a signal to Myc by an unknown pathway. In contrast to the transforming activity of Bcr-Abl, the normal c-Abl protein suppresses growth (figure 3). Knowledge of the signaling pathways utilized by Bcr-Abl and c-Abl should provide novel insight into potential therapeutic interventions for Philadelphia chromosome positive leukemias.

REFERENCES

Afar, D. E. H., Goga, A., McLaughlin, J., Witte, O. N., and Sawyers, C. L. (1994). Differential complementation of BCR-ABL point mutants with c-MYC. Science *264*, 424-426.

Cleveland, J. L., Dean, M., Rosenberg, N., Wang, J. Y., and Rapp, U. R. (1989). Tyrosine kinase oncogenes abrogate interleukin-3 dependence of murine myeloid cells through signaling pathways involving c-myc: conditional regulation of c-myc transcription by temperature-sensitive v-abl. Mol. Cell. Biol. *9*, 5685-5695.

Cogswell, P. C., Morgan, R., Dunn, M., Neubauer, A., Nelson, P., Poland-Johnston, N. K., Sandberg, A. A., and Liu, E. (1989). Mutations of the Ras protooncogenes in chronic myelogenous leukemia: A high frequency of Ras mutations in bcr/abl rearrangement-negative chronic myelogenous. Blood *74*, 2629.

Dikstein, R., Herretz, D., Ben-Neriah, Y., and Shaul, Y. (1992). c-abl has a sequence-specific enhancer binding activity. Cell 69, 751-757.

Franz, W. M., Berger, P., and Wang, J. Y. (1989). Deletion of an N-terminal regulatory domain of the c-abl tyrosine kinase activates its oncogenic potential. EMBO J 8(1), 137-147.

Gertler, F. B., Bennett, R. L., Clark, M. J., and Hoffman, F. M. (1989). Drosophila abl tyrosine kinase in embryonic CNS axons: A role in axonogenesis is revealed through dosage-sensitive interactions with disabled. Cell 58, 103-113.

Goga, A., McLaughlin, J., Pendergast, A. M., Parmar, K., Muller, A., Rosenberg, N., and Witte, O. N. (1993). Oncogenic activation of c-ABL by mutation within its last exon. Mol. Cell. Biol. 13, 4967-4975.

Henkemeyer, M., Gertler, F. B., Goodman, W., and Hoffman, F. M. (1987). The Drosophila Abelson homolog: identification of mutant alleles that have pleiotropic effects late in development. Cell 51, 821-828.

Jackson, P., and Baltimore, D. (1989). N-terminal mutations activate the leukemogenic potential of the myristoylated form of c-abl. EMBO J 8, 449-456.

Janssen, J. W. G., Steenvoorden, A. C. M., Lyons, J., Anger, B., Bohlke, J. U., Bos, J. L., Seliger, H., and Bartram, C. R. (1987). RAS gene mutations in acute and chronic myelocytic leukemias, chronic myeloproliferative disorders, and myelodysplastic syndromes. Proc. Natl. Acad. Sci. USA 84, 9228.

Kipreos, E. T., and Wang, J. Y. J. (1990). Differential phosphorylation of c-ABL in cell cycle determined by cdc2 kinase and phosphatase activity. Science 248, 382-385.

Kipreos, E. T., and Wang, J. Y. J. (1992). Cell cycle-regulated binding of c-ABL tyrosine kinase to DNA. Science 256, 382-385.

Kruh, G. D., Perego, R., Miki, T., and Aaronson, S. A. (1990). The complete coding sequence of arg defines the Ableson subfamily of cytoplasmic tyrosine kinases. Proc. Natl. Acad. Sci. USA 87, 5802-5806.

Kurzrock, R., Gutterman, J. U., and Talpaz, M. (1988). The molecular genetics of Philadelphia chromosome positive leukemias. New Engl. J. Med. 319, 990-998.

Levine, A. (1993). The tumor suppressor genes. Annu. Rev. Biochem 62, 623-651.

Lugo, T., and Witte, O. N. (1989). The BCR-ABL oncogene transforms Rat-1 cells and cooperates with v-myc. Mol. Cell. Biol. 9, 1263-1270.

Pendergast, A. M., Muller, A. J., Havlik, M. H., Clark, R., McCormick, F., and Witte, O. N. (1991). Evidence for regulation of the ABL tyrosine kinase by a cellular inhibitor. Proc. Natl. Acad. Sci. USA 88, 5927-5931.

Pendergast, A. M., Quilliam, L. A., Cripe, L. D., Bassing, C. H., Dai, Z., Li, N., Batzer, A., Rabun, K. M., Der, C. J., Schlessinger, J., and Gishizky, M. L. (1993a). BCR-ABL-induced oncogenesis is mediated by direct interaction with the SH2 domain of the GRB-2 adaptor protein. Cell 75, 175-185.

Pendergast, A. M., Gishizky, M. L., Havlik, M. H., and Witte, O. N. (1993b). SH1 domain autophosphorylation of P210 BCR-ABL is required for transformation but not growth factor-independence. Mol. Cell. Biol. 13, 1728-1736.

Puil, L., Liu, J., Gish, G., Mbamalu, G., Bowtell, D., Pelicci, P. G., Arlinghaus, R., and Pawson, T. (1994). BCR-ABL oncoproteins bind directly to activators of the Ras signalling pathway. EMBO J. 13(4), 764-773.

Sawyers, C. L. (1992). The bcr-abl gene in chronic myelogenous leukemia. Can. Surveys 15, 37-51.

Sawyers, C. L., Callahan, W., and Witte, O. N. (1992). Dominant negative MYC blocks transformation by ABL oncogenes. Cell 70, 901-910.

Sawyers, C. L., McLaughlin, J., Goga, A., Havlik, M., and Witte, O. (1994). The nuclear tyrosine kinase c-ABL negatively regulates cell growth. Cell 77, 1-20.

Sawyers, C. L., McLaughlin, J., and Witte, O. N. (1994). Genetic requirement for Ras in the transformation of fibroblasts and hematopoietic cells by the Bcr-Abl oncogene. J. Exp. Med., in press.

Schlessinger, J. (1993). How receptor tyrosine kinases activate Ras. Trends Biol. Sci. 18, 273.

Schwartzberg, P. L., Stall, A. M., Hardin, J. D., Bowdish, K. S., Humaran, T., Boast, S., Harbison, M. L., Robertson, E. J., and Goff, S. P. (1991). Mice homozygous for the abl^m1 mutation show poor viability and depletion of selected B and T cell populations. Cell 65, 1165-1175.

Tanaka, N., Ishihara, M., Kitagawa, M., Harada, H., Kimura, T., Matsuyama, T., Lamphier, M. S., Aizawa, S., Mak, T. W., and Taniguchi, T. (1994). Cellular commitment to oncogene-induced transformation or apoptosis is dependent on the transcription factor IRF-1. Cell 77, 829-839.

Tybulewicz, V. L. J., Crawford, C. E., Jackson, P. K., Bronson, R. T., and Mulligan, R. C. (1991). Neonatal lethality and lymphopenia in mice with a homozygous disruption of the c-abl proto-oncogene. Cell 65, 1153-1163.

Urbanao-Ispizua, A., Gill, R., Matutes, E., Levi, S., Wiedemann, L. M., Catovshy, D., and Marshall, C. J. (1992). Low frequency of Ras oncogene mutations in Philadelphia-positive acute leukemia and report of a novel mutation H61 Leu in a single case. Leukemia 6, 342.

Van Etten, R. A., Jackson, P., and Baltimore, D. (1989). The mouse type IV c-abl gene product is a nuclear protein, and activation of transforming ability is associated with cytoplasmic localization. Cell *58*, 669-678.

DISCUSSION

D. Livingston

A few things: Jean Wang, as you know, has reported that RB and C-Abl form a complex. I would be interested in your thoughts on the potential significance of that, and, secondly, these DNA binding deletion mutants which seem to transform, transform a variety of cells and if so, do they form complexes with RB?

C. Sawyers

Jean Wang reported that Abl and RB bind and that the binding site within Abl is in the kinase domain. We have tried with this functional assay to ask the question: "Is RB required for Abl toxicity or growth suppression. Tyler Jacks sent us fibroblasts from RB knockout mice and using the antibiotic selection assay, we have seen no effect of loss of RB on growth suppression. In other words, Abl is a growth suppressor in the absence of RB. That is what we know about that connection so far. I do not know if the Delta LX mutation (Delta DNA binding domain mutation) binds RB or not. We have not tested it. The prediction is that it probably still would bind because the binding domain for Abl that Jean identified is still intact. We have to test it. The Delta DNA binding domain mutation has only been shown to transform fibroblasts. Andrei has tested it in a pre-B lymphoid transformation assay and has not seen transformation. But the Delta SH-3 Abl does not transform in that assay in our hands either.

D. Livingston

Have you tried to make the Bcr Delta DNA binding domain mutant and then ask where it transforms normally?

C. Sawyers

It transforms normally in fibroblasts. We have not put it through the whole battery of hematopoietic cell assays.

R. Perlmutter

Having competing growth suppressive and growth promoting effects complicates the interpretation of these sorts of experiments. The first question is with Bcr Abl, can you perform an intra-genic complementation where two transformation defective Bcr-Abl mutants will complement one another and give a transforming phenotype?

C. Sawyers

This is trying to get at the two-signal model. We did the following experiment. We asked if the SH-2 domain mutant is just missing one signal and, say, the Grb-2 mutant is missing the other signal, if we add the two together, could we restore the whole phenotype?

We did not see that. We thought that they were acting in a dominant negative fashion against each other. However, we did see that the SH2 domain mutant inhibited transformation by wild-type Bcr-Abl.

R. Perlmutter

OK, but that leads to the second question, which is: How you interpret exactly the effect of FLVRES myc on the mutation. The straightforward interpretation which is satisfying is that in one fashion or other, Bcr Abl communicates with the myc pathway. You get rid of the SH-2 binding of phosphotyrosine, and you can complement with myc. An alternative interpretation is that mutation is, in fact, behaving as a dominant negative on some other pathway and revealing the ability of myc to promote growth. You have now isolated the ability of this Bcr Abl mutant to pull out a growth suppressor function, which is ordinarily part of its activity. Does that make sense?

C. Sawyers

We have tossed about that idea and not come up with a clean way of testing it. This may not really test it formally but we went ahead and made combinatorial constructs where an SH-2 mutant was combined with a Grb-2 mutant or an auto-phos mutant, all in the same molecule. Our thinking was that if it was one pathway we may be able to complement the double mutant vs. two pathways. We found that a double mutant could not be complemented. But there are arguments against why that methodology is not really the best way to test that idea. That is all I can say I think at this point.

R. Perlmutter

There are difficulties in saying directly that SH-2 in one fashion or other impinges on a pathway that ordinarily activates myc, for example.

L. Degos

You report that Dexamethasone induces an increase of expression of normal Abl. Did you try to see what happens in the malignant cells from patients if it increases the normal alleles of Abl either in short term culture or in long term culture?

C. Sawyers

We have not done that, but that experiment was done with a cell line which was engineered with an MMTV promoter to express Abl in an inducible fashion. We have no evidence that the endogenous level of Abl is affected at all by Dexamethasone. So that was just a special case with a specific construct.

D. Cantrell

Can you tell me what is known about how Ras contributes to transformation in these Abl transformed cells. Has anybody seen activation of Raf or MAP kinase in the cells, particularly if you take the fresh leukemia cells?

C. Sawyers

We have not done that and I am not aware of other people who have. Ras is clearly activated if you do the GDP-GVTP loading type experiments in experimental models. I do not know that that has been done in fresh CML cells and I do not know about MAP kinase activation. Jim, do you have any ideas on this?

J. Griffin

We have looked in stable phase CML neutrophils for ras activation, raf phosphorylation, MAP kinase tyrosine phosphorylation and none of them are activated in the stable phase. However, stable phase CML cells are less "transformed" than blast phase CML cells or cell lines transformed by Bcr-Abl oncogene. Stable phase CML cells for one thing are not factor independent. They are always factor dependent. So it is really a different disease in some ways.

D. Baltimore

Two questions: One is: Does the DNA binding deletion change the intra-nuclear localization of the protein?

C. Sawyers

We see localization in both the nucleus and the cytoplasm. The way we have done nuclear localizations is transfection of COS cells and immunofluorescence. That is what we get to work best. C Abl wild type is almost exclusively nuclear in that assay but the Delta DNA binding has a cytoplasmic portion as well, which is definitely different from wild type. We then took the Delta DNA binding deletion and mutated the nuclear localization signal, and now we saw no nuclear protein in loss of transformation. So that was really the key experiment that showed that this transformation is through a nuclear pathway.

D. Baltimore

The other question is whether you looked at DNA binding by the DNA binding deletion?

C. Sawyers

We did DNA cellulose over two years ago when Ann Marie Pendergast was in the lab. I did not do it, and I cannot tell you what the final result was. It seemed to be consistent with what everyone else was saying, but we should go back and check that.

C. Schneider

I wonder if c-Abl is still suppressive in p53 null cells?

C. Sawyers

I am wondering the same thing and I do not have the answer yet. We are doing the experiments. We are asking Tyler Jacks for about every knockout cell line we can come up with. But that is obviously another possibility.

C. Brugnara

After two weeks of G418 selection, there were some additional bands in your C Abl blots in the K(290) R and delta SH-3 lanes. Do you have any explanation for that?

C. Sawyers

I do not have a clean explanation. The antibody we are using is to the very end of the protein. They could be degradation products. I do not really know. I think they are degradation products because when we look at cells overexpressing viral Abl at high levels, we see similar bands. If we get high level expression I think we just begin to see that. It might be just the way we are lysing the cells.

D. Metcalf

I was interested in your observation that c-Abl seemed to be arresting cells in G-1, because the characteristic of CML cells *in vivo* is their prolonged G-1 period. Has anyone measured whether the Bcr-Abl in any of the transformation systems actually prolongs cell cycle times?

C. Sawyers

We have not done that and I do not know if anyone else has done precisely that experiment but, as you probably know, a number of groups have been reporting that Bcr-Abl can block apoptosis in a lot of assays which would suggest prolonged cell survival. Whether or not that is a specific effect of prolonging cell cycle intervals or if it is just the number of cell cycles, is open to question.

H. Beug

I would be interestd in the multi-step nature of this disease because if I am right, the Bcr-Abl mainly acts in the early phases, pre-leukemic or after treatment of one of the acute phases. What Bcr-Abl seems to do is to keep the particular clone of stem cells alive and allows it to overgrow all the other clones. Therefore, you essentially obtain normal progeny of essentially all lineages. Only in blast crisis, where apparently secondary oncogenic changes occur that may cooperate with Bcr-Abl you get the full blown leukemia. Is there any way you could comment on that? Are other oncogenes, like myc, affected in the disease that would relate to the oncogene cooperation you mentioned.

C. Sawyers

All I can say is that there is clear cytogenetic abnormalities during progression of blast crisis and they are consistent and implicate certain areas of the chromosome. What those mutations specifically are, I think, is a very open question. There is a good bit of evidence that p53 mutations occur in blast crisis and not in chronic phase and depending on whose series, it is somewhere between 10 to 30% of patients. But I think the majority of patients have an unidentified abnormality. Is myc overexpressed? Maybe. It is hard to know. It is difficult to really do a myc Northern blot in a precise way comparing blast cells to mature myeloid cells. Trisomy 8 is one of the abnormalities but whether or not it specifically implicates myc I think is unknown.

D. Baltimore

V Abl does increase the cell cycle time of cells. V-Abl transformation is a very complicated business. Jean Wong found years ago that using NIH 3T3 cells as an assay, that you can pick sub-lines in NIH 3T3 cells that will either respond by G-1 arrest or transformation to V Abl. I do not know what Jean said it was that she was looking at but presumably there is some natural variation in the expressions of something that leads to the differential response. But just in culture, you actually get selection against V Abl transformed cells in favor of normal cells because of a slower cell cycle. One question I had for you is in fact: Do you know whether the DNA binding mutation increases myc. Is it a myc inducing mutation?

C. Sawyers

We do not know. We have not done it. It is a good idea.

J. Griffin

One of the most striking differences that you see between stable phase and transition to blast phase in patient samples is the change in the level of kinase activity, at least as assessed by the number and amount of phospho tyrosine containing proteins per cell. Do you have any thoughts on what happens, either from a model system or actually looking at Bcr-Abl kinase between stable phase and blast phase.

C. Sawyers

I think there are problems in interpreting that because the cells you are comparing are pretty different in a stable phase patient when you are doing the assay on mature neutrophils vs. blasts in the blast crisis phase.

J. Griffin

Actually a couple of groups have purified myeloblasts from stable phase peripheral blood and they really have a very simple pattern of protein tyrosine phosphorylation. Obviously it is hard to get enough normal myeloblasts to do a proper comparison. So it looks like there is at least some difference in the kinase activity. Perhaps not as much as you would assess from looking at neutrophils.

C. Sawyers

I do not know the way to really do that in a model system. The only chronic phase CML model is the animal model which is, as we have heard, a difficult model to work with. There are difficulties in transplantation of the model to secondary mice and so forth and we have not tried to look at that. I cannot think of a molecular change that might be the explanation for that either. Dr. Ihle suggests phosphatases could be deleted and that could be an explanation.

D. Livingston

If, after transfecting or infecting with the C Abl retro in the neo assay, look for cells that finally grow out, do you ever find cells which can grow in agar? And, if so, are there spontaneous deletions in any region of the protein?

C. Sawyers

Yes, actually Andrei Goga did essentially that experiment a year or so ago and found colonies which would grow and then cloned them up and used serologic testing to find mutations in Abl or deletions. That is how the first deletion which he called delta LX was found. There was a huge deletion in the last exon which went from approximately the nuclear localization signal to beyond the DNA binding domain. There is a number of other clones that have not even been characterized yet that behaved similarly in that assay. In fact the way these mutations were created was using this transient transfection system to make retroviruses. In our system we have not separated Gag and envelop proteins. This might be an assay to screen for oncogenic mutations in normally non-oncogenic alleles. I know other groups are trying the system to do precisely that.

R. Perlmutter

Just to return to the issue raised by Don Metcalf and David Baltimore. The growth suppressive effects of the non receptor protein tyrosine kinases that are in other respects transforming is a fairly uniform observation. For example, the activated mutants of hck and lck in fibroblasts anyway, have a growth suppressive effect and are strongly selective against it. It is very difficult to get retroviruses which express those genes, to express high levels of the protein in cells until ultimately transformed foci arise. And if you do an experiment in which you couple metallothionine promoter to lck or hck precisely the same effect is seen. That is, as you increase the level of protein, you lengthen the cell cycle time.

J. Adams

I wonder about the mechanism of the G1 arrest. Have you explored whether it might involve induction of the messenger RNA for the p21 inhibitor of cyclin-dependent kinases?

C. Sawyers

We are doing it. We do not know. We are trying also to map where in G1 using some temperature sensitive mutants the growth arrest occurs.

J. Adams

On a different matter, has anyone followed up on this DNA binding to determine whether it is sequence specific?

C. Sawyers

I think people have followed up on it but no one has reported much on it. We have not found a specific site.

E. Mihich

Do you see a consistency in your domain deletion experiments. In the constructs that you are testing you said you found each domain was required in relation to the transformation that you were checking. But did you see any consistency, any parallel or discordance between the different functions, that is growth retardation, anti-apoptotic effect, transformation etc. in the importance of the domain. Are they all going together or can one dissociate effects?

C. Sawyers

We have just compared growth suppression to transformation. We have not looked at apoptosis. The auto-phosphorylation site within the kinase domain which is required for transformation is not required for growth suppression of Abl. The DNA binding deletion is not required for transformation of Bcr-Abl but is for c Abl. Nuclear localization is required for c Abl and not for Bcr-Abl.

12

ADVANCES IN THE UNDERSTANDING OF THE MOLECULAR PATHOGENESIS OF AGGRESSIVE B CELL LYMPHOMAS

Katarina Cechova, Wei Gu, Bihui H. Ye, Francesco Lo Coco,
Chih-Chao Chang, Jiandong Zhang, Anna Migliazza, Wilfredo Mellado,
Huifeng Niu, and Riccardo Dalla-Favera

Division of Oncology, Department of Pathology and Department of
  Genetics and Development
College of Physicians and Surgeons
Columbia University
New York, New York 10032

INTRODUCTION

Non-Hodgkin lymphoma (NHL) include neoplasms originating from lymphoid cells and characterized by a high degree of biological and clinical heterogeneity (for review see Magrath, 1990). Most NHL derive from the B-cell lineage, in particular from mature B-cells characterized by rearranged immunoglobulin (Ig) heavy and light chain genes and by the expression of cell surface Ig and B-cell associated markers. The wide clinico-pathological heterogeneity of NHL correlates with distinct genetic lesions, particularly chromosomal translocations, associated with its pathogenesis (Table 1; Gaidano and Dalla-Favera, 1993). Among low-grade NHL, "mantle zone" lymphoma are associated in 50% of cases with the t(11;14) translocation involving the juxtaposition of the IgH locus to the BCL-1/PRAD-1/cyclin D1 gene coding for a protein involved in the control of cell cycle progression (Tsujimoto et al., 1984; Motokura et al., 1991; Raffeld et al., 1991). In follicular-type NHL(FL), the t(14;18) translocation juxtaposes the IgH locus to BCL-2 (Bakhshi et al., 1985; Tsujimoto et al., 1984; Cleary et al., 1985), a gene coding for a protein that prevents programmed cell death or apoptosis (Korsmeyer, 1992). After years of indolent course, a significant fraction of FL undergoes histologic transformation and clinical progression into Diffuse Large Cell Lymphoma (DLCL), an event which is associated with loss or mutations of the p53 tumor suppressor gene (Lo Coco et al., 1993). "De novo" DLCL are associated with rearrangements and deregulation of the BCL-6 gene, which codes for a zinc-finger transcription factor (Ye et al., 1993a, 1993b; Kerckaert et al., 1993). In Burkitt Lymphoma (BL), the t(8;14), t(8;22), and t(2;8) chromosomal translocations lead to the deregulated expression of the c-Myc proto-oncogene by juxtaposition to one of the Ig loci (Dalla-Favera et al., 1982; Taub et al.,

Normal and Malignant Hematopoiesis, Edited by Enrico Mihich and Donald Metcalf
Plenum Press, New York, 1995

Table 1. Distribution of molecular lesions in B-NHL

NHL type	Molecular lesions				
	BCL-1	BCL-2	c-MYC	p53	BCL-6
Mantle-zone	50 %	—	—	—	—
Follicular	—	70-90 %	—	—	6%
Diffuse	—	20 %	10 %	—	30-40%
Burkitt	—	—	100 %	30 %	—

1982; Dalla-Favera et al., 1983; Dalla-Favera, 1991). A sizable fraction (35%) of BL are also associated with loss or mutations of the p53 gene (Gaidano et al., 1991).

This chapter will focus on recent advances on the elucidation of the molecular pathogenesis of DLCL and BL, collectively termed aggressive lymphomas. In particular, we will review the mechanism by which chromosomal translocations contribute to the altered regulation of the c-Myc proto-oncogene in BL and recent results on the identification of a novel gene, BCL-6, which is frequently altered in DLCL.

MECHANISM OF C-MYC ONCOGENE ACTIVATION IN BURKITT LYMPHOMA

Normal Function of c-Myc as a Transcription Factor Controlling Cell Proliferation

The c-myc proto-oncogene codes for a nuclear phosphoprotein ubiquitously expressed in somatic cells where it is involved in the control of proliferation and differentiation (reviewed by Marcu et al., 1992). Substantial experimental evidence supports the notion that the c-Myc gene product may be a transcription factor. The c-myc protein can bind DNA in vitro and a specific sequence (CACGTG) has been identified as its specific binding site (Blackwell et al. 1990; Halazonetis and Kandil, 1991; Prendergast and Ziff, 1991). In addition, the c-myc protein contains domains which are typical of several types of transcription factors including a carboxy-terminal basic, helix-loop-helix and leucine-zipper (b-HLH-LZ) domain which can mediate the formation of oligomeric complexes capable of specific DNA binding (Landschulz et al., 1988) and an amino-terminal domain capable of transcriptional transactivation when assayed in experimental chimeric constructs (Kato et al., 1990; 1992). A second b-HLH-LZ protein, Max, can specifically associate with c-Myc in vitro to form heterodimeric complexes capable of specific DNA binding with higher affinity than either c-Myc or Max homodimers (Blackwood and Eisenman, 1991; Prendergast et al., 1991). The existence of c-Myc/Max complexes has been documented in vivo (Blackwood et al., 1992). Experiments involving the transient cotransfection of c-Myc and/or Max expression vectors with a reporter gene linked to a c-Myc/Max binding site demonstrate a direct role of c-Myc and Max as transcriptional regulators and show that c-Myc/Max heterodimers activate whereas Max represses transcription. In cells stably transfected with c-Myc and/or Max, overexpression of c-Myc and increased formation of c-Myc/Max complexes accelerate proliferation, whereas Max overexpression leads to inhibition of growth (Gu et al., 1993).

The opposite transcriptional and biological activities of c-Myc/Max heterodimers versus Max homodimers, together with the pattern of regulation of c-Myc and Max gene

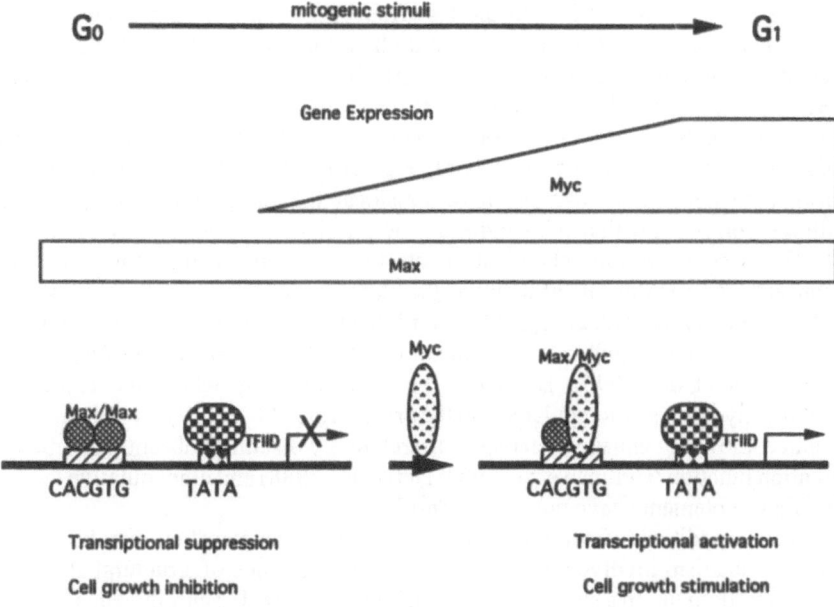

Figure 1. A model for the role of c-Myc and Max in the control of transcription and cell proliferation. Resting cells, which synthetize Max but not Myc, contain Max-Max homodimers which suppress the transcription of target genes and cell proliferation. Upon mitogenic stimulation, Myc is synthetized and Myc-Max complexes are formed which go to displace Max-Max complexes from DNA-binding sites. Myc-Max complexes can activate transcription and stimulate proliferation.

expression (Marcu et al., 1992), have suggested a model for the role of these transcriptional factor in the control of cell proliferation and differentiation. Through continued gene expression and due to the extreme long half life of the protein, Max is constitutively expressed during cell quiescence, proliferation and terminal differentiation (Blackwood et al, 1992); conversely, through a specific pattern of gene transcription and synthesis of an extremely labile RNA and protein, c-Myc expression is tightly regulated during proliferation and differentiation, being absent in quiescent cells, rapidly induced by mitogenic stimulation (G0 to G1 transition) and rapidly suppressed after induction of differentiation (reviewed by Luscher and Eisenman, 1990). Thus, resting cells are under the effects of the transcriptional and growth suppressive activities of Max, while mitogenically stimulated cells contain high levels of the oppositely acting c-Myc/Max heterodimers. Cell proliferation and differentiation may be induced in vivo by varying c-myc levels which, in turn, would control the levels of the differentially acting c-Myc/Max and Max/Max transcriptional complexes (Gu et al., 1993).

Consequences of c-Myc Translocation in Burkitt Lymphoma: Transcriptional Deregulation and Mutations in the Transcriptional Transactivation Domain

Three types of reciprocal chromosomal translocations have been shown to involve the c-myc locus on chromosome 8q24 and one of the Ig loci in BL (for review see Dalla-Favera, 1991). In 80% of cases a t(8;14)(q24;q32) is detectable in which breakpoints located centromeric to c-myc lead to its translocation into the IgH locus on chromosome

14q32 (Dalla-Favera et al., 1983). In the less frequent variant (2;8)(p11;q24) (15%) and (8;22)(q24;q11) (5%) translocations, an Ig light-chain locus is translocated telomeric to the c-myc locus, which remains on chromosome 8 (Hollis et al., 1984; Davis et al., 1984).

Several experimental observations have indicated that one common consequence of these translocations is the loss of the normal regulation of the c-Myc gene which becomes expressed constitutively under the control of Ig regulatory elements. First, studies in somatic cell hybrids indicated that the expression of translocated alleles, but not the one of normal c-myc alleles, correlates with cell backgrounds permissive for Ig gene expression (Haluska et al., 1987). Second, various chemical agents that are capable of influencing Ig gene expression are also capable of influencing the expression of Ig gene-linked translocated c-myc alleles (Eick and Bornkamm, 1989). Third, the translocated c-myc allele is unresponsive to the negative autoregulatory mechanism regulating the expression of normal c-myc alleles (Grignani et al., 1990). However, the mechanism by which the expression of translocated c-myc genes is deregulated and the precise role of Ig regulatory domains remains unclear, since known Ig enhancer elements are relatively distant or absent from the Ig/myc recombination junction (Gellman et al., 1983; Neri et al., 1988) and other putative long-range active regulatory elements have not been identified.

A second additional feature of translocated c-myc genes which may be relevant to explain the mechanism involved in deregulation is the presence of structural alterations in the 5' regulatory domain. In a significant fraction of BL, the t(8;14) chromosomal breakpoint decapitates the c-myc gene of its first exon (Pelicci et al., 1986). In all the remaining cases carrying breakpoints either 5' or 3' to the c-myc gene, a ~ 400 bp. region spanning the first exon/first intron junction is selectively and consistently mutated in translocated c-myc alleles (Pelicci et al., 1986; Cesarman et al., 1987). This region contains potentially important regulatory domains suggesting that these structural alterations may contribute to deregulated c-Myc expression. However, despite their consistent presence and clustering in functionally important sequences, no direct evidence has been obtained for a functional role of these mutations.

More recently, a second cluster of mutations within the domain coding for the amino-terminus of the c-Myc protein has been identified in approximately 70% of BL (Bhatia et al., 1993; Yano et al., 1993; Fig.2). The region involved in the mutations is part of the region coding for the c-Myc transcriptional transactivation domain which is necessary for the activity of c-Myc in stimulating transcription and cell proliferation. The next section summarizes recent results (Gu et al., 1994) showing that mutations within these domain may have functional significance and contribute to deregulated c-Myc activity in BL .

c-Myc Mutants Escape Suppression by RB-Related p107

Based on the hypothesis that mutation within the transcriptional transactivation of c-Myc may alter its activity by interfering with other proteins binding to this domain, we screened a cDNA expression library from a B cell lymphoma cell line with a ^{32}P-labelled Myc fusion-protein representative of the Myc NH_2-terminal 210 amino acids. One positive clone displaying a strong interaction with the c-Myc probe was characterized and shown to contain sequences identical to a portion of the gene coding for p107, a growth-suppressor protein related to Rb (Ewen *et al.*, 1989; 1991). Like Rb, p107 associates with viral transforming proteins, including adenovirus E1A and SV40 large T antigen (Ewen et al., 1989; 1991). In particular, the cDNA sequence corresponded to almost the entire "pocket" domain of p107, a region essential for viral protein binding as well as for cell cycle regulation and growth suppression (Zhu *et al.*, 1993; Hu ey al., 1990; Qin et al. 1992). To determine whether p107 and Myc associate in vivo, we performed co-immunoprecipitation experiments aimed at determining whether p107 was present in immunoprecipitates generated by an

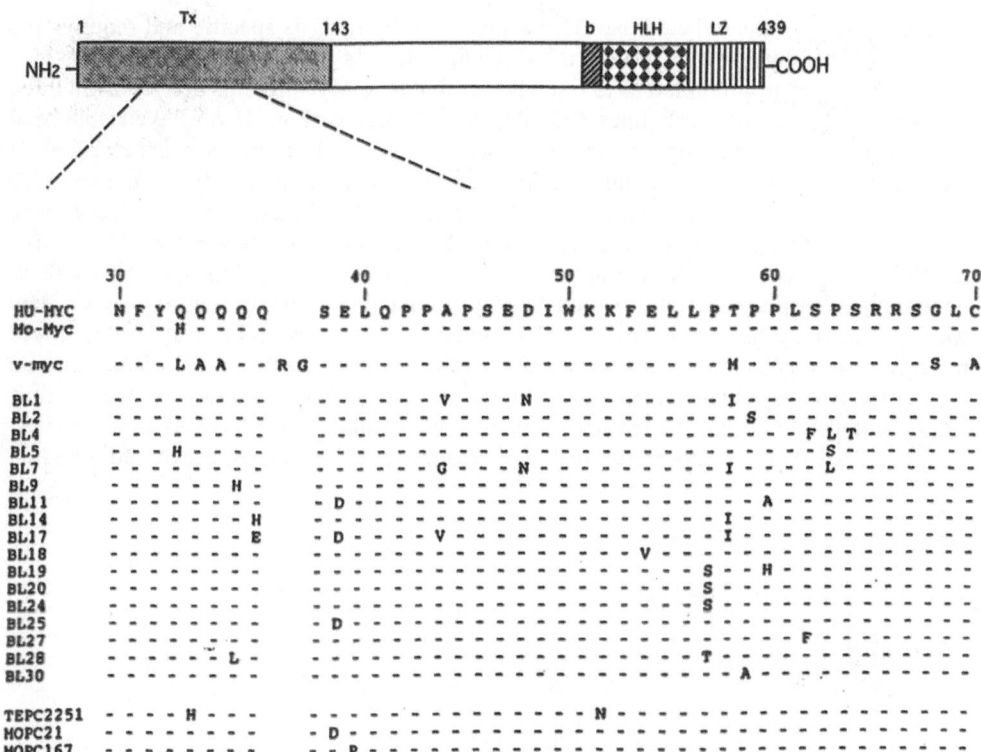

Figure 2. Mutations in the c-Myc transcriptional transactivation domain in Burkitt lymphoma. Schematic representation of the c-Myc protein and its functional domains: transcriptional transactivation domain (Tx) and basic/helix-loop-helix/leucine zipper (b/HLH/LZ). The sequences below show the aminoacid residues involved in the mutations in Burkitt lymphoma (BL) and mouse plasmocytoma (TEPC2251, MOPC21, MOPC167) cell lines as compared to human (HU-MYC) and mouse (MO-HU) wild-type sequences.

antibody to Myc and viceversa. This analysis (see Fig. 3 for representative results) indicated that p107 was detectable in immunoprecipitates obtained with antibodies to Myc, but not in those obtained with an antibody to Rb or with control serum. Notably, p107 was also detectable in immunoprecipitates obtained with the antibody to Max suggesting that p107 can bind Myc when Myc is complexed with Max. The detection of p107 in immunoprecipitates generated by an antibody to Max (fig. 3), together with the facts that no direct interaction has been detectable between p107 and Max in vitro and most Myc molecules are complexed with Max in vivo, suggested that at least some p107 is bound, via Myc, to Myc-Max complexes.

To determine the functional consequences of p107 binding, we studied whether p107 affected the transactivation activity of Myc. We cotransfected a p107 expression plasmid with a plasmid carrying NH_2-terminal Myc sequences fused to the yeast GAL4 DNA-binding sequences into NIH3T3 cells, and assayed transcription from a reporter gene linked to GAL4-binding sites Cotransfection of the p107 plasmid lead to a dose-dependent suppression of the activity of the c-Myc transactivation domain (Fig. 4). The p107 "pocket" domain was both necessary and sufficient for suppression, since mutants carrying point mutations or deletions of the entire domain are impaired in their suppressive activity, while a deletion mutant retaining only the "pocket" domain is active. We also found that Rb was completely inactive in suppressing Myc transactivation while it could suppress the E2F transactivation domain. These results indicate that Myc suppression by p107 is specific and requires the same functional domain (pocket) of 107 which mediates its growth suppressive effects.

These results prompted us to test whether mutant c-Myc proteins derived from three Burkitt lymphoma (BL) cell lines (P3HR1, ST486, CA46; Fig. 2) displayed abnormal behavior in p107 binding or suppression. Since the Myc-p107 complex was detectable in these three BL lines by the immunoprecipitation assay and because only the mutant Myc protein is expressed in these cell lines, this result indicated that the mutations do not eliminate Myc-p107 binding. In contrast, marked differences were observed between wild-type Myc and the BL-derived mutant Myc proteins in the transactivation assay (Fig. 4B). Although all three mutants displayed transactivation activity comparable to that of wild-type Myc, they were substantially less responsive to p107-induced suppression (Gu et al, 1994). More recently, these results have been corroborated by the observations that p-107 could suppress the transactivation activity of wild-type, but not mutant, c-Myc plasmids expressing full-length protein and co-transfected with a reporter gene linked to c-Myc/Max binding sites. Furthermore, c-Myc vectors were capable of rescuing p-107-mediated growth suppression

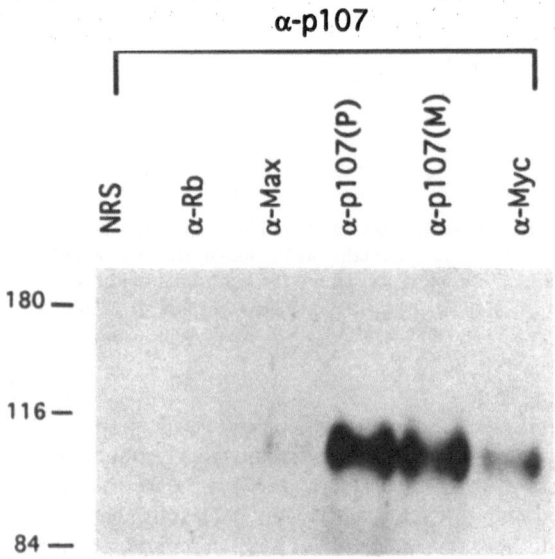

Figure 3. Association between MYC and p107 in vivo. Cell lysates obtained from a lymphoblastoid cell line transfected with a c-Myc expression vector were immuno-precipitated with normal rabbit serum (NRS), an antibody to Rb (α-Rb), an anti-body to Max [α-Max)], a polyclonal anti-body to p107 [α-p107(P)], a monoclonal antibody to p107 [α-p107(M)], or an anti-body to Myc (α-Myc) and then analyzed by immunoblot using [α-p107(P)].

Figure 4. p107 suppresses the activity of wild-type, but not mutant, c-Myc transactivation domain. Panel (A): Suppression of Myc transactivation by p107. In the lanes marked G5-CAT and G0-CAT, NIH 3T3 cells were transfected with a target [G5-CAT: five copies of the Gal-4 binding site preceding a minimal promoter and the chloramphenicol acetyltransferase gene (CAT)] or control-target plasmid (G0-CAT: as in G5-CAT but without the Gal-4 binding site) to show the basal levels of transcription from G_5 in the absence of transactivation. In the lanes marked Gal-MYC(1-210), G5-CAT was cotransfected with a vector expressing the Gal-4 DNA binding domain linked the amino terminal portion of Myc (Gal-Myc(1-210), together with a control vector (control), a vector expressing Rb (pRb), or vectors expressing wild-type (p107) or mutant (p107F846, p107DE, p107N385) p107. In the lanes marked Gal-E2F1 and Gal-VP16, G5-CAT was cotransfected with Gal-E2F1 or Gal-VP16 together with control, Rb (pRb) or p107 vector plasmid. Panel (B): Resistance of lymphoma-derived MYC mutants to suppression by p107. NIH 3T3 cells were transfected with G_5-CAT, a wild-type (WT) or a mutant (P3HR1, ST486, CA46) Myc expression vector, and a control (-) or a CMV-p107 (+) plasmid.

when transfected into SAOS cells, and mutant c-Myc alleles were significantly more active in this assay (Gu et al., in preparation).

These observations point to a functionally important activity of p-107 in suppressing c-Myc activity during cell-cycle progression in mammalian cells and suggest an additional mechanism contributing to deregulated c-Myc activity in BL. Mutations within the transcrip-

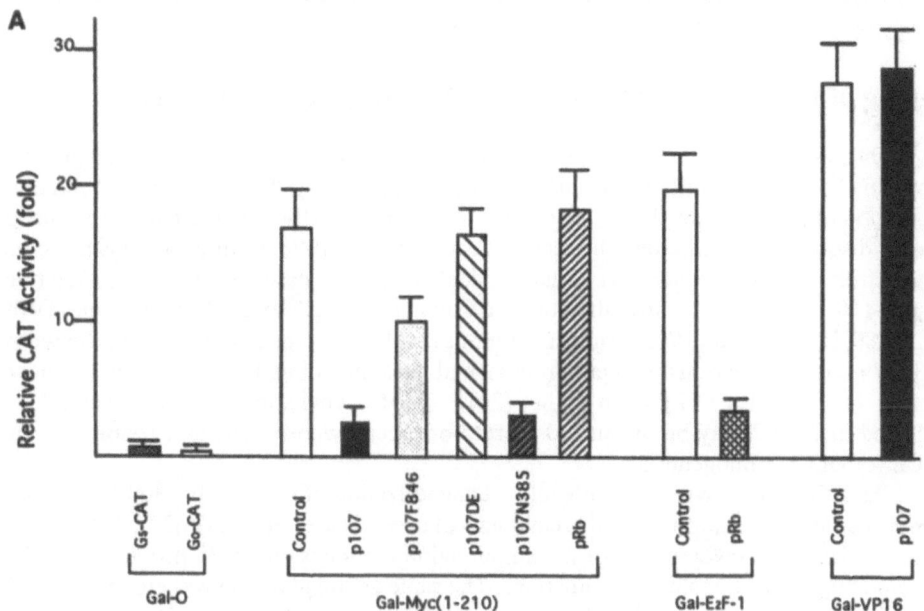

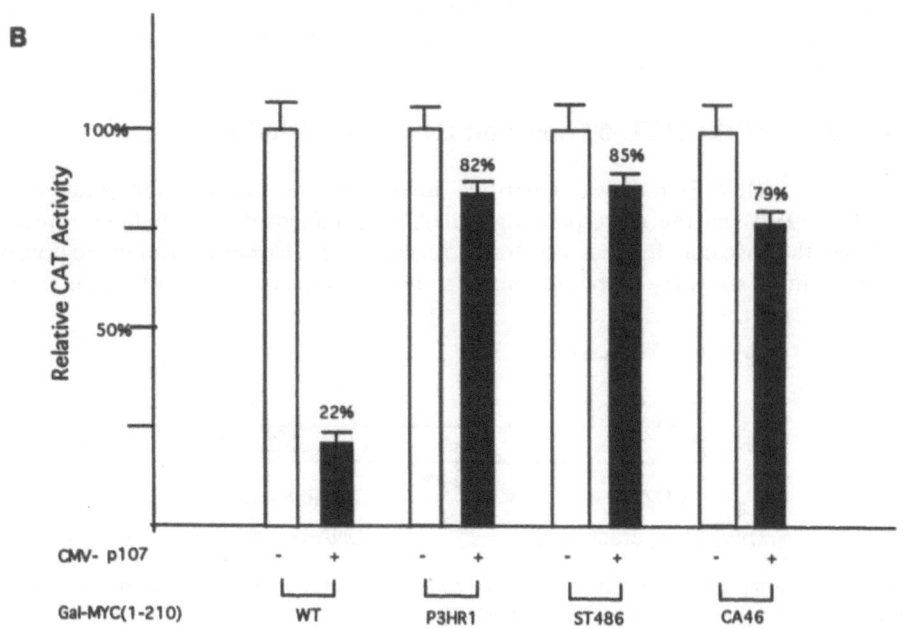

tional activation domain may further deregulate the activity of contitutively transcribed, Ig-driven c-Myc alleles by enabling them to escape p107-mediated modulation. The precise mechanisms by which p-107 suppresses c-Myc activity and mutant c-Myc protein escape this suppression remain to be elucidated.

ALTERATIONS OF THE BCL-6 GENE IN DIFFUSE LARGE CELL LYMPHOMA

Cloning of Chromosomal Breakpoints Affecting 3q27 in DLCL

Despite the fact that DLCL represents the most frequent and lethal human lymphoma, accounting for ~40% of initial NHL diagnoses and representing the common final stage of progression of FL (Magrath, 1990), no genetic lesion specifically or consistently associated with the disease had been found till recently (Ladanyi et al. 1991, Offit et al. 1989). A clue derived from cytogenetic analysis of large panels of DLCL cases have revealed relative frequent (10-12%) chromosomal alterations affecting band 3q27 in this NHL subtype (Offit et al., 1989; Bastard et al., 1992). These alterations involve reciprocal translocations between the 3q27 region and various partner chromosomal sites including, but not limited to, those carrying the Ig heavy-(14q32) or light- (2p12, 22q11) chain loci. These observations suggested that 3q27 may be the site of a proto-oncogene whose structural lesion may be critical for DLCL pathogenesis.

As a first step toward the molecular characterization of the putative 3q27 proto-oncogene, we cloned the chromosomal breakpoints of several cases of (3;14)(q27;q32) translocations in which the involvement of the immunoglobulin locus on 14q32 provided a probe for the cloning of the translocation junctions. The same genomic region was cloned from all the t(3;14)(q27;q32) cases analyzed (Ye et al., 1993b; Baron et al., 1993; Deweindt et al., 1993; Kerckaert et al., 1993; and Miki et al., 1994). Furthermore, DNA probes from this region identified rearrangements in this locus in 13 of 17 cases carrying 3q27 alterations, irrespective of the partner chromosomes involved in the translocations (Ye et al., 1993a). These results indicated that the chromosomal breakpoints clustered within a restricted genomic region in various DLCL cases further suggesting that the affected genetic locus may be important for lymphomagenesis.

Identification of the BCL-6 Gene and Its Protein Product

A gene, called BCL-6, was identified adjacent to the translocation breakpoints (Ye et al., 1993a) and the corresponding cDNA was sequenced. Its predicted protein product has the essential features illustrated in Fig.5. The carboxy-terminal region of BCL-6 containins six C_2H_2 zinc-finger motifs and a conserved stretch of seven amino

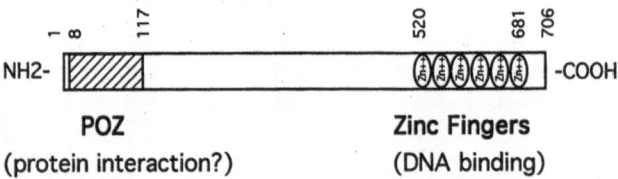

Figure 5. Schematic representation of the functional domains of the BCL-6 protein. The approximate positions of the zinc-finger motifs (Zn++) and the NH_2-POZ domain are indicated.

acids (the H/C link) connecting the successive zinc-finger repeats as typical of various members of the *Krüppel*-like subfamily of zinc-finger proteins (Rosenberg et al., 1989). Zinc-finger encoding genes usually act as nuclear, DNA binding transcription factors and represent plausible candidate oncogenes as they have been shown to participate in the control of cell proliferation, differentiation, and organ pattern formation (for review see El-Baradi and Pieler, 1991). In addition, alterations of zinc-finger genes have been detected in a variety of tumor types including WT-1 in Wilm's tumor (Haber et al., 1990), PLZF and PML in acute promyelocytic leukemia (Chen et al., 1993; de The' et al., 1991; Kakizuka et al., 1991; and Pandolfi et al., 1991), EVI-1 in mouse and human myeloid leukemia (Morishita et al., 1988; Fichelson et al., 1992), TTG-1 in T-cell ALL (McGuire et al., 1989), and HTRX in acute mixed-lineage leukemia (Djabali et al., 1992; Tkachuk et al., 1992; and Gu et al., 1992).

The NH_2-terminal region of BCL-6 is devoid of the FAX (Knochel et al., 1989) and KRAB (Dellefroid et al., 1991) domains often present in *Krüppel*-related zinc-finger proteins, but it does have homologies (Fig. 1) with other zinc-finger transcription factors including the human ZFPJS protein, a putative human transcription factor that regulates the major histocompatibility complex II promoter (Sugawara, M.M., unpublished), the *Tramtrack* (*ttk*) and *Broad-complex* (*Br-c*) proteins in *Drosophila* that regulate transcription during development (Harrison and Travers., 1990; Dibello et al., 1991), the human KUP protein (Chardin et al., 1991), and the human PLZF protein. Such homology is also found in proteins (*e.g.* VA55R) of the poxvirus family (Koonin et al., 1992) as well as in the *Drosophila kelch* protein involved in nurse cell-oocyte interaction (Xue and Cooley, 1993). This homology domain, recently named POZ (for POX/Zinc finger), has been found in an additional zinc-finger protein, ZID, and shown to act as a specific protein-protein interaction domain capable of regulating DNA binding by the zinc-finger domain (Bardwell and Treisman, 1994).

Recent preliminary results show that BCL-6 is localized in the nucleus, contains a transcriptional repressor domain, and is able to bind to specific DNA sequences (Chang et al., in preparation; Ye et al., in preparation). Taken together these observations further suggest that BCL-6 may function as a DNA-binding transcription factor possibly involved in the control of cell differentiation and tissue development.

The BCL-6 Gene Is Expressed in Multiple Organs and Is Regulated during B Cell Differentiation

A single 3.8 kb RNA species was detectable in a number of normal human tissues tested (see Fig. 6 A) although at different abundance. Within the B cell lineage, high levels of BCL-6 RNA were found in all cell lines derived from mature B-cells except for EBV-immortalized lymphoblastoid cell lines (LCL) (see Fig. 6B for representative results), low or undetectable amounts of BCL-6 mRNA were found in cells preceding (pre-B) or following (plasma cells) mature B cells in the B cell differentiation pathway. Cell lines derived from T cells, from other hematopoietic cell lineages, or from other normal tissues, lacked notable levels of BCL-6 mRNA. Normal B cells derived from human peripheral blood also contained detectable levels of BCL-6 RNA both under resting or mitogen-stimulated conditions (Zhang, J. et al., unpublished). Western blot analysis of BCL-6 protein levels confirmed this pattern of expression. These results suggest that BCL-6 may have a function in multiple tissues and, in particular, may play a role in the control of normal B-cell differentiation and lymphoid organ development.

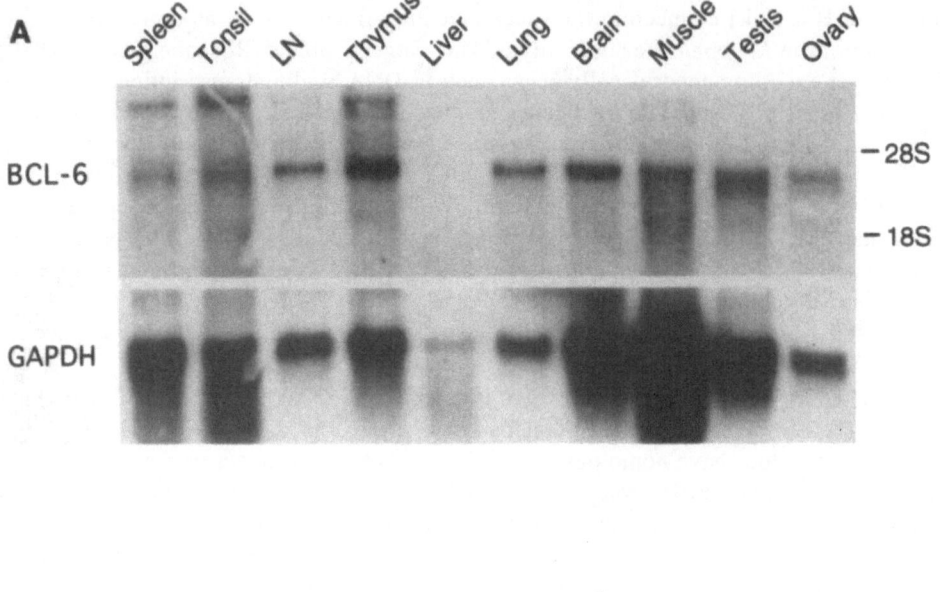

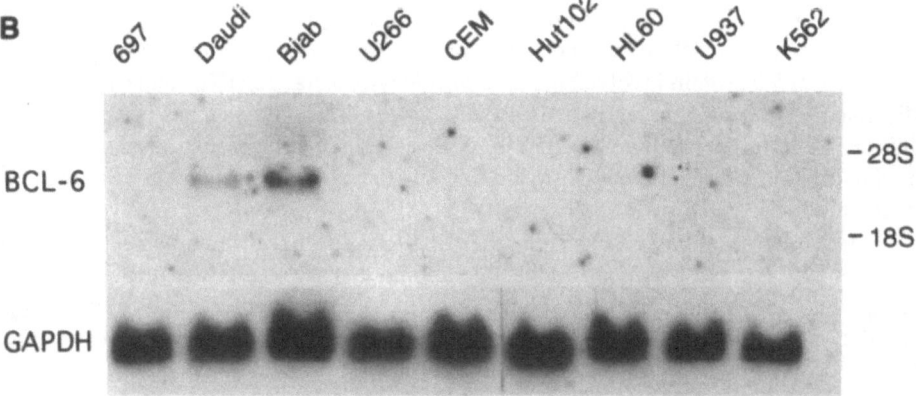

Figure 6. Analysis of BCL-6 RNA expression in normal tissues and tumor derived cell lines. Northern blot analysis of BCL-6 and GAPDH (as control for RNA amounts) RNA expression in normal human tissues (panel A) and hematopoietic tumor cell lines (panel B). The results indicate that the BCL-6 gene is widely expressed in differnt human tissues, but is regulated during hematopoietic differentiation with the highest levels of expression in cell lines displaying a mature B cell phenotype (Daudi, Bjab). BCL-6 RNA is undetectable in cell lines corresponding to pre-B cells, plasmacells U266), immature (CEM) or mature (Hut102) T cells, promyelocytes (HL60), promonocytes (U937), or early erythroid precursors (K562).

The BCL-6 Gene Is Frequently Altered in its 5' Non-Coding Region in DLCL

In order to define the incidence of BCL-6 rearrangements in various lymphoprolif-erative diseases, the entire BCL-6 genomic locus spanning approximate 26 kb of genomic DNA was cloned (Fig.7). Using various probes from this locus, we analyzed a panel of cases not previously selected on the basis of 3q27 alterations but representative of the major subtypes of NHL as well as of other lymphoproliferative diseases including acute lym-

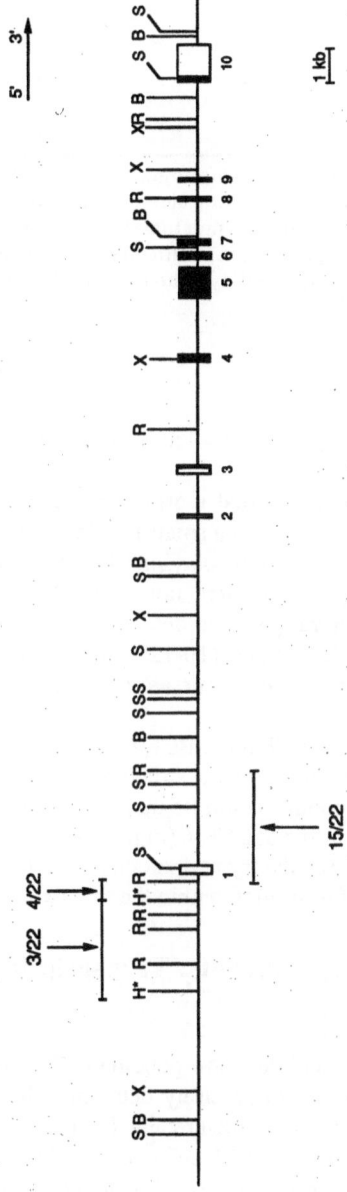

Figure 7. Exon–intron organization of the BCL-6 gene and mapping of breakpoints detected in DLCL. Coding and non-coding exons are represented by filled and empty boxes, respectively. Arrows indicate the breakpoint position for major breakpoint clusters as determined by restriction digestion/hybridization analysis. Numbers above arrows indicate numbers of rearranged/tested cases. Restriction enzyme symbols are: S, Sac I; B, Bam HI; X, Xba I; H, Hind III; R, Eco RI.

Table 2. Rearrangements of the BCL-6 gene in
lymphoid tumors

Tumor[a]	Rearranged/tested	%
NHL		
SML	0/8	0
FL	2/31	6
DLCL	16/45	35
BL	0/22	0
ALL	0/45	0
CLL	0/51	0
MM	0/23	0

[a]NHL, non-Hodgkin's lymphoma; ALL, acute
lymphoblastic leukemia; CLL, chronic lymphocytic
leukemia; MM, multiple myeloma; SML, small
lymphocytic lymphoma; FL, follicular lymphoma;
DLCL, diffuse large cell lymphoma; BL, Burkitt
lymphoma.

phoblastic leukemia (ALL), chronic lymphocytic leukemia (CLL), and multiple myeloma
(MM) (Lo Coco et al., 1994).

This analysis showed (see Table 2) that rearrangements of the BCL-6 gene were
detectable only in DLCL (35% of cases) and in a small fraction of FL (6%) (Lo Coco et al.,
1994). All except two of the DLCL cases displaying BCL-6 alterations lacked BCL-2
rearrangements. Although cytogenetic data were not available for the panel of tumors
studied, the frequency of BCL-6 rearrangements detected by Southern Blot analysis far
exceeded that expected for 3q27 aberrations (10-12% in DLCL) at cytogenetic level,
suggesting that BCL-6 rearrangements can occur as a consequence of submicroscopic
chromosomal aberrations.

The positions of the breakpoints within the BCL-6 locus have been mapped allowing
for a preliminary understanding of their effect on the structure/function of the BCL-6 gene (Ye
et al., 1993a). All the observed breakpoints could be mapped within the 5' flanking region,
within the first exon or within the first intron of BCL-6 (Fig. 7). As the result, the coding domain
of BCL-6 is left intact whereas the 5' regulatory region containing the promoter sequences is
either completely removed (in case of truncation within the first exon or intron) or truncated.

Consequences of BCL-6 Rearrangements: Transcription Driven by Heterologous Promoters

The rearranged genomic locus and the corresponding cDNA were cloned and characterized from a number of cases in order to precisely determine the consequences of these
rearrangements on BCL-6 mRNA and protein structure. Fig. 8 shows a schematic representation
of these clones derived from a representative DLCL cell line, called Ly-8. This cell line carries a
t(3;14)(q27;q32) translocation and, as a result, the BCL-6 gene is truncated within its first intron
and linked to the IgH locus within the Switch-γ (Sγ) recombination region. The corrsponding
cDNA shows that the BCL-6 first exon sequences are replaced by sequences derived from the I
region of γ3 locus (Kuze et al., 1991). The I region is located upstream from Sγ and its promoter is
normally responsible for expression of the sterile transcripts in germ-line IgH genes. These sterile
transcripts contain I region sequences spliced to the downstream sequences from the Ig constant
region. In Ly-8 cells, I sequences are spliced to BCL-6 exon 2 sequences resulting in a fusion

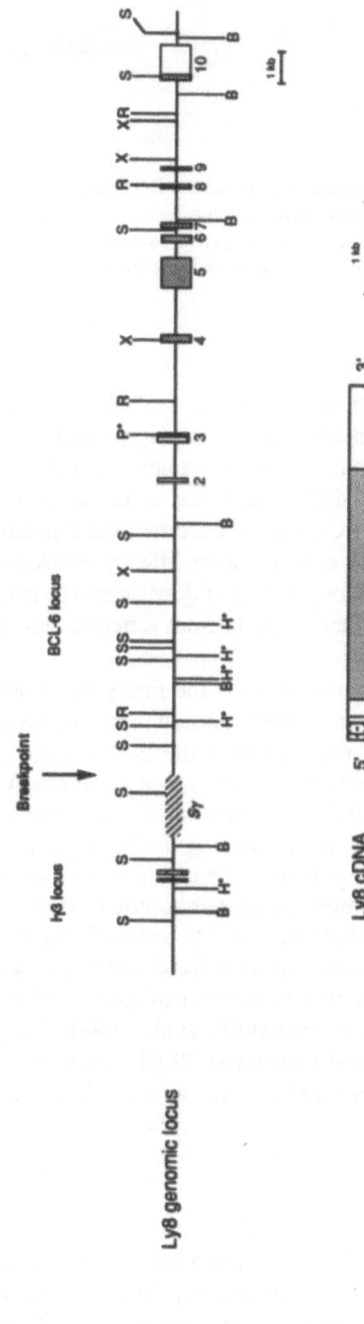

Figure 8. Schematic representation of the rearranged BCL-6 allele spanning the t(3;14) translocation breakpoint cloned from Ly-8 genomic DNA (top panel) and of its corresponding cDNA cloned from a Ly-8 cDNA library. Ig region exons (stippled boxes), Sγ sequences, non-coding (empty boxes) and coding (filled boxes) BCL-6 exons are also indicated. Restriction enzyme symbols are: S, Sac I; B, BamH I; X, Xba I; H, Hind III; R, EcoR I; P, Pst I; G, Bgl II. Restriction sites marked by an asterisk have been only partially mapped.

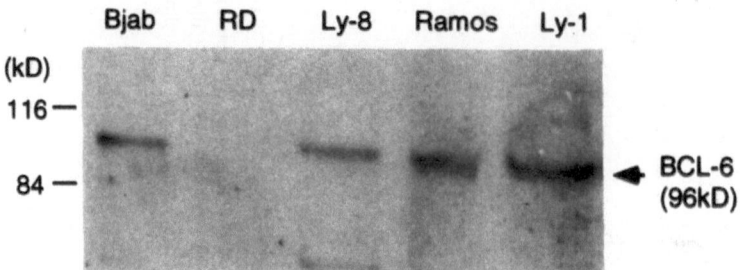

Figure 9. Western blot analysis of BCL-6 protein expression in a B cell line carrying a rearranged BCl-6 gene (Ly-8, see Fig. 8) and comparison with cell lines carrying unrearranged BCL-6 genes and either express (Bjab, Ly-1, Ramos; positive controls) or do not express (RD; negative control) BCL-6 RNA. A rabbit antiserum (N-70-6) raised againsts a synthetic KLH-conjugated oligopeptide corresponding to the amino-terminal region of BCL-6 was used.

transcript which retains the entire BCL-6 coding domain. Analysis of this cDNA predicts that the functional consequence of this BCL-6 rearrangement is the expression of a normal BCL-6 protein under the control of a heterologous promoter sequence leading to the loss of its normal pattern of regulation. Fig. 9 shows that a normally sized protein was, in fact. detectable in Ly-8 cells. This protein must be coded by the rearranged BCL-6 allele since the second unrearranged BCL-6 allele is silent in these cells. These observations suggest that in different translocations involving 3q27 the BCL-6 coding domain (exon 3-10) may be linked downstream to heterologous sequences which, based on cytogenetic analysis, can originate from different chromosomes in different cases.

In conclusion, the results presented identify the first genetic lesion associated with DLCL, the most clinically relevant form of NHL. Although no proof yet exists of a role for these lesions in DLCL pathogenesis, the feature of the BCL-6 gene product, its specific pattern of expression in B cells and the clustering of lesions disrupting its regulatory domain strongly suggest that deregulation of BCL-6 expression may contribute to DLCL development. In addition to contributing to the understanding of DLCL pathogenesis, the identification of BCL-6 lesions may have relevant clinical implications. DLCL represent a heterogeneous group of neoplasms which are treated homogeneously despite the fact that only 50% of patients experience long-term disease free survival (Schneider et al., 1990). A recent study has shown that BCL-6 rearrangements were noted in a statistically significant subset of cases characterized by primary involvement of extranodal tissues, lack of bone marrow involvement and favorable prognosis (Offit et al., 1994). Thus, BCL-6 rearrangements may identify a biologically distinct subset of DLCL and may serve as a diagnostic and prognostic marker in the management of patients with this disease.

ACKNOWLEDGMENTS

We are grateful to a number of collaborators whose work has been instrumental to generate many of the results described in this chapter, in particular to I. Magrath, D.M. Knowles and R.S.K. Chaganti and to their respective collaborators. The work described has been supported by N.I.H. grants CA37165 and CA-44029. F.L.C. is presently at Sezione di Ematologia, Dipartimento di Biopatologia Umana of the University 'La Sapienza' of Rome, and was partially supported by Associazione Italiana contro la Leucemia (AIL). B.H.Y. is a

Leukemia Society of America Fellow. C.C. Chang is partially supported by the Lymphoma Research Foundation.

REFERENCES

Bakhshi, A., J.P. Jensen, P. Goldman, J.J. Wright, O.W. McBride, A.L. Epstein, and S.J. Korsmeyer. 1985. Cloning the chromosomal breakpoint of t(14;18) human lymphomas: clustering around J_H on chromosome 14 and near a transcriptional unit on 18. *Cell* 41: 889.

Bardwell, V.J. and R. Treisman. 1994. The POZ domain: A conserved protein-protein interaction motif. *Genes Dev.* 8:1664.

Baron, B.W., G. Nucifora, N. McCabe, R. Espinosa III, M.M. Le Beau, and T.W. McKeithan. 1993. Identification of the gene associated with the recurring chromosomal translocations t(3;14)(q27;q32) and t(3;22)(q27;q11) in B-cell lymphomas. *Proc. Natl. Acad. Sci. USA* 90:5262.

Bastard, C., H. Tilly, B. Lenormand, C. Bigorgne, D. Boulet, A. Kunlin, M. Monconduit, and H. Piguet. 1992. Translocations involving band 3q37 and Ig gene regions in non-Hodgkin's lymphoma. *Blood* 79: 2527.

Bhatia, K. *et al.* 1993. Point mutations in the c-Myc transactivation domain are frequent occurence in Burkitt's lymphoma and mouse plasmocytoma. *Nature Genet.* 5:56

Blackwell, T. K., Kretzner, l., Blackwood, E. M., Eisenman, R. N., and Weintraub. 1990. Sequence-specific DNA binding by the c-Myc protein. *Science* 250: 1149

Blackwood, E. M., and Eisenman, R. N. 1991. Max: A helix-loop-helix zipper protein that forms a sequence-specific DNA binding complex with Myc. *Science* 251: 1211-1217.

Blackwood, E. M., Luscher, B., and Eisenman, R. N. 1992. Myc and Max associate in vivo. *Genes & Dev.* 6: 71-80.

Cesarman, E., R. Dalla-Favera, D. Bentley, M. Groudine. 1987. Mutations in the first exon are associated with altered transcription of c-myc in Burkitt lymphoma. *Science* 238: 1272.

Chaganti, R.S.K., L.A. Doucette, K. Offit, D.A. Filippa, G.J. Allen, M.R. Condon, S.C. Jhanwar, B.D. Clarkson, and P.H. Lieberman. 1989. Specific translocationsin non-Hodgkin's lymphoma: incidence, molecular detection, and histological and clinical correlations. *Cancer Cells* 7: 33. Cold Spring Harbor Laboratories, Cold Spring Harbor, New York.

Chardin, P., G. Courtois, M.-G. Mattei, and S. Gisselbrecht. 1991. The KUP gene, located on human chromosome 14, encodes a protein with two distant zinc fingers. *Nucleic Acid Res.* 19:1431

Chen, Z., N.J. Brand, A. Chen, S. Chen, J.-H. Tong, Z.-Y. Wang, S. Waxman, and A. Zelent. 1993. Fusion between a novel *Krüppel*-like zinc finger gene and the retinoic acid receptor-α locus due to a variant t(11;17) translocation associated with acute promyelocytic leukaemia. *EMBO J.* 12: 1161.

Cleary, M.L. and J. Sklar. 1985. Nucleotide sequence of a t(14;18) chromosomal breakpoint in follicular lymphoma and demonstration of a breakpoint-cluster region near a transcriptionally active locus on chromosome 18. *Proc Natl Acad Sci USA* 82:7439.

Dalla-Favera, R. 1991. Chromosomal translocations invoving the c-myc oncogene and their role in the pathogenesis of B-cell neoplasia. in The Origins of Human Cancer , J. Brugge, T. Curran, E. Harlow, and F. McCormick, eds. (Cold Spring Harbor Laboratories Press), pp. 543-551.

Dalla-Favera, R. 1991. Chromosomal translocations involving the c-myc oncogene and their role in the pathogenesis of B cell neoplasia. *Origin of Human Cancer* (eds J. Brugge, T. Curran, E. Harlow, and F. McCormick). pp. 543-541. Cold Spring Harbor Laboratory (publ.).

Dalla-Favera, R., M. Bregni, J. Erickson, D. Patterson, R.C. Gallo, and C.M. Croce. 1982. Human c-myc oncogene is located on the region of chromosome 8 that is translocated in Burkitt lymphoma cells. *Proc. Nat. Acad. Sci. USA* 79: 7824.

Dalla-Favera, R., S. Martinotti, R.C. Gallo, J. Erikson, and C.M. Croce. 1983. Translocation and rearrangements of the c-myc oncogene locus in human undifferentiated B-cell lymphomas. *Science* 219: 963.

Davis, M., S. Malcolm, T.H. Rabbitts. 1984. Chromosome translocation can occur on either side of the c-myc oncogene in Burkitt lymphoma cells. *Nature* 308: 286.

de The' H, C. Lavau, A. Marchio, C. Chomienne, L. Degos, and A. Dejean. 1991. The PML-RARα fusion mRNA generated by the t(15;17) translocation in acute promyelocytic leukemia encodes a functionally altered RAR. *Cell* 66:675.

Dellefroid, E.J., D.A. Poncelet, P.J. Lecocq, O. Revelant, and J.A. Martial. 1991. The evolutionarily conserved Krüppel-associated box domain defines a subfamily of eukaryotic multifingered proteins. *Proc. Nat. Acad. Sci. USA* 88:3608.

Deweindt, C., J-P. Kerckaert, H. Tilly, S. Quief, V.C. Nguyen, and C. Bastard. 1993. Cloning of a breakpoint cluster region at band 3q27 involved in human non-Hodgkin's lymphoma. *Genes, Chrom, & Cancer* 8:149.

Dibello, P.R., D.A. Withers, C.A. Bayer, J.W. Fristrom, and G.M. Guild. 1991.The *Drosophila Broad-Complex* encodes a family of related proteins containing zinc fingers. *Genetics* 129:385

Djabali, M., L. Selleri, P. Parry, M. Bower, B.D. Young, and G.A. Evans. 1992. A trithorax-like gene is interrupted by chromosome 11q23 translocations in acute leukemias. *Nature Genet.* 2:113.

Eick, D., G.W. Bornkamm. 1989. Expression of normal and translocated c-myc alleles in Burkitt's lymphoma cells: evidence for different regulation. *EMBO J.* 8: 1965.

El-Baradi T and Pieler T. 1991. Zinc finger proteins: what we know and what we would like to know. *Mech .Dev.* 35:155.

Ewen, M. E. *et al.* 1989. An N-terminal transformation-governing sequence of SV40 large T antigen contributes to the binding of both p110Rb and second cellular protein, p120.*Cell* 58: 257

Ewen, M.E, Y. Xing, J. B. Larence, D. M. Livingston. 1991. Molecular cloning, chromosomal mapping, and expression of the cDNA for p107, a retinoblastoma gene product-related protein. *ibid.* 66: 1155.

Fichelson, S., F. Dreyfus, R. Berger, J. Melle, C. Bastard, J.M. Miclea, and S. Gisselbrecht 1992. Evi-1 expression in leukemic patients with rearrangements of the 3q25-q28 chromosomal region. *Leukemia* 6:93.

Gaidano, G. and R. Dalla-Favera. 1993. Biologic and molecular characterization of non-Hodgkin's Lymphoma. *Current Opinion in Oncology* 5:776.

Gaidano, G., P. Ballerini, J. Gong, A. Neri, E.W. Newcomb, I.T. Magrath, D.K. Knowles, and R. Dalla-Favera. 1991. p53 mutations in human lymphoid malignancies: association with Burkitt lymphoma and chronic lymphocytic leukemia. *Proc. Natl. Acad. Sci. USA* 88:5413.

Gellman, E.P., M.C. Psallidopoulos,T.S. Papas, R. Dalla-Favera. 1983. Identification of reciprocal translocation points within the c-myc and immunoglobulin m loci in a Burkitt lymphoma. *Nature* 306: 799.

Gribben, J.G., D. Neuberg, A.S. Freedman, C.D. Gimmi, K.W. Pesek, M. Barber, L. Saporito, S.D. Woo, F. Coral, N. Spector, S.N. Rabinowe, M.L. Grossbard, J. Ritz, and L. Nadler. 1993. Detection by polymerase chain reaction of residual cells with the BCL-2 translocation is associated with increased risk of relapse after autologous bone marrow transplantation for B-cell lymphoma. *Blood* 81:3449.

Grignani, F., L. Lombardi, G. Inghirami, L. Sternas, K. Cechova, R. Dalla-Favera. 1990. Negative autoregulation of c-myc gene expression is inactivated in transformed cells. *EMBO J.* 9: 3913.

Gu, W., Cechova, K., Tassi, V., Dalla-Favera, R. 1993. Opposite regulation of gene transcription and cell proliferation by c-Myc and Max. *Proc. Natl. Acad. Sci. 90* : 2935

Gu, Y., T. Nakamura, H. Alder, R. Prasad, O. Canaani, G. Cimino, C.M. Croce, and E. Canaani. 1992. The t(4;11) chromosome translocation of human acute leukemias fuses the *ALL*-1 gene, related to Drosophila *trithorax*, to the AF-4 gene. *Cell* 71:701.

Gu. W., Bhatia, K., Magrath, I.T., Dang, C.V., Dalla-Favera, R. 1994. Binding and suppression of the Myc transcriptional activation domain by p107. *Science* 264: 251

Haber, D.A., A.J. Buckler, T. Glaser, K.M. Call, J. Pelletier, R.L. Sohn, E.C. Douglass, and D.E. Housman. 1990. An internal deletion within an 11p13 zinc finger gene contributes to the development of Wilms' tumor. *Cell* 61:1257.

Halazonetis, T., and Kandil, A. N. 1991. Determination of the c-Myc DNA-binding site. *Proc. Natl. Acad. Sci.* 88: 6162-6166.

Haluska, F.G., Y. Tsujimoto, C.M. Croce. 1987. Oncogene activation by chromosomal translocation in human malignancy. *Ann. Rev. Genet.* 21: 321.

Harrison, S.D. and A.A. Travers. 1990. The tramtrack gene encodes a Drosophila finger protein that interacts with the ftz transcriptional regulatory region and shows a novel embryonic expression pattern. *EMBO J.* 9:207.

Hollis, G.F., K.F. Mitchell, J. Battey, H. Potter, R. Taub, G.M. Lenoir, P. Leder. 1984. A variant translocation places the lambda immunoglobulin genes 3' to the c-myc oncogene in Burkitt's lymphoma. *Nature* 307: 752.

Hu, Q.J., N. Dyson, E. Harlow. 1990. The regions of the retinoblastoma protein needed for binding to adenovirus E1A or SV40 large T antigen are common sites for mutations, *EMBO* 9:1147.

Kakizuka, A., W.H. Miller, Jr, K. Umesono, R.P. Warrell Jr., S.R. Frankel, V.V. Murty, E. Dmitrovsky, and R.M. Evans. 1991. Chromosomal translocation t(15;17) in human acute promyelocytic leukemia fuses RARα with a novel putative transcription factor, PML. *Cell* 66:663.

Kato, G. J., Barret, J., Villa -Garcia, M., and Dang, C. V. 1990. An amino-terminal domain of c-myc required for transformation activates transcription. *Mol. Cell. Biol.* 10: 5914

Kato, G. J., Lee, W. M. F., Chen, L., and Dang, C. V. 1992. Max: functional domains and interaction with c-Myc. *Genes & Dev.* 6: 81.

Kerckaert, J-P., C. Deweindt, H. Tilly, S. Quief, G. Lecocq, and C. Bastard. 1993. LAZ3, a novel zinc-finger encoding gene, is disrupted by recurring chromosome 3q27 translocations in human lymphoma. *Nature Genet.* 5:66.

Knochel, W., A. Poting, M. Koster, T. el Baradi, W. Nietfeld, T. Bouwmeester, and T. Pieler. 1989. Evolutionary conserved modules associated with zinc fingers in Xenopus laevis. *Proc. Natl. Acad. Sci. USA* 86:6097

Koonin, E.V., T.G. Senkevich, and V.I. Chernos. 1992. A family of DNA virus genes that consists of fused portions of unrelated cellular genes. *Trends Biochem.Sci.* 17:213.

Korsmeyer, SJ. 1992. Bcl-2 initiates a new category of oncogenes: Regulators of cell death. *Blood* 80:879-886.

Kuze, K., A. Shimizu, and T. Honju. 1991. Characterization of the enhancer region for germline transcription of the gamma 3 constant region gene of human immunoglobulin. *Int. Immunol.* 3:647.

Lo Coco, F., B.H. Ye, F. Lista, P. Corradini, K. Offit, D.M. Knowles, RSK. Chaganti, and R. Dalla-Favera. 1994. Rearrangements of the BCL-6 gene in diffuse large-cell non-Hodgkin lymphoma. *Blood* 83:1757.

LoCoco, F., G. Gaidano, D.C. Louie, K. Offit, RSK. Chaganti, and R. Dalla-Favera. 1993. p53 mutations are associated with histologic transformation of follicular lymphoma. *Blood* 82:2289.

Luscher, B.,and Eisenman, R. N. 1990. New light on Myc and Myb. Part I. Myc. *Genes & Dev.* 4: 2025.

Magrath, I. 1990. Lymphocyte ontogeny: a conceptual basis for understanding neoplasia of the immune system. In *The non-Hodgkin's lymphoma.* (ed. I. Magrath) pp. 29-48.Williams & Wilkins, Baltimore.

Marcu, K.B., Bossone, S.A., Patel, A.J. 1992. Myc function and regulation. *Ann. Rev. Biochem.* 61:809-860.

McGuire, E.A., R.D. Hockett, K.M. Pollock, M.F. Bartholdi, S.J. O'Brien, and S.J. Korsmeyer. 1989. The t(11;14)(p15;q11) in a T-cell acute lymphoblastic leukemia cell line activates multiple transcripts, including Ttg-1, a gene encoding a potential zinc finger protein. *Mol. Cell Biol.* 9:2124.

Miki, T., N. Kawamata, S. Hirosawa, and N. Aoki. 1994. Gene involved in the 3q27 translocation associated with B-cell lymphoma, BCL-5, encodes a Kruppel-like Zinc-finger protein. *Blood* 83:26.

Morishita, K., D.S. Parker, M.L. Mucenski, N.A. Jenkins, N.G. Copeland, and J.N. Ihle. 1988. Retroviral activation of a novel gene encoding a zinc finger protein in IL3-dependent myeloid leukemia cell lines. *Cell* 54:831.

Motokura, T., T. Bloom, K.H. Goo, H. Juppner, J.V. Ruderman, H.M. Kronenberg, and A. Arnold. 1991. A novel cyclin encoded by a bcl-1 linked candidate oncogene. *Nature* 350:512.

Neri, A., F. Barriga, D.M. Knowles, I.T. Magrath, R. Dalla-Favera. 1988. Different regions of the immuno-globulin heavy-chain locus are involved in chromosomal translocations in distinct pathogenetic forms of Burkitt lymphoma. *Proc. Natl. Acad.* 85:2748

Offit, K., F. LoCoco, D.C. Louie, N.Z. Parsa, D. Leong, C. Portlock, B.H. Ye, F. Lista, D.A. Filippa, A. Rosenbaum, M. Ladanyi, R. Dalla-Favera, and RSK. Chaganti. 1994. Rearrangement of the BCL6 gene as a prognostic marker in diffuse large cell lymphoma. *N. Engl.J . Med.* 331:74

Offit, K., S. Jhanwar, S.A. Ebrahim, D. Filippa, B.D. Clarkson, RSK. Chaganti. 1989. t(3;22)(q27;q11): A novel translocation associated with diffuse non-Hodgkin's lymphoma. *Blood* 74: 1876.

Pandolfi, P.P., F. Grignani, M. Alcalay, A. Mencarelli, A. Biondi, F. Lo Coco, F. Grignani, and P.G. Pelicci. 1991. Structure and origin of the acute promyelocytic leukemia myl/RARα cDNA and charac-terization of its retinoid-binding and transactivation properties. *Oncogene* 6:1285.

Pelicci, P.G., D.K. Knowles, I. Magrath, R. Dalla-Favera. 1986b. Chromosomal breakpoints and structural alterations of the c-myc locus differ in endemic and sporadic forms of Burkitt lymphoma. *Proc. Natl. Acad. Sci.U.S.A.* 83: 2984.

Prendergast, G. C., and Ziff, E. B. 1991. Methylation-sensitive sequence-specific DNA binding by the c-Myc basic region. *Science* 251: 186.

Prendergast, G. C., Lawe, D., and Ziff, E. B. 1991. Association of Myn, the murine homolog of Max, with c-Myc stimulates methylation-sensitive DNA binding and Ras cotransformation. *Cell* 65: 395.

Qin, X.Q., T. Chittenden, D. Livingston, W. G. Kaelin. 1992. Identification of a growth suppression domain within the retinoblastoma gene product. *Genes Dev.* 6: 953.

Raffeld, M., and E.S. Jaffe. 1991. Bcl-1, t(11;14), and mantle zone lymphomas. *Blood* 78: 259

Rosenberg UB, Schroder C, Preiss A, A. Kienlin, S. Côté, I. Riede, and H. Jäckle. 1989. Structural homology of the product of the Drosophila Kruppel gene with Xenopus transcription factor IIIA. *Nature* 319:336.

Schneider, A.M., D.J. Straus, A.E. Schluger, D.A. Lowenthal, B. Koziner,. B.J. Lee, G. Wong, and B.D. Clarkson. 1990. Treatment results with an aggressive chemotherapeutic regimen (MACOP-B) for intermediate and some high grade non-Hodgkin's lymphomas. *J .Clin. Oncol.* 8:94.

Taub R, Kirsch I, Morton C, G.Lenoir, D. Swan, S. Tronick, S. Aaronson, and P. Leder. 1982. Translocation of c-myc gene into the immunoglobulin heavy chain locus in human Burkitt lymphoma and murine plasmacytoma cells. *Proc Natl Acad Sci USA* 79:7837.

Tkachuk, D.C., S. Kohler, and M.L. Cleary. 1992. Involvement of a homolog of Drosophila Trithorax by 11q23 chromosomal translocations in acute leukemias. *Cell* 71:691.

Tsujimoto, Y., J. Yunis, L. Onorato-Showe, J. Erikson, P.C. Nowell, and C.M. Croce. 1984. Molecular cloning of the chromosomal breakpoint on chromosome 11 in human B-cell neoplasms with the t(11;14) chromosome translocation. *Science* 224:1403.

Xue, F. and L. Cooley. 1993. *Kelch* encodes a component of intercellular bridges in *Drosophila* egg chambers. *Cell* 72: 681.

Yano, T. *et al..* Clustered mutations in the second exon of the MYC gene in sporadic Burkitt's lymphoma. *Oncogene* 8, 2741 (1993)

Ye, B.H., F. Lista, F. Lo Coco, D.M. Knowles, K. Offit, RSK. Chaganti, and R. Dalla-Favera. 1993a. Alterations of a zinc finger-encoding gene, BCL-6, in diffuse large cell-lymphoma. *Science* 262:747.

Ye, B.H., P.H. Rao, RSK. Chaganti, and R. Dalla-Favera. 1993b. Cloning of bcl-6, the locus involved in chromosomal translocations affecting band 3q27 in B-cell lymphoma. Cancer Res. 53:2732.

Zhu, L. *et al.* 1993. Inhibition of cell proliferation by p107, a relative of the retinoblastoma protein. *Genes & Dev.* 7: 1111.

DISCUSSION

M. Goyns

Does the mutation of the myc gene produce proteins with altered half life?

R. Dalla Favera

No, the mutated proteins are still short-lived.

M. Goyns

Have you looked for mutations of myc in non Burkitt lymphomas?

R. Dalla Favera

We have not looked. Ian Magrath has looked and, as far as I know, there are very rare, very few cases of myelomas, a few cases of solid tumors. The exact frequency is not clear to me. I believe it is very low.

L. Nadler

One of the important questions in understanding the biology of lymphoid tumors is why these malignancies lock in phenotypic states of differentiation. I wonder if you have any thoughts about the two types of large cell lymphomas and their translocations. Where the neoplastic event happens, is it really at the pre-B cell stage as people have thought or in these two translocations? These are different tumors at different stages of differentiation, giving different phenotypes and different clinical behavior. Are they really two different cancers, that happen to grow fast?

R. Dalla Favera

This is a very good question. When we see lesions correlating with two different types of tumors, we do not know whether it is a genetic event or is it the target cells in which it occurs that determines most of the phenotype, or a combination of the two. In the case of translocations involving immunoglobulin genes, we can have some educated guesses. The ones that involved the joining region of immunoglobulin, are believed, I think with reason, but with no direct proof, to be occurring early during cell differentiation, at the time of VDJ joining; whereas the one involving myc in sporadic Burkitt lymphomas, and those involving bcl-6 when involving immunoglobulins, involve mostly switch recombinations occurring at the time of isotype switching, a much later event in B-cell differentiation. So that is the only clue that we have about the timing of occurrence of those lesions during differentiation. Whether the freezing in the differentiation stage is due to the activity of any these genes, e.g. bcl-6 and myc, we do not know. Myc is believed to arrest differentiation and therefore may arrest a cell whenever the translocation occurs, for bcl-6 we do not know.

L. Nadler

When the tumors transform, generating the transformed lymphoma, they do not behave like diffused large cell lymphomas. They do not classically invade extra lymphoid organs, enter the brain etc. They seem like different cancers. Do they seem like different cancers to you?

R. Dalla Favera

They seem different especially to me because I know that they have completely different sets of lesions; bcl-2 and p53 in transformed lymphoma and bcl-6 in the *de novo* lymphoma.

D. Baltimore

Is bcl-6 involved in diffuse large cell lymphomas? Did you say they are extra-nodal? And, did they involve a switch region translocation or not?

R. Dalla Favera

They are extra-nodal in presentation. Actually, in our series, I think 22 of 23 cases presented outside the lymph nodes.

D. Baltimore

And where did they present?

R. Dalla Favera

In all kinds of organs. Brain, lungs, stomach, gut, skin, all over the place. Regarding your second question, when bcl-6 rearranges with the immunoglobulin genes, which is in a minority of cases, it is a switch mistake.

I. Weissman

I will describe briefly which homing receptors are on normal B cell development and then see if there is some explanation we can come up with in terms of the extra-nodal site. Normally pre B and pre T cells are not expressing high levels of the homing receptors that take lymphocytes to lymph nodes or Payer's patches. Upon maturation, the early B cells, within the bone marrow, or the earliest T cell within the thymus, begin to express homing receptors. Upon activation by antigen, B cells and T cells lose Beta 7, so they are no longer

Payer's patch homing, and lose L selection expression. The B cells that are activated normally do not express either of them. They express Alpha-4 -Beta-1. Alpha-4 Beta-1 is a high affinity receptor for a ligand present in inflamed endothelium, V-CAM-1. There are also receptors for C-peptide and fibronectin. So it could be that the extra-nodal sites of location of the tumor correlates with particular inflammatory sites in those patients although it is hard to think of many inflammatory cells in the brain, commonly. Elsewhere you might expect it, in places like the lungs and so on. So I think it would be important to look at the homing receptor profile of these cells and ask whether they are constitutively activated homing receptors by simple *in vitro* binding to V-CAM. The only other point I would make is that very early in thymic development, in fetal life, the cells are mainly expressing not Alpha 4 Beta 1, but what is called Alpha Epsilon Beta-1, which promotes the localization of those cells to the epithelial sites in the body, skin, intestinal epithelium, reproductive epithelium and tongue and maybe there will be a recapitulation of this in some of these lymphomas. I do not know what the homing receptor profiles are in human tumors. What is it?

L. Nadler

The bcl-2 positive transformed large cell lymphomas are Alpha 4 positive, and the other ones are the opposite, and so it is exactly what you would have predicted. But I cannot explain the dramatic differences in a curable malignancy and an incurable malignancy based on their homing patterns. One is an incurable disease and one is a curable disease, and I cannot explain that.

I. Weissman

Have you, or anybody, looked for Alpha E vs. Alpha 4 expression?

R. Dalla Favera

A number of people are looking in detail to every single adhesion and homing molecules in these tumors. Moreover, there is extensive literature preceding bcl-6 in which people have found no difference between the normal lymphoma, extra-nodal or nodal. A couple of points need to be noted. Bcl-6 is rearranged only in B cell derived tumors, never in T cell derived tumors. The second is that there is something different on the adhesion, homing profile of these cells, not only because they are extra-nodal in presentation but also because they never go to the marrow, which is the very common fate of most of these tumors lacking BCL-6 rearrangements.

L. Nadler

There is one other marker. CD-44 and its isoforms are very important in this cancer also.

I. Weissman

CD-44 expressed at high levels turns out to be the most relevant homing receptor on activated T cells that go to the brain and cause allergic encephalomyelitis more than Alpha 4 Beta 1. And we have shown that antibodies to CD-44 completely block the development of that disease. So it could be there is something specific now with CD-44 expression.

D. Metcalf

In Australia there is an iron water tank mounted on wheels, manufactured by a fellow called Furphy. And it was around the water tank in country towns in the desert that people gathered to gossip and exchange rumors. So there is a term in Australia called a "furphy" for rumors without much factual basis. I have enjoyed Irv Weissman's discussion of thymic homing and it is very interesting. I am not sure that it has much to do with leukemia, but it is interesting. The reason I am a little bit skeptical is that if you look at the AKR mouse there is a difference between the common thymic lymphomas that arise relatively early in and apparently similar lymphoid tumors that arise late in life in non thymic organs. Leukemias arising in the marrow or the thymus may be a little more aggressive than those arising elsewhere. However, if you take a series of twenty or thirty primary AKR thymic lymphomas and ask what is their characteristic behavior on transplantation: are they more or less aggressive; and where do the cells localize, you find a very interesting phenomenon. Each leukemic population has its own characteristic pattern of localization. One may produce transplanted tumors in the ovary, for example. Others will commonly produce large transplanted tumors in peripheral lymph nodes. This is not correlated either with their initial latent period of development or the time taken for the first generation transplant. It is an individual characteristic of each one of those different tumors so I cannot see a correlation that would apply universally to these types of behavior pattern.

I. Weissman

Just to extend "Furphyism", Jean Butcher and I did an analysis of frozen sections of AKR lymphomas arising in the thymus. Both the primary site and the secondary site in the host of origin, that is we implied the secondary site because they had a thymic lymphoma and they needed a Pyrus patch or a peripheral node or both or neither. And they correlated exactly with the expression of Beta 7 for Pyrus patch and L selection for lymph nodes. Now that is not to say they were more or less aggressive. I was not going to try to combine it with David's suggestion but that is in fact how we found the homing receptors, made antibodies to the gene products and then characterized normal cells that were expressing it. On transplantation they had a predictable metastatic behavior depending on which of the homing receptors they expressed.

D. Metcalf

The only point I was making was that there is not an overt correlation between aggressiveness of a malignancy and the homing pattern of its cells.

J. Ihle

A relatively simpler question. You talked about bcl-6 expression in mature B cells, the question is whether or not it is specifically expressed and activated in mature B cells and what is the possibility that, in fact, it is the transcription factor that is regulating some of the expression of these homing antigens.

R. Dalla Favera

Maybe. We are trying hard to accumulate data on where exactly bcl-6 is regulated during the normal fate of a B cell. It is expressed right to the point where the B cell has some important decisions to make including whether to become a plasma cell, to proliferate, or to die. We want to see whether bcl-6 is involved in the decision. We do not have yet antibodies working in immunohistochemistry allowing us to determine where it is expressed and where it is not. These are the critical questions. Not only the adhesion profile but also the fate of the B cell, and at which time it switches off bcl-6.

D. Livingston

Can you make a point mutation in the transactivation domain of myc that still transactivates but does not bind 107 and, if you can do that, what is its biology?

R. Dalla Favera

So far anything that knocks out transactivation, knocks out p107 binding. So there is coincidence between the two.

D. Livingston

You mean any mutant you have made or any naturally occurring mutant?

R. Dalla Favera

Any mutant that we made.

D. Livingston

Is that hundreds of mutants or just a few?

R. Dalla Favera

A limited series of deletion mutants.

D. Livingston

I think you have to look at lots of points right?

R. Dalla Favera

I agree. Regarding the naturally occurring mutants, we analyzed 3 Chi Dang another half a dozen. All of them bind p107, all of them fail to be repressed.

D. Livingston

And are they wild-type in transactivation by all assays?

R. Dalla Favera

They are all transactivation wild type. They all transactivate the same way in a GAL-4 fusion transactivation assay.

D. Livingston

And can you see complexes by gel shift assay in extracts of these tumors?

R. Dalla Favera

As you may know, you cannot perform gel shifts with myc and this is a major problem. We have been able, recently, to get myc soluble and to obtain complexes with recombinant myc and max *in vitro* and obtain DNA binding and shifts with those. We have not been able to analyze the effects of p107 in this assay so far. When we have p107 the whole complex stays in the well, it does not enter the gel and we cannot study the effects.

D. Livingston

What about 107 E-2F complexes or 107/cyclin-A/E-2F complexes? Do you detect them in these tumors compared, say for example, to other lymphoid tumors?

R. Dalla Favera

We are addressing those questions.

C. Sawyers

Another question on the myc story. As you know there are several phosphorylation sites in the N terminus of myc, threonine 58 and serine 62 which I think have been involved and when mutated, have effects on transactivation as well as transformation, sometimes very confusing effects. Are those sites involved in some of your naturally occurring mutants and what do they do with p107 binding or transformation assays?

R. Dalla Favera

They may or may not be involved. It is not a general rule that you have phosphory-lation mutants. As you correctly mentioned, the data on the ability of various phosphorylation mutants to transactivate is highly controversial. We do not see any difference in the ability of phosphorylation mutants to transactivate.

C. Sawyers

Have you tested any of your naturally occurring mutants in a traditional ras plus myc transformation assay to see if they are more active than wild type myc?

R. Dalla Favera

No, but we do not see much difference in a Rat-1 transformation assay.

J. Adams

I wonder if I could return to your interesting earlier remarks about the differences between hematopoietic tumors and solid tumors, namely, that the solid tumors display more evidence of genetic instability. I wonder if this difference could reflect the need for a smaller number of genetic alterations to generate hematopoietic malignancies? The argument for that has been made, as you know, from the age incidence of different types of malignancies and we would argue from our transgenic studies that it probably takes only about three mutations to get a full blown lymphoid malignancy, at least in mice. A larger number of mutations may be required to produce most solid tumors, as has been suggested for certain human tumors such as colon cancer. If so, I think it would follow that tumors of that type would tend to exhibit early alterations in genes that would themselves affect mutability, such as a breakdown in checkpoint control or in DNA repair. And that as the consequence would provide an accumulation of other types of alterations, some of which would be relevant and some irrelevant to the actual tumor development. Thus, genetic instability would be more prevalent in solid tumors because a larger number of independent events is necessary to produce a malignancy than in hematopoietic cells.

R. Dalla Favera

There is a difference between generalized genome instability and the specific instability observed in hematologic malignancies. There may also be differences among various hematological malignancies. I do not think, however, that lymphoma for instance, may require a smaller number of lesions than a colon cancer. We had some cases of AIDS associated lymphomas, admittedly an atypical case, in which we found half a dozen lesions. These tumors actually developed quite fast because they develop a few years after the patients are infected by HIV, so we have some sort of time frame in which they developed. I think that the data on the transgenic mice may be somehow biased biologically because the situation is very artificial. In these mice you have a whole tissue with a lesion as opposed to the history of naturally occurring tumors where a single cell has to make it in the context of an essentially normal tissue. So less lesions may be necessary to make a tumor in transgenic mice where the target is greatly expanded than in a natural situation where the target is minimal.

E. Mihich

What we are discussing these days essentially is the approach of identifying gene products, making constructs and seeing what the products of these constructs will do, in terms of their function, with respect to the original function of the unmodified gene. And all of this is looking at the product of gene functions and alterations. There is another approach, namely, to identify a protein and its configuration, and get to the medicinal chemist and try to develop synthetic fragments of that protein and see what the minimum structure would be for function and what alteration might be applied to that structure to modify the function. And the question is: In addition to the approaches that we have discussed these days, what is being done in studying some of these molecules by looking at the synthetic structures that can be these days rather readily made and see what the function would be of those structures. Is this approach ready for us to be pursued or is it premature, with some structures such as p107 for instance.

R. Dalla Favera

Well, I would make some distinctions in terms of target. Those that are immunogenic for instance, all the fusion proteins in the immature malignancies seem to be a more amenable target for a number of therapeutic approaches. If you take instead the example of interaction between p107 and myc, it is one of many that are present in every tissue and thus, they seem to be more problematic to me in terms of targeting by any kinds of means.

L. Nadler

You can find bcl-2 translocations in virtually every tonsil of every child. The incidence is low and you can find them in pretty much everyone's bone marrow in this room. What about bcl-6? We probably all, in this room, have every one of those translocations and I would argue our immune system is handling them or possibly not handling them and we live in symbiosis with that.

R. Dalla Favera

It is not clear that everybody has bcl-2 translocation, although I agree a good fraction of normal people have bcl-2 translocations.

L. Nadler

If you look, you find them.

R. Dalla Favera

I do not think there are other examples, although we may not have the sensitivity to look at them. I do not know anything about bcl-6 in this respect and it is a particularly difficult case because the translocations are promiscuous so you do not have an easy PCR assay, and therefore, are not amenable to a simple PCR assay.

13

STRUCTURE AND EXPRESSION PATTERN OF THE PML GENE

Marta Fagioli and Pier Giuseppe Pelicci[*]

Istituto di Medicina Interna e Scienze Oncologiche
Policlinico Monteluce
Perugia University
Perugia, Italy

INTRODUCTION

APL is characterized by a reciprocal translocation that involves chromosomes 15 and 17. As a consequence of this translocation two recombinant chromosomes are formed: 15q+ and 17q- . Expression of the t(15;17) is strictly limited to neoplastic promyelocytes and is often (70-90%) the only chromosomal anomaly seen in the neoplastic metaphases. Since the t(15;17) is associated exclusively with a promyelocytic phenotype and there are no other known chromosome alterations able to express this phenotype, it is suggested that the t(15;17) is implicated in the pathogenesis of APL (1-2).

The chromosome 15 and 17 break sites were isolated by three groups using distinct experimental approaches (3-6). The chromosome 15 breakpoint falls within a previously unknown gene, which was originally designated myl, but then renamed PML, for promyelocytes. The chromosome 17 breakpoint is located in the locus that encodes the retinoic acid receptor α (RARα) (7-8). Two fusion genes are formed as a consequence of the translocation: the PML/RARα gene on the recombinant 15q+ chromosome and its reciprocal RARα/PML on the recombinant 17q- chromosome (9). The chimeric PML/RARα and RARα/PML transcripts have the potential to encode PML/RARα and RARα/PML fusion proteins, respectively (9-14).

The candidate transforming gene is PML\RARα, since it is expressed in 100% cases, whereas RARα\PML is expressed in only 70-80% (9). We have recently identified APL cases with submicroscopic 15;17 recombinations leading to the formation of non-reciprocal PML\RARα or RARα\PML fusion genes. Each of the two reciprocal translocation products may, therefore, be independently formed and selected by the leukemic phenotype, implying that both are involved in tumorigenesis (15).

[*] Corresponding author: P. G. Pelicci, Istituto Clinica Medica I, Policlinico Monteluce, 06100 Perugia, Italy. Tel. (39) 75-578 3255; FAX. (39) 75-578 3444 or (39) 75-572 6264.

Direct proof for a role of PML/RARα protein in the pathogenesis of APL is still lacking (e.g. ectopic expression of the fusion protein into normal hematopoietic precursors). Indirect evidence has, instead, been obtained from expressing PML/RARα in hematopoietic precursor cell lines. We expressed the PML/RARα protein in the U937 myeloid precursor cells and showed that they: i) lose the capacity to differentiate when induced by vitamin D_3 and TGFβ1; ii) acquire enhanced sensitivity to the differentiation action of RA; iii) exhibit a higher growth rate that is accounted for by a reduction in apoptotic cell death (16).

On the basis of its structure and preliminary biochemical data it appears that PML/RARα may act as multifunctional protein with the potential to interfere with the endogenous activation pathways of RARα and other members of the nuclear receptor family, such as thyroid hormone receptors (TR) and VD receptors (VDR). A first step in understanding the mechanism(s) of action of PML/RARα is, therefore, to define which RARα regions are retained in the fusion protein and to establish whether their corresponding functions are maintained or modified when they fuse with PML.

RARα is a member of the retinoic acid receptor family and is physiologically implicated in the regulation of embryonic and myeloid differentiation (17). It acts as a retinoic acid (RA) dependent enhancer factor to regulate activity of RA-target genes. Important biochemical properties of RARα are, therefore, to specifically bind retinoids and DNA. RARα directly bind to specific cis-acting elements (RA responsive elements, RAREs) located in the promoter region of RA-target genes (18-20). High affinity binding of RARα to RARE takes places only if another member of the retinoic acid receptor family, RXR, is present and the molecular basis of the RAR/RXR synergism is their dimerization (21). The biochemical properties of RARs have been mapped to specific regions of the RAR molecule. Primary RAR sequences can be divided into six regions (A-F) on the basis of their alignement with the other members of the nuclear receptor superfamily (22). The DNA binding domain is made up of two zinc fingers and corresponds to region C. The ligand-binding domain corresponds region E. Region E and D respectively contain the domains for dimerization and nuclear localization. The A/B region has transcriptional activation function. As the chromosome 17 break always occurs in RARα intron 2, the RARα component of the PML/RARα protein is the same in all cases of APL and corresponds to regions B to F (22-26). One would expect, therefore, PML/RARα to retain the DNA binding, ligand binding, RXR dimerization and nuclear localization properties of RARα. PML/RARα forms homodimers which bind to RAREs with a different specificity than RAR/RXR heterodimers and heterodimers with RXR. Formation of PML/RARα-RXR heterodimers can indirectly influence the functional activity of other nuclear receptors like RARs, TR and DR, which physiologically dimerize with RXR allowing efficient formation of stable RE-bound complexes.

Little is known about PML and its contribution to the transformng activity of the PML/RARα fusion protein. We will summarize the information we have accumulated on the structure, expression pattern and function of PML.

RESULTS AND DISCUSSION

Organization of the PML Locus

The PML locus contains 9 exons. Since portions of PML exons 7 and 8 are differently assembled into the PML cDNAs (see below) they are individually designated 7a, 7b, 8a (Fig. 1). The ATG codon is found in exon 1 and alternative TGA termination codons in exons 7, 8 and 9 (see below). A limited restriction map of the human PML locus is shown in Fig.1. The PML-derived transcripts proved to be extremely complex and, due to alternative splicings of some exons, a single primary transcript can form many different mature transcripts. The alternative splicings can cut out entire or portions of exons. As exon 7 possesses internal splice sites, the whole exon or only part of it (7a) can be assembled into the mature transcripts together with exon 8 or exon 8a and 9. Another splicing involves the alternative assembly of one retained intron. 641 bp of exon 7 are spliced out and exon 7a abutts exon 7b (Fig. 1). Similar phenomena have been described for other genes, for example the rat fibronectin and the rat γ-fibrinogen gene.

An important aspect of these alternative splicings is that they always involve coding exons and, in consequence, the variously spliced PML transcripts encode different PML isoforms. The exons implicated in the alternative splicings fall into two groups: central exons (4, 5 and 6) and 3' exons (7, 8 and 9). Therefore, all PML isoforms share the N-terminal aminoacid region, but differ in their central and C-terminus portions.

Modular Organization of the PML Protein

The N-terminal aminoacid portion common to all PML isoforms (AA 1-394), which is encoded by exons 1, 2 and 3, contains three clusters rich in cysteine and histidine residues which are retained in all PML and PML/RARα isoforms (Fig.2). The first cluster corresponds to a new zinc finger motif (RING motif) that defines a recently recognized family of proteins with functions, where known, that require DNA binding (27-30). These functions include: i) regulation of development, for example the XNF-7 gene in Xenopus; ii) regulation of gene expression, for example RPT-1, which affects the expression of the IL2 receptor gene; iii) repair of UV-damaged DNA (e.g. Rad18); iv) DNA recombination (e.g. RAG-1). RING genes other than PML have been implicated in tumorigenesis; examples are T18 and Rfp, two transforming proteins that result from the fusion between a RING protein and the B-Raf and ret proteins, respectively, and Bmi-1, a gene that cooperates with myc in lymphoma development (30 and references therein). That RING is a zinc finger domain with DNA binding activity is supported by the facts that a RING peptide binds zinc with tetrahedral co-ordination to cysteines and DNA in a zinc-dependent fashion (30).

A group of RING genes contains a cysteine/histidine rich region, termed the B box, which corresponds to an additional putative zinc finger domain (28-29). PML belongs to the RING + B box family, which also includes XNF7, T18, rfp, rpt1 and the human RNA-binding autoantigen SS-A/Ro. PML contains two B boxes, which correspond to the second and third cysteine/histidine rich cluster (Fig. 2).

Immediately C-terminal to the finger-like region is a region that can form multiple α-helices. A portion of the α-helix has the potential to assume a coiled-coil configuration and contains four clusters of heptad repeats with hydrophobic aminoacids at first, fourth and eighth positions (31-32). Similar repeats are found in the ligand binding domain of TR, RAR and VDR and are considered to be dimerization interfaces (32). The α-helix without the heptad repeats is variably retained in the different PML isoforms, while the finger-like and coiled-coil regions are retained in all, suggesting that they all have the potential to bind DNA and form homo- and/or hetero-dimers (33).

The next region, that spans AA 502-570, is encoded by the 3' portion of exon 6 and the entire exon 7a and corresponds to the serine/proline-rich domain. This region is retained in all PML isoforms but the 3-4-7 and the 3-7 which only contain a portion of it (Fig. 2B). It has been proposed that, as this domain could be phosphorylated at the serine residues and so be the target of kinases or phosphatases, it could be involved in regulating the molecule's activity (12-13). An analogous serine/proline-rich region has been identified in other molecules, for example in the rabbit muscle glycogen synthase where it is probably involved in the hormonal control of the enzymatic activity. Furthermore, the PML serine/proline-rich domain contains four repeats of the X-S-P-X consensus (X: any aminoacid; S: serine; P: proline) that has been identified as the minimum recognition sequence of a new ser-

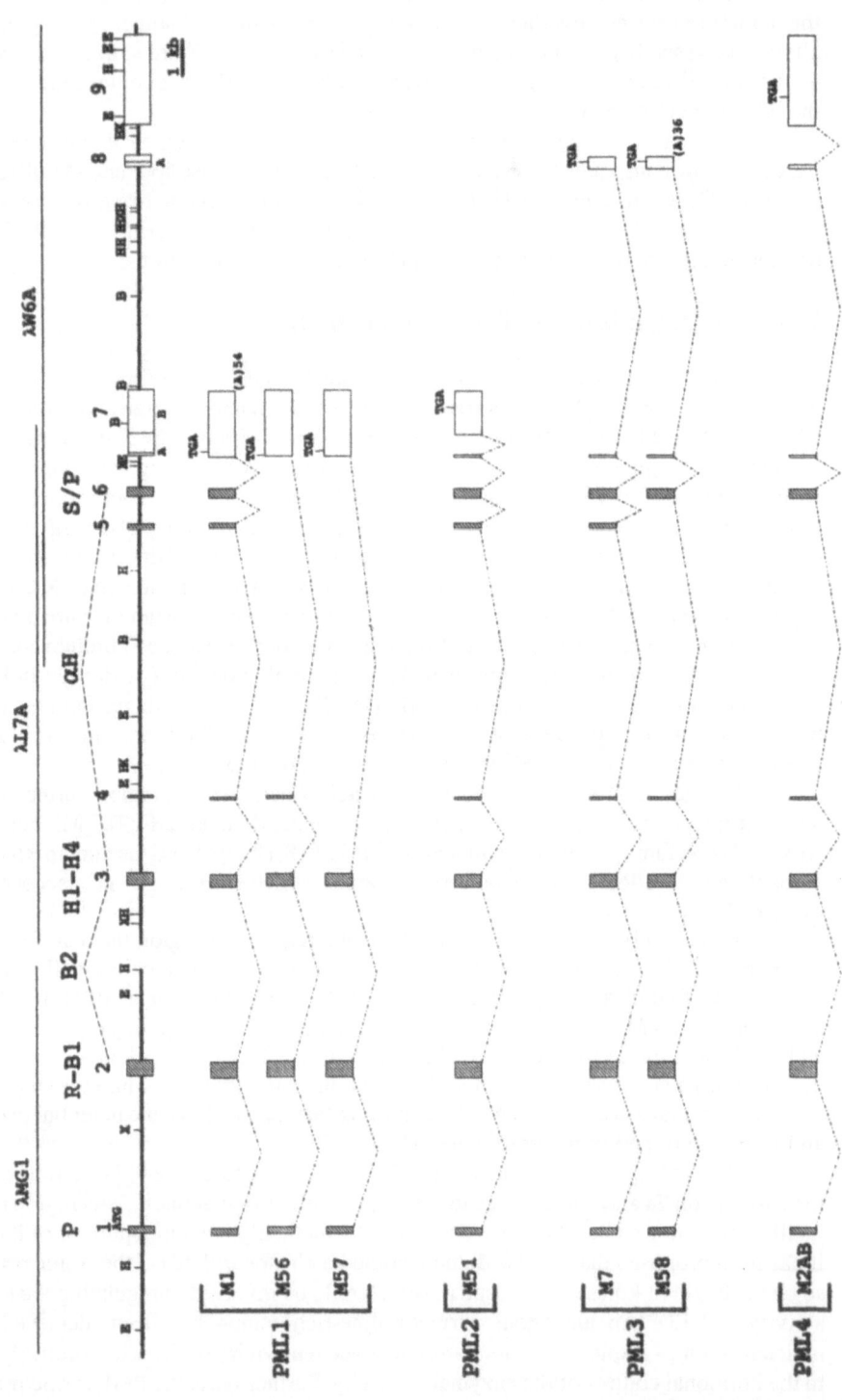

Figure 1. Organization of the PML locus and exon assembly of PML cDNAs. A limited restriction enzyme map of the PML locus is shown at the top with the corresponding PML λ-phages also indicated. Boxes indicate PML exons and the exon numbers are given above the boxes (numbering is temporary since the PML cap site has not been mapped). There is a gap within the PML locus map because the λ-MG1 and λ-L7A PML genomic clones do not overlap. The regions coding for the PML proline-rich motif (P), the three cysteine-rich clusters (R, B1, and B2), the α-helix domain (αH), the the four heptads (H1-H4) and the serine/proline-rich motif (S/P) are indicated above the corresponding coding exons (regions encoded by more than one exon are shown by broken lines). B = BamHI, E = EcoRI, H = HindIII, X = XbaI. A schematic representation of the PML exons assembly in the indicated cDNAs is given below. Exons are indicated by boxes and the splicing events are shown by the broken zig-zag line. The position of the TGA termination codon is given for each cDNA. Clones M51 and M2ab were shorter than shown, and initiated from nucleotide 655 and 1151 with respect to the reported PML M1 cDNA sequence (11), respectively.

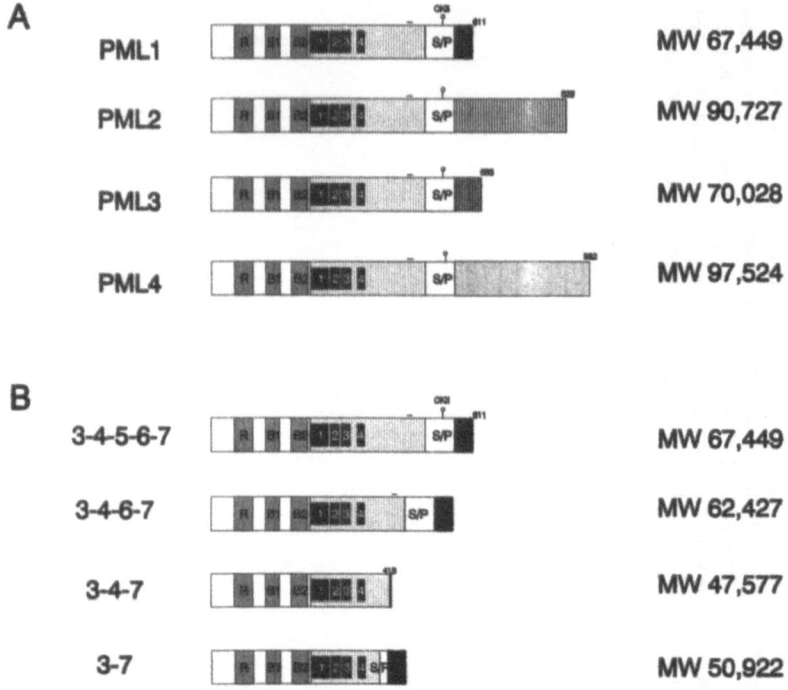

Figure 2. Modular organization of PML isoforms. The putative PML domains (designated as in Figure 1) are illustrated for each predicted PML isoform (PML1 to 4, 3-4-5-6-7, 3-4-6-7, 3-4-7, 3-7 as indicated and detailed in the text). Regardless the exon combinations actually identified by cDNA cloning or predicted by PCR analysis (see text) only the exon 3-4-5-6-7 combination is shown in panel A and only the N-terminus of the PML1 isoform is shown in panel B. The predicted molecular weight of each putative isoform is given on the right. The alternative C-termini are shown by difference in shading.

ine/threonine kinase that acts on tyrosine hydroxylase and numerous other substrates *in vitro*. It contains a casein kinase II (CKII) phosphorylation site (our unpublished results) (Fig. 2). Phosphorylation by CKII is a post-translational modification shared by many transcription factors and usually associated with modifications in their biological activity (34).

Expression Pattern of PML Isoforms

The expression pattern of all PML isoforms was determined by analyzing a series of human RNAs with RNA probes representative of the various isoforms in RNase protection experiments (33). The samples analyzed included cells derived from the myeloid lineage (HL-60, KG1, KG1a, U937), the T-lymphoid lineage (CEM, Molt4, Fro2.2), the B-lymphoid lineage (CB33, Daudi, UH1), lung (WI38, L132, Calu1, H69), breast (MCF7), liver (HepG2), pancreas (MiaPaca), uterus (Skov), brain (SKNMC), bone (U-2 OS), teratocarcinoma (Ntera2). Results were comparable for the cell lines analyzed. RNA probes spanning each alternative PML cDNA 3' end were used to measure the expression levels of the PML1, 2, 3 and 4 isoforms. These experiments demonstrated that the PML1 isoform was the least represented in the majority of cell lines and so confirmed the data obtained in the Northern blot experiments (not shown). Analogous experiments were carried out to evaluate the expression of the isoforms derived by the alternative assemblage of PML exons 4, 5 and 6. The 3-4-6 and the 3-4-5-6 transcripts were expressed at comparable levels, whereas 3-7 expression was very low. The expression of the 3-4-7 transcript could not be evaluated in this analysis because its protected fragment comigrates with the 3-4-5-6 fragment. Finally, we excluded the existence of additional PML isoforms that involve exons 1 and 2 by RNase protection experiment using an exon1-exon2 representative RNA probe. In synthesis, the results indicated that: i) PML transcripts are present in all cell lines examined, but the levels vary slightly from one line to another; ii) all isoforms, but the 3-7, are more or less equally expressed in all cell lines examined.

Cell Localization of PML Proteins

To study the expression and cell localization of PML proteins we have generated a monoclonal antibody (PG-M3) directed against the PSPSPTERAPASEEE(C)-NH2 synthetic peptide, corresponding to aminoacids 37-51 of the PML protein (35). At immunocytochemistry, both mock- and PML-transfected U937 cells displayed intense, speckled nuclear staining for the PG-M3 mAb. However, PML nuclear dots were much larger in PML- than mock-transfected cells (Fig.3A and B).

All human myeloid cell lines tested (HL60, KG1, TF1, K-562, U937) by the APAAP technique exhibited the same staining pattern, i.e. speckled positivity predominantly located within the nuclei (not shown). In particular, the number of speckles varied and ranged from 5 to 20 (average 10) in individual cells, while the size of the dots differed both from cell to cell and whithin single cells. The dots were usually distributed throughout the nucleus with no association to nucleoli. There was no significant staining of the nucleus between the dots. A few PML dots were also noted in the cytoplasm of 10%-20% of cells (Fig.1A and B).

In contrast, a micropunctate staining pattern consisting of hundreds of intranuclear dots, sometimes with weak labeling of the nuclear interspaces was typical of the NB4 promyelocytic cell line bearing the t(15;17) (a gift from M. Lanotte) (Fig.3C). These dots were so tiny that they were usually very difficult to count accurately. The percentage of NB4 cells that manifested cytoplasmic staining for PG-M3 was similar to that of other myeloid and non-myeloid cell line (10%-20%).

Cellular Localization of Transfected Pml Isoforms and the PML/RARα Protein

From the above results, it appears that PML and PML/RARα proteins are localized in both the nucleus (virtually all cultured cells) and the cytoplasm (10%-20% cells). Because many PML and PML/RARα isoforms have been identified that might differ in their cellular localization, we analyzed the staining pattern of different, individually transfected PML isoforms.

Transfection of NIH-3T3 cells with cDNAs encoding the PML2 3-4-5-6-7 or PML3 3-4-6-7 isoforms resulted in a strong nuclear staining of all cells with the typical speckled pattern. The PML-stained structures within the nuclei of NIH-3T3 cells were round, large and often had a vesicular-like appearance (not shown). Moreover, numerous positive dots were evenly distributed throught the cytoplasm of most cells.

Positivity for PML was observed in the cytoplasm, but never within the nuclei, of NIH-3T3 cells transfected with cDNA encoding PML3 3-7 (Fig.3D). The cytoplasmic staining pattern was mainly microgranular with 1-2 large, sometimes ring-like, spots (Fig.3D). The staining pattern of mouse NIH-3T3 cells transfected with the cDNA encoding PML 3-4-7 was predominantly, but not exclusively, cytoplasmic and was limited to 1-2 large dots (not shown).

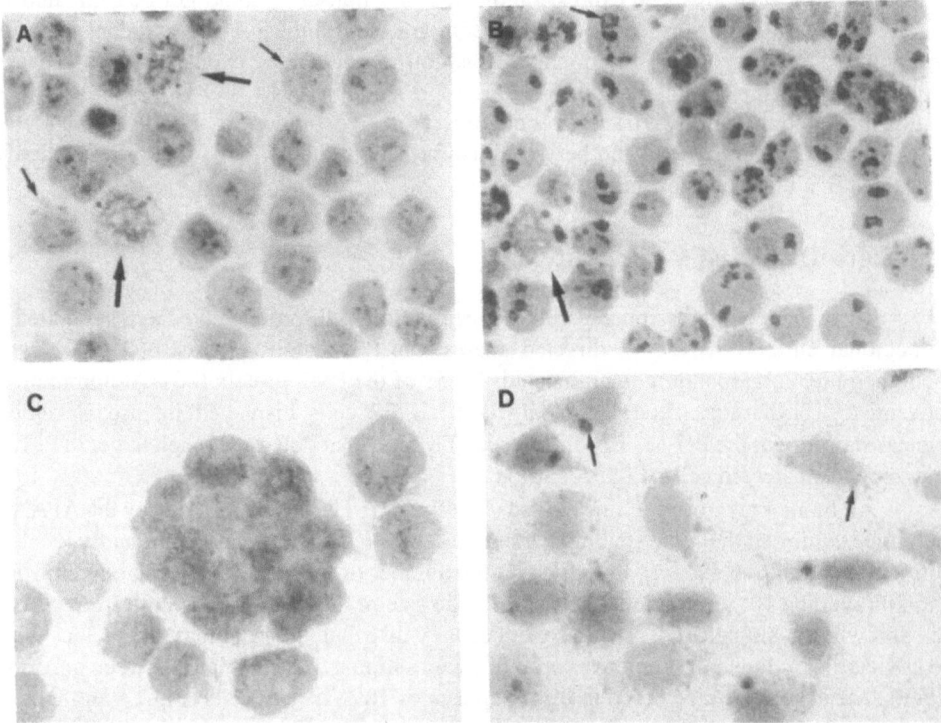

Figure 3. Localization of PML proteins. **A and B.** Reactivity of the PG-M3 anti-PML monoclonal antibody with the human U-937 cell line in basal condition (A) and after transfection with cDNA encoding PML protein (Fig 2; isoform PML3 3-4-6-7) (B). Labeling of endogenous PML produces a speckled pattern (average 10 dots/per cell). Only occasional cells (short arrows, A) show a few positive granules in the cytoplasm. The long arrows (A and B) point to mitotic figures that display only cytoplasmic dots not associated with metaphase chromosomes. The short arrow in B indicates a typical "ring-like" nuclear structure. **C.** Reactivity of PG-M3 with the human APL cell line NB4. The acute promyelocytic cell line NB4 shows a micropunctate staining pattern of PML. **D.** Reactivity of anti-PML antibody with mouse NIH-3T3 cells transfected with cDNAs encoding the PML3 3-7 isoform of human PML. PML expression is mostly restricted to the cytoplasm where a few very large dots are observed (arrows). APAAP technique; hematoxylin counterstain; x 1,000.

In summary, staining was cytoplasmic in PML isoforms that lacked the PML region 422-553 (PML 3-4-7 and PML3 4-7) while it was both nuclear and cytoplasmic in all other PML isoforms. Since the 418-466 region is not present in the nuclear PML3 3-4-6-7 isoform, the PML nuclear localization signal (NLS) must be located within the region spanning aminoacids 466 to 553. Furthermore, these data suggest that each of the different PML isoforms contributes to the cytoplasmic localization of the endogenous PML proteins.

Expression of the PML Protein in Normal Tissues

In normal peripheral blood, the PG-M3 mAb reacted with all mononuclear cells, monocytes showing a higher number of nuclear dots (3-5/cell) than neuthrophils and lymphocytes (2-3 dots/cell) (data not shown). Bone marrow myeloid cells at different stages of maturation were PG-M3 positive. Cells of the erythroid lineage only occasionally displayed nuclear positivity for PML and megakaryocytes were usually negative (data not shown). Almost all cortical thymocytes were PML negative; only rare, usually larger, cortical thymocytes reacted with PG-M3. The thymic cortex also contained isolated non-lymphoid elements that displayed intense speckled staining for PML. The amount of PML was greater in medullary than cortical thymocytes; the most strongly PML positive cells in the medulla were thymic epithelial cells, which had a staining pattern similar to that of tonsil epithelium.

PML protein expression was variable in normal human tonsil cell populations. 2-3 small dots per cell were seen in mantle B- lymphocytes, while germinal center B cells usually stained weakly (1 small nuclear dot/cell) or not at all for PML. In contrast, germinal center macrophages exhibited intense nuclear PML staining. The second most strongly labeled cell population in the tonsil were endothelial and squamous epithelium cells.

The presence of PML protein was also demonstrated by immunohistology on nuclei of many non-lymphohemopoietic cells, including hepatocytes, cells of kidney tubules, different kind of epithelia (data not shown). This indicates that PML is quite ubiquitous protein.

PML Expression during Macrophage Differentiation and Activation

The reported strong positivity for PML of tissue macrophages prompted us to analyze the expression of PML in promonocytic U937 cells following stimulation with agents able to induce macrophage activation or differentiation.

The number of PG-M3 positive nuclear dots was greatly increased (peak at day 1) in U937 cells treated with IFN-γ, compared to unstimulated U937 cells (not shown). PML expression was not significantly changed by incubation with vitamin D3 alone, whilst both the number and the size of PG-M3 positive nuclear dots were increased following incubation with vitamin D3 plus TGFβ1, as compared with negative controls (not shown). However, maximum nuclear expression of PML was observed when U937 cells were both activated and induced to mature following stimulation with IFN-γ plus vitamin D3 and TGFβ1. Taken togheter, the above results suggest that PML is upregulated during both monocyte maturation and activation.

CONCLUSIONS

The chromosome 15 breakpoint of the t(15;17) is variably located in three regions of the PML locus. In 90-95% of cases, it is equally distributed between intron 6 (breakpoint cluster region 1; bcr1) and intron 3 (bcr3). In the remaining 5-10% cases, it is located within exon 6 (bcr2) (10). Regardless of the extreme variability of the PML break sites, PML/RARα genes which have the potential to encode for a fusion PML/RARα protein are consistently selected by the leukemia. In bcr1 or bcr3 cases, the 5' portion of PML intron 6 or 3 respectively fuses with the 3' portion of RARα intron 2. During assemblage of the PML/RARα junction in the fusion transcript the chimeric intron is spliced out and the longest PML and RARα open reading frames ORFs become aligned. The operative mechanism of bcr2 is more complex: a cryptic donor site of the retained portion of PML exon 6 and the RARα intron 2 physiological acceptor site take part in the assemblage (10).

Due to the chromosome 15 breaksite heterogeneity, the retained PML portion differs in each PML/RARα. If the various PML/RARα proteins are compared, it will be seen that the only portions of PML consistently retained in the fusion protein are the finger-like and the coiled-coil region (10).

The localization of PML/RARα differs from that of PML and RARα. PML/RARα has micropunctated nuclear pattern, PML has a speckled nuclear pattern, whereas RARα is finely dispersed in the nucleus (14, 31, 36, 37). Anti-PML antibodies reveal that APL cells, which express all three proteins, display the PML/RARα-like micropunctated nuclear pattern, indicating that the fusion protein localization dominates over the two wild type proteins (31). Strikingly, the treatment of APL blasts with RA converts the micropunctated nuclear pattern to the speckled, PML-like, pattern (36).

Analysis of the RA binding proteins in PML/RARα expressing cells has demonstrated that the fusion protein is part of multiple nuclear complexes with molecular weights of 600 and 1200 kDa (the apparent molecular weight of PML/RARα is 110 kDa) (38). Although little is known of the proteins that in vivo belong to these complexes, it has been demonstrated *in vitro* that PML/RARα dimerizes with itself, with PML and, as already discussed, with RXR (31).

Since it is unlikely that the PML/RARα-PML heterodimers perform the functions of wild type PML, it is conceivable that PML/RARα interferes with PML functions. This is supported by recent experiments showing that PML/RARα mutants lacking different portions of the PML protein did not retain their transforming activity *in vitro*. Further studies are required to define the function of PML proteins and the role of the PML signaling protein in the pathogenesis of APL.

ACKNOWLEDGMENTS

This research was supported by grants from the "Associazione Italiana per la Ricerca sul Cancro" (A.I.R.C.); the Italian Council of Research (ACRO project); E.C. (Biotech and Biomed).

REFERENCES

1. Grignani F., Fagioli M., Alcalay M., Pandolfi P.P., Donti E., Biondi A., Lo Coco F., Grignani F., Pelicci P.G.: Acute promyelocytic leukemia: from genetics to treatment. Blood, 83:10, 1
2. Warrell RP Jr, Frankel SR, Miller WH Jr, Scheinberg DA, Itri LM, Hittelman WN, Vyas R, Andreeff M, Tafuri A, Jakubowski A, Gabrilove J, Gordon MS, Dmitrovsky E: Differentiation therapy for acute promyelocytic leukemia with tretinoin (all-trans-retinoic acid). N Engl J Med 324:1385, 1991.
3. de Thè H, Chomienne C, Lanotte M, Degos L, Dejean A: The t(l5;17) translocation of acute promyelocytic leukaemia fuses the retinoic acid receptor α gene to a novel transcribed locus. Nature 347:558, 1990.
4. Borrow J, Goddard AD, Sheer D, Solomon E: Molecular analysis of acute promyelocytic leukemia breakpoint cluster region on chromosome 17. Science 249:1577, 1990.
5. Longo L, Pandolfi PP, Biondi A, Rambaldi A, Mencarelli A, Lo Coco F, Diverio D, Pegoraro L, Avanzi G, Tabilio A, Zangrilli D, Alcalay M, Donti E, Grignani F, Pelicci PG: Rearrangements and aberrant expression of the retinoic acid receptor α gene in acute promyelocytic leukemia. J Exp Med 172:1571, 1990.
6. Alcalay M, Zangrilli D, Pandolfi PP, Longo L, Mencarelli A, Giacomucci A, Rocchi M, Biondi A, Rambaldi A, Lo Coco F, Diverio D, Donti E, Grignani F, Pelicci PG: Translocation breakpoint of acute promyelocytic leukemia lies within the retinoic acid receptor α locus. Proc Natl Acad Sci U S A 88:1977, 1991.
7. Petkovich M, Brand NJ, Krust A, Chambon P: A human retinoic acid receptor which belongs to the family of nuclear receptors. Nature 330:444, 1987.
8. Giguere V, Ong ES, Segui P, Evans RM: Identification of a receptor for the morphogen retinoic acid. Nature 330:624, 1987.
9. Alcalay M, Zangrilli D, Fagioli M, Pandolfi PP, Mencarelli A, Lo Coco F, Biondi A, Grignani F, Pelicci PG: Expression pattern of the RARα-PML fusion gene in acute promyelocytice leukemia. Proc Natl Acad Sci USA 89:4840, 1992.
10. Pandolfi PP, Alcalay M, Fagioli M, Zangrilli D, Mencarelli A, Diverio D, Biondi A, Lo Coco F, Rambaldi A, Grignani F, Rochette-Egly C, Gaube MP, Chambon P, Pelicci PG: Genomic variability and alternative splicing generate multiple PML/RARa transcripts that encode aberrant PML proteins and PML/RARα isoforms in acute promyelocytic leukaemia. EMBO J 1:1397, 1992.

11. Pandolfi PP, Grignani F, Alcalay M, Mencarelli A, Biondi A, Lo Coco F, Grignani F, Pelicci PG: Structure and origin of the acute promyelocytic leukemia myl/RARα cDNA and characterization of its retinoid-binding and transactivation properties. Oncogene 6:1285, 1991.

12. Kakizuka A, Miller WH Jr, Umesono K, Warrel RP Jr, Frankel SR, Murty VVVS, Dmitrovsky E, Evans RM: Chromosomal translocation t(15;17) in human acute promyelocytic leukemia fuses RARα with a novel putative transcription factor, PML. Cell 66:663, 1991.

13. de Thè H, Lavau C, Marehio A, Chomienne C, Degos L, Dejean A: The PML-RARa fusion mRNA generated by the t(15;17) translocation in acute promyelocytic leukemia encodes a functionally altered RAR. Cell 66:675, 1991.

14. Kastner P, Perez A, Lutz Y, Rochette-Egly C, Gaub MP, Durand B, Lanotte M, Berger R, Chambon P: Structure, localization and transcriptional properties of two classes of retinoic acid receptor α fusion proteins in acute promyelocytic leukemia (APL): structural similarities with a new family of oncoproteins. EMBO J 11:629, 1992.

15. Lafage-Pochitaloff M., Alcalay M., Brunel V., Longo L., Sainty D., Simonetti J., Birg F., Pelicci P.G.:Acute promielocytic leukemia cases with non-reciprocal PML/RARα or RARα/PML fusion protein. Blood, in press. 1995.

16. Grignani Fr, Ferrucci PF, Testa U, Talamo G, Fagioli M, Alcalay M, Mencarelli A, Grignani F, Peschle C, Nicoletti I, Pelicci PG: The acute promyelocytic leukaemia specific PML\RARα fusion protein inhibits differentiation and promotes survival of myeloid precursor cells. Cell 74:423, 1993.

17. Evans RM: The steroid and thyroid hormone receptor superfamily. Science 240:889, 1988.

18. Leroy P, Nakshatri H, Chambon P: Mouse retinoic acid receptor α2 isoform is transcribed from a promoter that contains a retinoic acid response element. Proc Natl Acad Sci USA 88:10138, 1991.

19. de Thè H, Vivanco-Ruiz MDM, Tiollais P, Stunnenberg H, Dejean A: Identification of a retinoic acid responsive element in the retinoic acid receptor ß gene. Nature 343:177, 1990.

20. Smith WC, Nakshatri H, Leroy P, Rees J, Chambon P: A retinoic acid response element is present in the mouse cellular retinol binding protein I (mCRBPI) promoter. EMBO J 10:2223, 1991.

21. Mangelsdorf DJ, Ong ES, Dyck JA, Evans RM: Nuclear receptors that identifies a novel retinoic acid response pathway. Nature 345:224, 1990.

22. Krust A, Green S, Argos P, Kumar V, Walter P, Bornet JM, Chambon P: The chicken oestrogen receptor sequence: homology with v-erbA and the human oestrogen and gluococorticoid receptors. Embo J 5:891, 1986.

23. Diverio D, LoCoco F, D'Adamo F, Biondi A, Fagioli M, Grignani Fr, Rambaldi A, Rossi V, Avvisati G, Petti MC, Testi AM, Liso V, Specchia G, Fioritoni G, Recchia A, Frassoni F, Ciolli S, Pelicci PG: Identification of DNA rearrangements at the RARα locus in all patients with acute promyelocytic leukemia (APL) and mapping of APL breakpoints within the RARα second intron. Blood 79:1, 1992.

24. Chen SJ, Zhu YJ, Tong JH, Dong S, Huang W, Chen Y, Xiang WM, Zhang L, Song Li X, Qian GQ, Wang ZY, Chen Z, Larsen CJ, Berger R: Rearrangements in the second intron of the RARα gene are present in a large majority of patients with acute promyelocytic leukemia and are used as molecular marker for retinoic acid induced leukemic cell differentiation. Blood 78:2696, 1991.

25. Tong JH, Dong S, Geng JP, Huang W, Wang ZY, Sun GL, Chen SJ, Chen Z, Larsen CJ, Berger R: Molecular rearrangements of the MYL gene in acute promyelocytic leukemia (APL, M3) define a breakpoint cluster region as well as some molecular variants. Oncogene 7:311, 1992.

26. Chen Z, Chen SJ, Tong JH, Dong S, Huang W, Chen Y, Xiang WM, Zhang L, Song X, Qian XS, Wang ZY, Chen Z, Larsen CJ, Berger R: The retinoic acid α receptor gene is frequently disrupted in its 5' part in Chinese patients with acute promyelocytic leukemia. Leukemia 5:288, 1991.

27. Goddard AD, Borrow J, Freemont P, Solomon E: Characterization of a zinc finger gene disrupted by the t(15;17) in acute promyelocytic leukemia. Science 254:1371, 1991.

28. Freemont PS, Hanson IM, Trowsdale J: A novel cysteine-rich sequence motif. Cell 64:483, 1991.

29. Reddy BA, Etkin LD, Freemont PS: A novel zinc finger coiled-coil domain in a family of nuclear proteins. TIBS 17:344, 1992.

30. Lovering R, Hanson IM, Borden KLB, Martin S, O'Reilly NJ, Evan GI, Rahman D, Pappin DJC, Trowsdale, Freemont P: Identification and preliminary characterization of a protein motif related to the zinc finger. Proc Natl Acad Sci U.S.A. 90: 2112, 1993.

31. Perez A, Kastner P, Sethi S, Lutz Y, Reibel C, Chambon P: PMLRARα homodimers: distinct DNA binding properties and heterodimeric interaction with RXR. EMBO J 12:3171, 1993.

32. Forman BM, Samuels HH: Interaction among a subfamily of nuclear hormone receptors: The regulatory Zipper model. Mol Endocr 4:1293, 1990.

33. Fagioli M, Alcalay M, Pandolfi PP, Venturini L, Mencarelli A, Simeone A, Acampora D, Grignani F, Pelicci PG: Alternative splicing of PML transcripts predicts coexpression of several earboxy-terminally different protein isoforms. Oncogene 7:1083, 1992.

34. Hunter T, Karin M: The regulation of transcription by phosphorylation. Cell 70:375, 1992.

35. Flenghi L., Fagioli M., Tomassoni L., Pileri S., Pacini R., Grignani F., Casini T., Ferrucci P.F., Martelli M. F., Pelicci P.G., Falini B.: Cellular localization and tissue distribution of PML proteins as defined by a new monoclonal antibody PG-M3 directed against the aminoterminal portion of the PML gene product. Preferential expression of PML in post-mitotic cells. Blood, in press. 1995.

36. Daniel MT, Koken M, Romagne O, Barbey S, Bazarbachi A, Stadler M, Guillemin MC, Degos L, Chomienne C, de Thé H: PML protein expression in hematopoietic and acute promyelocytic leukemia cells. Blood 82:1858, 1993.

37. Gaub MP, Lutz Y, Ruberte E, Petkovich M, Brand N, Chambon P: Antibodies specific to the retinoic acid human nuclear receptors α and β. Proc Natl Acad Sci USA 86:3089, 1989.

38. Nervi C, Poindexter CE, Grignani F, Pandolfi PP, Lo Coco F, Avvisati G, Pelicci PG, Jetten AM: Characterization of the PML/RARα chimeric product of the acute promyelocytic leukemia-specific t(15;17) translocation. Cancer research 52:3687, 1992.

DISCUSSION

S. Landolfo

Is the PML/RAR Alpha fusion protein maintaining the PML promoters or does it undergo regulation as PML or what?

P. Pelicci

Yes. The PML/RAR Alpha expression is driven by the PML promoter. We do not know the regulation of the PML promoter to say whether or not it is maintained after the fusion with RAR Alpha.

S. Landolfo

You showed that PML is inducible by interferon Gamma so presumably it contains a gas-driven element. Is it also regulated by Alpha interferon?

P. Pelicci

Yes, it is regulated by Alpha interferon. However, its promoter does not appear to be regulated by interferon in *in vitro* experiments..

L. Nadler

With retinoic acid the APL cells die under those conditions. What mechanism do they die by?

P. Pelicci

Apoptosis.

L. Nadler

Is anything known about what drives them to apoptosis? Is this by a direct mechanism from the retinoic acid or is it secondary to an immune response?

P. Pelicci

This phenomenon can be easily monitored *in vitro* by using the APL cell line. NB-4, suggesting that there is no immune involvement in this process. NB4 cells undergo terminal differentiation induced by retinoic acid and, neutrophils eventually die. Cell death is associated with increased hypodyploid DNA content and DNA degradation. So I think this would suggest that they die by apoptosis.

A. D'Andrea

It is interesting that the addition of retinoic acid changes the nuclear localization pattern of the PML RAR Alpha. I was wondering if the nuclear localization pattern also changes or differs at different times during the cell cycle?

P. Pelicci

We do not have any evidence to say that the localization of PML/RAR Alpha changes during the cell cycle in fibroblasts. This is difficult to assess in APL cells because it is difficult to synchronize cells.

L. Degos

First, a comment: *In vitro,* the malignant promyelocytic cells behave better when retinoic acid is added. They die without retinoic acid. There is maturation, and then they die. It seems that the process is not a direct trigger of apoptosis. Then I have a question about the U9 3 7 cells which are monocyte derived cells. Perhaps the absence of complete maturation is due to their lineage. Moreover, we know that after treatment a monocytopenia occurs in leukemic patients and in vitro experiments disclose a decrease of monocyte colonies with retinoic acid treatment of normal bone marrow. So, perhaps your results are due to the cell line characteristics and not to the treatment of normal bone marrow.

P. Pelicci

I do agree. The choice of U937 cells as a model system at the beginning of our studies was forced by the fact that, for example, we were never able to express the fusion proteins in HL60 cells, using different promoters and different approaches. I agree that pro-monocytes are not the best model system to express the PML/RAR Alpha protein.

R. Dalla Favera

So, the PML/RAR Alpha maintains both the respective homodymerization sites and it is physically displaced versus the wild type proteins. And now you are finding the RAR Alpha PML protein in the absence of the reciprocal. So, is it possible that PML/RAR Alpha or the reciprocal work by simply sucking out all the wild type proteins to this different structure and that consequently there is no PML/RAR Alpha bound to DNA in APL.

P. Pelicci

Yes, this is consistent with the data we have so far. This is one of the working hypotheses. You are suggesting that PML/RAR Alpha may have the function of displacing the PML and RARα in the nuclear bodies, thus sequestering RXR. This is possible. For example, you can rescue the block of differentiation by vitamin D3 by overexpressing RXR, which strongly suggests that indeed one of the mechanisms by which PML/RAR Alpha works is sequestering RXR for vitamin D3 receptor.

R. Dalla Favera

So, one consequence would be that you should not see biological effects, or you should miss some biological effects in cells that do not contain the RAR Alpha, because there is no sequestration. You could not see biological effects of the fusion protein in cells lacking RAR Alpha. I guess PML is all over the place.

P. Pelicci

Yes, this is possible but my impression is that the sequestration of RXR is only one of the mechanisms by which PML/RAR Alpha acts and I can give you results of some experiments for this thought. For example, overexpression of RXR does not rescue the block of differentiation of PML/RAR Alpha during TPA induced differentiation. We know that vitamin D3 requires RXR but we know very little about the signaling pathway triggered by TPA. These experiments suggest that there are multiple mechanisms by which PML/RAR Alpha interferes with terminal differentiation.

I. Weissman

The localization of PML/RAR Alpha in the nuclear bodies looked like it included the nucleolus. Am I mistaken by what I was looking at?

P. Pelicci

Those are 15 to 20 small dots and the nucleolus is clearly not associated with the nuclear bodies.

M. Caligiuri

It is my understanding that the majority of patients treated with retinoic acid without chemotherapy relapse with the disease?

P. Pelicci

Yes.

M. Caligiuri

Do you still get the localization in patients re-treated with retinoic acid who were found to be resistant?

P. Pelicci

There are only a few patients which indeed relapse because the standard treatment is now combination of retinoic acid plus chemo. So we did not have the chance to study relapsing patients or resistant patients. Staining with anti PML antibodies the NB4 resistant cell lines *in vitro* the localization differs from that of the sensitive cell lines.

L. Degos

In the resistant cell line the resistance is due to an additional genetic effect. However, in patients the resistance is reversible. So after six months to one year after discontinuation of RA treatment the patients become again sensitive to retinoic acid. So something else occurs in patients which is not a genetic defect. Is it an induced hypercatabolytic state? I shall discuss this hypothesis in my talk.

D. Livingston

Do you know if the fusion protein, like the wild type, will interact *in vitro* even, with a pair of 140-160Kd polypeptides recently described by Myles Brown at Dana Farber? These two proteins interact with the ligand binding domains of ER and RAR and possibly other nuclear small molecule receptor family members as well.

P. Pelicci

I do not know. We have not done the experiments.

D. Livingston

Those are possible sequestration targets.

P. Pelicci

A good experiment that has to be done.

C. Sawyers

I was a bit confused about the way you interpret your myc results. If I understood it correctly, the fusion protein activates myc in the reporter construct assays and retinoic acid increases that activation. So how do you put that together with the biological block in differentiation?

P. Pelicci

Probably I was not clear on that. We see two different effects in myeloid cells and fibroblasts. What we do see in fibroblasts is activation of myc expression and superactivation by retinoic acid, while in myeloid cells, if you express the fusion protein, you see shutting off the myc oncogene.

14

CONTRIBUTION OF THE CYCLIN D1 GENE TO LYMPHOMAGENESIS

Jerry M. Adams,[*] Beverley J. Warner, Sharon E. Bodrug,[†] Mary L. Bath, Geoffrey J. Lindeman, and Alan W. Harris

The Walter and Eliza Hall Institute of Medical Research
P.O. Royal Melbourne Hospital
Victoria 3050, Australia

INTRODUCTION

Recent years have witnessed a revolution in our perception of the cell cycle. Pioneering genetic studies in yeast uncovered the central role of kinases in governing the key steps in the cycle. Later, biochemical studies launched in *Xenopus* and in clams revealed that the activity of the kinase required not only a catalytic subunit, structurally related to the *cdc*-2 gene product of fission yeast, but also a regulatory subunit. Because the latter often oscillated in abundance during the cycle, it was denoted a cyclin and its catalytic partner eventually came to be designated as a cyclin-dependent kinase or Cdk. In the current view of the cell cycle (reviewed by Norbury and Nurse 1992; Reed 1992), each step in cycle progression is orchestrated by a particular Cdk and associated cyclin. One key unresolved issue for multicellular organisms is how mitogenic and growth inhibitory signals from the cellular environment impinge on this basic cycling machinery. Another is how it is perturbed by oncogenic changes. Finally, the major issue addressed below is whether altered expression of a cycle component can itself contribute to oncogenesis.

It is during the G1 phase of the cycle that the integration of extracellular signals dictates the critical decisions to replicate the cellular genome, to enter a quiescent state or to differentiate (Pardee 1989; Reed 1992). Hence the recently identified G1 cyclins (Reed 1992; Sherr 1993), denoted cyclin C, D1, D2, D3 and E, have attracted considerable interest. Little is known about cyclin C, but cyclin E in association with Cdk-2 probably governs the G1/S transition (Ohtsubu and Roberts 1993). The role of the three D cyclins remains enigmatic (Sherr 1993). Whereas nearly all other basic cycle components are expressed in every cell, various combinations of the D cyclins are found in different cell types. Within the hematopoietic system, for example, myelomonocytic cells express D1 and D3 and

[*] Address correspondence to Dr. Adams.

[†] Present address: La Jolla Cancer Research Foundation, 10901 North Torrey Pines Road, La Jolla, CA 92037, USA.

lymphoid cells D2 and D3. It is tempting to speculate that particular D cyclins are made in response to particular signal transduction pathways from growth factor receptors. At least D1 and D2 appear to be synthesized before cyclin E, and therefore they may govern one or more earlier transitions in G1, such as the 'restriction point', the last point at which growth factor signals are required for progression into S phase and through mitosis (Pardee 1989). Cyclin D1 is required by fibroblasts for progression through G1 (Baldin et al., 1993; Quelle et al., 1993), and its enforced overexpression shortened that phase (Quelle et al., 1993), as did overexpression of D-type cyclins in a myeloid cell line (Ando et al., 1993; Kato and Sherr 1993).

The cyclin gene most strongly implicated in neoplasia is the D1 gene, which is overexpressed due to genetic alteration in several types of human tumors (Motokura and Arnold 1993). In some parathyroid adenomas, a chromosome 11 inversion has linked it to an active promoter (Motokura et al., 1991), while the gene is amplified in many cancers of the breast and upper body (Lammie and Peters 1991; Motokura and Arnold 1993). The tumor where it is most consistently implicated however is centrocytic or mantle-zone B-cell lymphomas; in up to 70% of cases, a translocation has juxtaposed the 11q13 band bearing the cyclin D1 gene with the IgH locus on chromosome 14 (Motokura and Arnold 1993; Withers et al., 1991; Rosenberg et al., 1991). The analogy of the *myc*/IgH translocations that hallmark Burkitt's lymphoma led to the initial proposal that the chromosome 11 breakpoints fell near an oncogene, provisionally designated *bcl*-1 (Tsujimoto et al., 1984). Although the cyclin D1 gene lies 120 kb from the first 11q13 breakpoints described (Figure 1), closer breakpoints have subsequently been found and cyclin D1 now appears to be the nearest gene. Its consistent high expression in these lymphomas and minimal expression in diverse others makes cyclin D1 the prime candidate for the relevant oncogene (Withers et al., 1991; Rosenberg et al., 1991; Motokura and Arnold 1993).

D cyclins have also been implicated in oncogenesis in rodents. Several T lymphomas induced by retroviruses contain proviral insertions near the D1 gene (Lammie et al., 1992) or the D2 gene (Hanna et al., 1993). Moreover, cyclin D1 expression is markedly elevated during the late stages of skin carcinogenesis (Bianchi et al., 1993).

To date, evidence for the oncogenic action of cyclins has been largely circumstantial. A definitive test is enforced expression of a candidate oncogene in the relevant cell type of transgenic mice (Adams and Cory 1991). For example, mice having the *myc* gene coupled to the immunoglobulin heavy chain enhancer (Eμ), mimicking the chromosome junction in Burkitt's lymphoma, contain an expanded population of cycling pre-B lymphocytes and eventually succumb to B-lymphoid malignancy as a consequence of somatic mutation (Adams et al., 1985; Langdon et al., 1986; Harris et al., 1988b; Alexander et al., 1989).

To explore how constitutive cyclin D1 expression affects lymphocytes, we have generated analogous mice having a murine cyclin D1 cDNA (Matsushime et al., 1991) regulated by this enhancer, which can promote transgene expression in both the B- and T-cell lineages (Adams and Cory 1991). As we have reported recently (Bodrug et al., 1994), lymphoid differentiation is modulated in the young mice, but the cell cycle behaviour of the

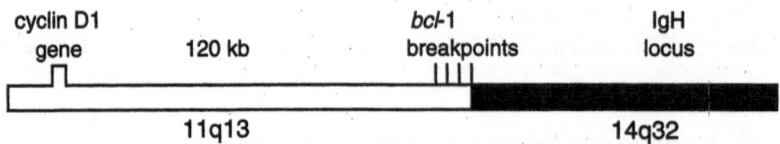

Figure 1. Relation of the cyclin D1 gene to the bcl-1 breakpoint cluster in the translocation between 11q13 and 14q23 that typifies centrocytic lymphoma.

lymphocytes appears unperturbed. Surprisingly, tumorigenesis in these mice appears to depend on the concomitant action of the *myc* gene.

RESULTS

We constructed eight strains of Eμ-cyclin D1 mice. Each showed abundant transgenic mRNA in the spleen and thymus and in mitogenically activated B- and T-cell populations. Moreover, immunoblots with a monoclonal antibody specific for murine cyclin D1 (Quelle et al., 1993) revealed that both B and T cells contained abundant cyclin D1 polypeptide. In accord with other studies with lymphocytes (Matsushime et al., 1991; Ajchenbaum et al., 1993), the non-transgenic B and T cell populations exhibited no endogenous cyclin D1 mRNA or protein. Hence, the transgene generated strong ectopic cyclin D1 expression in both lymphoid lineages.

Lymphomagenesis Is Greatly Enhanced by myc and Modestly by ras

Spontaneous tumors of Eμ-cyclin D1 mice have been uncommon. Only five (2%) have developed a tumor. All five were thymic T-cell lymphomas. It is uncertain whether the transgene contributed to their development because approximately 1% of our non-transgenic mice develop T-cell lymphoma.

In transgenic models, certain well studied oncogenes (e.g. *pim*-1, *bcl*-2) produce only a very low incidence of spontaneous tumors themselves and another oncogene is required to reveal their impact (Adams and Cory 1991). To determine whether cyclin D1 could cooperate in tumorigenesis with a deregulated c-*myc* gene, the Eμ-cyclin D1 mice were crossed with Eμ-*myc* mice. Mice inheriting both transgenes rapidly developed lymphoma (Figure 2), most by 8 weeks of age. This marked acceleration in tumor onset is highly significant (P=0.0002). These mice had developed disseminated malignant lymphoma like that in Eμ-*myc* animals (Adams et al., 1985; Harris et al., 1988a). Most were of pre-B or B-cell type, like those bearing only the *myc* transgene, but two were CD4$^+$8$^+$ T-cell

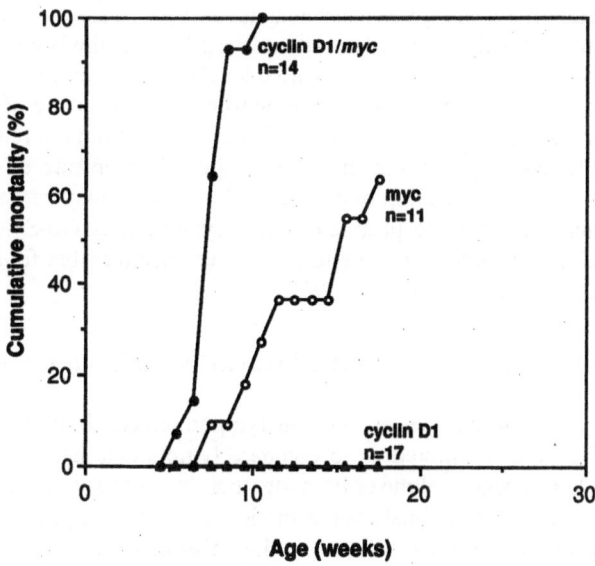

Figure 2. Cumulative mortality in progeny of a cross of Eμ-cyclin D1 mice with Eμ-*myc* mice. Both strains are hemizygous for the transgene, and the four types of progeny were identified by polymerase chain reaction (PCR) using primers specific to each transgene. The mice were monitored twice a week and sacrificed when terminally ill. All had lymphoma. Adapted from Bodrug et al (1994).

lymphomas. Hence, cyclin D1 and *myc* probably can collaborate in the development of T as well as B lymphomas.

To test whether cyclin D1 could also cooperate in tumorigenesis with an activated mutant N-*ras* transgene, an analogous cross was made with Eμ-N-*ras* mice. These mice, in which the transgene is expressed predominantly in the T-cell lineage, are predisposed at 3 to 6 months of age to develop thymic T-cell lymphoma, whereas the older mice often develop histiocytic sarcoma (Harris et al., 1988a; Haupt et al., 1992). In this cross the bi-transgenic progeny tended to develop thymoma somewhat earlier than those inheriting only the Eμ-N-*ras* transgene (P=0.088). All the tumors analysed comprised CD4$^+$8$^+$ T lymphoblasts. Although the acceleration was modest, the results suggest that the cyclin D1 gene may collaborate in oncogenesis with a *ras* as well as a *myc* oncogene.

Lymphocyte Cell Cycle Behaviour and Mitogen Responsiveness Appear Normal

In some but not all transgenic models of tumorigenesis, a preneoplastic phenotype is discernible (Adams and Cory 1991). We therefore investigated whether the lymphocytes of Eμ-cyclin D1 mice were altered. In adult Eμ-cyclin D1 mice, this compartment appeared unperturbed. Flow cytometric analysis revealed normal numbers of pre-B and B cells in the bone marrow and spleen, as well as of the major T-cell subsets in the thymus and mature T cells in the spleen. The size of cells in the thymus and spleen was also unaltered. Cell cycle analysis of thymocytes revealed normal percentages of cells within the S plus G2/M phases (7-9%). Similarly, no change was observed in the proportion of cycling cells among total spleen cells (4-6%) or purified splenic B and T cell populations. This finding is in accord with evidence in other cell types that cyclin D1 is insufficient to drive quiescent cells into cycle (Quelle et al., 1993; Ando et al., 1993).

D cyclins are made in response to mitogenic stimuli (Matsushime et al., 1991; Ajchenbaum et al., 1993; Won et al., 1992; Sewing et al., 1993; Ando et al., 1993), and overexpression of cyclin D1 in fibroblast cell lines reduced their requirement for serum and contracted the G1 interval (Quelle et al., 1993). We therefore evaluated whether constitutive cyclin D1 expression had affected lymphocyte growth properties in culture. The unstimulated lymphocytes failed to proliferate and died at a normal rate, so they were not rendered autonomous nor given a survival advantage. To assess whether their sensitivity to mitogens had been enhanced, we monitored responses of the B cells to graded doses of lipopolysaccharide and of the T cells to varied levels of three mitogenic regimes. The responses to either optimal or sub-optimal stimulation were normal, and the stimulated populations had a normal size and cell cycle distribution. Thus, cyclin D1 did not replace or augment any of the tested mitogenic signals, nor increase the proliferation rate of the stimulated lymphocytes. On mitogenic stimulation of quiescent B or T lymphocytes, the G0 to S period was not notably shortened, but the protracted and asynchronous G0-S interval in lymphoid populations (22-26 h) could obscure a few-hour decrease like that found in fibroblast cell lines (Quelle et al., 1993; Jiang et al., 1993).

Impeded Lymphocyte Maturation in Young Cyclin D1 and D1/*myc* Mice

The marked synergy in lymphomagenesis of the cyclin D1 and *myc* transgenes (Figure 2) prompted us to compare lymphocyte numbers in healthy young mice bearing either transgene alone or both together. In contrast to the adult D1 animals discussed above, the young D1 animals had a modest reduction in pre-B cells, and less than one-third the normal B-cell level (Figure 3); they also had just over half the normal levels of CD4$^+$ and

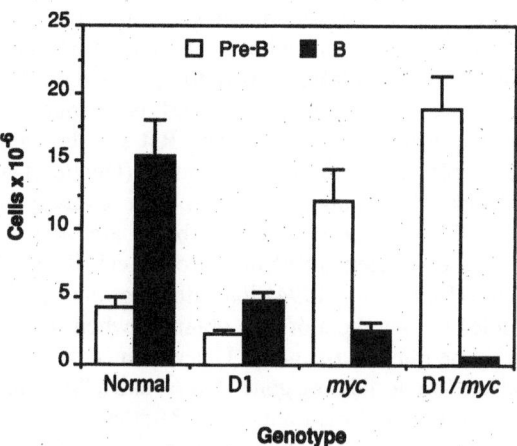

Figure 3. Effects of the cyclin D1 and *myc* transgenes on B-lymphoid cell numbers in the spleen of 12-d old progeny from crosses of Cyd 76 and Eμ-*myc* mice. Cell numbers were derived by counting total spleen cells and determining by flow cytometry (Langdon et al., 1986) the percentages of pre-B cells (CD45R(B220)$^+$, sIg$^-$) and B cells (B220$^+$, sIg$^+$). Error bars denote S.E.M. Adapted from Bodrug et al (1994).

CD8$^+$ T cells (not shown). Thus the cyclin D1 transgene appeared to impede the development of B cells and mature T cells. Moreover, the pre-B to B-cell ratio (Figure 3) for the D1 mice was somewhat higher than that for their normal littermates, suggesting that the transition from pre-B to B cells is somewhat inhibited. The young Eμ-*myc* mice had the expected (Langdon et al., 1986) several-fold increase in pre-B cells and a drop in B cells (Figure 3). Significantly, the bi-transgenic animals had almost no B cells (<5% the normal level) and even more pre-B cells than the Eμ-*myc* mice. These pre-B cells, like those of Eμ-*myc* mice (Langdon et al., 1986), had the high forward light scatter characteristic of a cycling lymphocyte population. Thus, the oncogenic synergy between *myc* and cyclin D1 is paralleled by cooperative effects on the differentiation of the target cell population.

DISCUSSION

The case that the cyclin D1 gene is the long-sought *BCL*-1 oncogene (Tsujimoto et al., 1984; Motokura et al., 1991; Motokura and Arnold 1993; Withers et al., 1991; Rosenberg et al., 1991) is greatly strengthened by our evidence, documented more extensively elsewhere (Bodrug et al., 1994), that its constitutive expression in lymphocytes can contribute to the development of both B- and T-lymphoid malignancy. Hence, cyclin D1 is likely to play a key role in the etiology not only of centrocytic lymphoma but also of the many cases of prolymphocytic leukemia, multiple myeloma and chronic lymphocytic leukemia bearing the 11q13 translocation (Motokura and Arnold 1993). Moreover, the likelihood is increased that cyclin D1 is a critical gene within the 11q13 amplicon found in a substantial minority of breast cancers and squamous cell carcinomas of the head, neck, esophagus and lung, although other genes within the large (1 Mb) amplified region could also contribute (Lammie and Peters 1991).

While the very low incidence of spontaneous lymphomas in Eμ-cyclin D1 mice initially seemed somewhat surprising, there are several precedents of transgenic models where well studied oncogenes have elicited few tumors except in the context of another genetic change (Adams and Cory 1991). Moreover, our evidence that the lymphomagenic action of the cyclin D1 transgene required other oncogenic events (Figure 2) is consistent with the complex karyotype of nearly all centrocytic lymphomas (Leroux et al., 1991). The strong synergy in lymphomagenesis observed here of cyclin D1 gene with the *myc* gene (Figure 2) has also been found recently with the N-*myc* and L-*myc* genes (Lovec et al.,

1994a). These observations raise the question whether genes of the *myc* family participate in any of the tumors with 11q13 alterations and, conversely, whether D cyclin genes contribute in neoplasms known to involve *myc*.

Tests of whether cyclin D1 is oncogenic in fibroblasts have yielded somewhat conflicting results. NIH-3T3 and Rat-2 cells overexpressing cyclin D1 remained contact inhibitable and anchorage dependent (Quelle et al., 1993), but overexpressing Rat-6 fibroblasts, at very high cell doses, did eventually give small tumors in nude mice (Jiang et al., 1993). These differences probably are related to the (unknown) genetic changes that allowed establishment of the parental cell lines. In primary rodent fibroblasts one recent study (Hinds et al., 1994) indicated that the cyclin D1 gene itself was inert but increased (3-fold) the numbers of foci obtained when it was cotransfected with a partially disabled E1A gene plus the *ras* gene. This evidence that cyclin D1 can collaborate to an extent with both an "immortalizing gene" (E1A) and a "transforming gene" (*ras*) is akin to our findings with the *myc* and *ras* transgenes. Another study with primary fibroblasts found that cyclin D1 plus *ras* sufficed to produce tumorigenic foci (Lovec et al., 1994b). Considered *in toto*, these studies suggest that the oncogenic effects of cyclin D1 are relatively subtle and may be affected by cell type as well as its abundance.

Our evidence that cyclin D1 can modulate lymphocyte maturation (Figure 3) may point to a role for the D cyclins in the control of differentiation. Curiously, the retarded maturation was evident only in young animals. Perhaps the effects of the transgene in the older animals are balanced by the homeostatic pressure to establish an effective immune system. A role for D-type cyclins in the control of differentiation has also been inferred recently by Kato and Sherr (1993) from the consequences of cyclin D overexpression in the factor-dependent myeloid line 32D. While untransfected 32D cells treated with granunolocyte-colony stimulating factor terminally differentiated, the cells overexpressing cyclin D2 or D3 instead arrested and died.

Since impeded differentiation is a critical step in leukemogenesis (Adams and Cory 1992), it is tempting to relate the marked synergy in oncogenesis of the cyclin D1 and *myc* transgenes (Figure 2) to the diminished B-cell maturation produced by each gene and the profound reduction elicited by their combined action (Figure 3). The influence of the two genes on self-renewal and maturation is sketched in Figure 4.

Perhaps Myc and cyclin D1 close windows in G1 during which differentiation can occur. Recent evidence indicates that the Myc protein, a transcription factor, can promote expression of the cyclin E and cyclin A genes (Jansen-Durr et al., 1993). Hence, an attractive hypothesis is that constitutive expression of cyclins D1, E and A in the bi-transgenic cells facilitates entry of many of the lymphocytes into S phase (Figure 5), and thereby keeps most of the cells in cycle and retards their maturation.

As well as the positively acting cyclins, important regulators of the cell cycle include several recently discovered inhibitors of the cyclin-dependent kinases (Hunter 1993). Some inhibitors mediate physiological growth inhibitory signals while others arrest the cell cycle in response to DNA damage to allow DNA repair. Hence at least some inhibitors almost

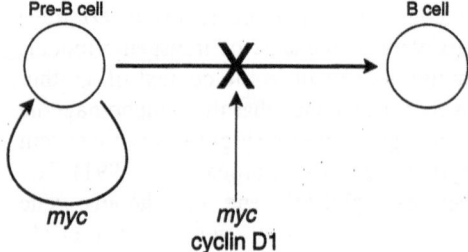

Figure 4. Proposed effects of cyclin D1 and Myc on B cell proliferation and maturation. Myc both enhances self renewal and partially impedes maturation. Cyclin D1 also retards maturation. Hence their concerted action favors an elevated level of cycling immature cells.

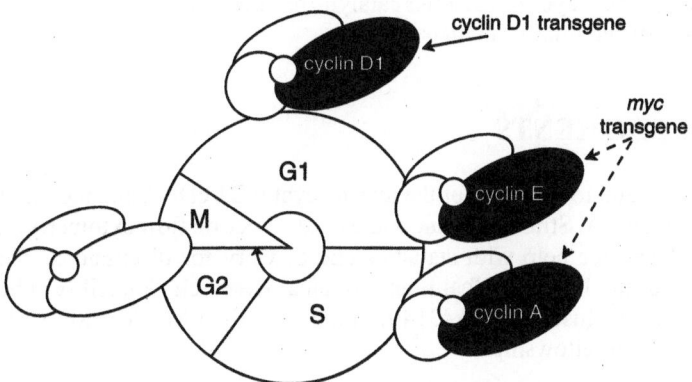

Figure 5. Model for the concerted action of cyclin D1 and *myc* transgenes.

certainly are tumor suppressors. Notably, the gene encoding the inhibitor p16, which associates preferentially with Cdk4 (Serrano et al., 1993), is very frequently deleted or disrupted in cell lines from diverse human tumors and in some primary tumors (Kamb et al., 1994). Because Cdk4 is the preferred catalytic partner of the D cyclins (Matsushime et al., 1992), the oncogenic role of cyclin D1 might be related to its ability to compete with p16 for Cdk4, or for its close relative Cdk6 (Figure 6). In other words, overexpressed cyclin D1 might keep Cdk4 (or Cdk6) in an active form, overriding the inhibitory action of p16.

The critical substrates of the cyclins and associated kinases operative in the G1 and S phases probably include the tumor suppressor RB and its relatives p107 and p130 (reviewed by Sherr 1993; Hamel et al., 1993; Wiman 1993). Until phosphorylated, RB-like molecules can sequester transcription factors such as E2F, which allow phase-restricted gene expression. The D cyclins can physically associate with the RB 'pocket' through their N-terminal motif (Dowdy et al., 1993; Kato et al., 1993). This association probably allows the cyclin to target a Cdk to RB in order to phosphorylate it, but it might also allow RB to regulate the activity of cyclin D. Hence overexpression of cyclin D1 might inactivate RB by targetting it for phosphorylation or allow cyclin D1 to escape from regulation by RB. RB also appears to be required for specific differentiation steps (Wiman 1993; Hamel et al., 1993), perhaps by a similar conditional interaction with lineage-specific transcription factors (Gu et al., 1993). Hence, by governing the interaction of RB family members with transcrip-

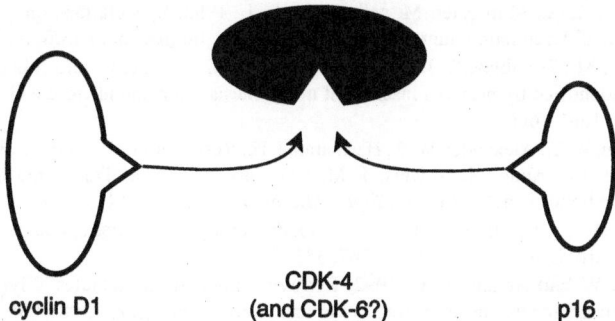

cyclin D1   CDK-4   p16
                    (and CDK-6?)

Figure 6. Competition of cyclin D1 and p16 for Cdk4. Cdk6 is very similar in structure to Cdk4 and might therefore also be a common partner of cyclin D1 and p16.

tional regulators, the G1 cyclins and their catalytic partners may represent the nexus between proliferative and differentiative signals.

ACKNOWLEDGMENTS

We are grateful to C. Sherr for the murine cyclin D1 cDNA and the murine cyclin D1 monoclonal antibody, A. Strasser for antibodies and advice on flow cytometry and mitogenesis assays, A. Kyne for help with statistics and S. Cory for discussions. This work was supported by both the National Health and Medical Research Council (Canberra) and the U.S. National Cancer Institute (CA43140) and S.E.B. by a Medical Research Council of Canada post-doctoral fellowship.

REFERENCES

Adams, J. M. and Cory, S. 1991. Transgenic models of tumor development. *Science*, 254, 1161-1167.

Adams, J. M. and Cory, S. 1992. Oncogene cooperativity in leukemogenesis. *Cancer Surveys*, 15, 119-141.

Adams, J. M., Harris, A. W., Pinkert, C. A., Corcoran, L. M., Alexander, W. S., Cory, S., Palmiter, R. D. and Brinster, R. L. 1985. The c-*myc* oncogene driven by immunoglobulin enhancers induces lymphoid malignancy in transgenic mice. *Nature*, 318, 533-538.

Ajchenbaum, F., Ando, K., DeCaprio, J. A. and Griffin, J. D. 1993. Independent regulation of human D-type cyclin gene expression during G_1 phase in primary human T lymphocytes. *J. Biol. Chem.*, 268, 4113-4119.

Alexander, W. S., Bernard, O., Cory, S. and Adams, J. M. 1989. Lymphomagenesis in Eμ-*myc* transgenic mice can involve *ras* mutations. *Oncogene*, 4, 575-581.

Ando, K., Ajchenbaum-Cymbalista, F. and Griffin, J. D. 1993. Regulation of G_1/S transition by cyclins D2 and D3 in hematopoietic cells. *Proc. Natl. Acad. Sci. USA*, 90, 9571-9575.

Baldin, V., Lukas, J., Marcote, M. J., Pagano, M. and Draetta, G. 1993. Cyclin D1 is a nuclear protein required for cell cycle progression in G_1. *Genes & Devel.*, 7, 812-821.

Bianchi, A. B., Fischer, S. M., Robles, A. I., Rinchik, E. M. and Conti, C. J. 1993. Overexpression of cyclin D1 in mouse skin carcinogenesis. *Oncogene*, 8, 1127-1133.

Bodrug, S. E., Warner, B. J., Bath, M. L., Lindeman, G. J., Harris, A. W. and Adams, J. M. 1994. Cyclin D1 transgene impedes lymphocyte maturation and collaborates in lymphomagenesis with the *myc* gene. *EMBO J*, 13, 2124-2130.

Dowdy, S. F., Hinds, P. W., Louie, K., Reed, S. I., Arnold, A. and Weinberg, R. A. 1993. Physical interactions of the retinoblastoma protein with human D cyclins. *Cell*, 73, 499-511.

Gu, W., Schneider, J. W., Condorelli, G., Kaushal, S., Mahdavi, V. and Nadal-Ginard, B. 1993. Interaction of myogenic factors and the retinoblastoma protein mediates muscle cell commitment and differentiation. *Cell*, 72, 1-20.

Hamel, P. A., Phillips, R. A., Muncaster, M. and Gallie, B. L. 1993. Speculations on the roles of *RB1* in tissue-specific differentiation, tumor initiation, and tumor progression. *FASEB J.*, 7, 846-854.

Hanna, Z., Jankowski, M., Tremblay, P., Jiang, X., Milatovich, A., Francke, U. and Jolicoeur, P. 1993. The vin-1 gene, identified by provirus insertional mutagenesis, corresponds to the G1-phase cyclin D2. *Oncogene*, 8, 1661-1666.

Harris, A. W., Langdon, W. Y., Alexander, W. S., Hariharan, I. K., Rosenbaum, H., Vaux, D., Webb, E., Bernard, O., Crawford, M., Abud, H., Adams, J. M. and Cory, S. 1988a. Transgenic mouse models for hematopoietic tumorigenesis. *Current Topics Microbiol. Immunol.*, 141, 82-93.

Adams, J. M. 1988b. The Eμ-*myc* transgenic mouse. A model for high-incidence spontaneous lymphoma and leukemia of early B cells. *J. Exp. Med.*, 167, 353-371.

Haupt, Y., Harris, A. W. and Adams, J. M. 1992. Retroviral infection accelerates T lymphomagenesis in Eμ-N-*ras* transgenic mice by activating c-*myc* and N-*myc*. *Oncogene*, 7, 981-986.

Hinds, P. W., Dowdy, S. F., Eaton, E. N., Arnold, A. and Weinberg, R. A. 1994. Function of a human cyclin gene as an oncogene. *Proc. Natl. Acad. Sci. USA*, 91, 709-713.

Hunter, T. 1993. Braking the cycle. *Cell*, 75, 839-841.

Jansen-Durr, P., Meichle, A., Steiner, P., Pagano, M., Finke, K., Fotz, J., Wessbecher, J., Draetta, G. and Eilers, M. 1993. Differential modulation of cyclin gene expression by *MYC*. *Proc. Natl. Acad. Sci. USA*, 90, 3685-3689.

Jiang, W., Kahn, S. M., Zhou, P., Zhang, Y.-J., Cacace, A. M., Infante, A. S., Doi, S., Santella, R. M. and Weinstein, I. B. 1993. Overexpression of cyclin D1 in rat fibroblasts causes abnormalities in growth control, cell cycle progression and gene expression. *Oncogene*, 8, 3447-3457.

Kamb, A., Gruis, N. A., Weaver-Feldhaus, J., Liu, Q., Harshman, K., Tavtigian, S. V., Stockert, E., Day, R. S. I., Johnson, B. E. and Skolnick, M. H. 1994. A cell cycle regulator potentially involved in genesis of many tumor types. *Science*, 264, 436-440.

Kato, J. and Sherr, C. J. 1993. Inhibition of granulocyte differentiation by G1 cyclins D2 and D3 but not D1. *Proc. Natl. Acad. Sci. USA*, 90, 11513-11517.

Kato, J.-y., Matsushime, H., Hieberg, S. W., Ewen, M. E. and Sherr, C. J. 1993. Direct binding of cyclin D to the retinoblastoma gene product (pRb) and pRb phosphorylation by the cyclin D-dependent kinase CDK4. *Genes Dev.*, 7, 331-342.

Lammie, G. A. and Peters, G. 1991. Chromosome 11q13 abnormalities in human cancer. *Cancer Cells*, 3, 413-420.

Lammie, G. A., Smith, R., Silver, J., Brookes, S., Dickson, C. and Peters, G. 1992. Proviral insertions near cyclin D1 in mouse lymphomas: a parallel for BCL1 translocations in human B-cell neoplasms. *Oncogene*, 7, 2381-2387.

Langdon, W. Y., Harris, A. W., Cory, S. and Adams, J. M. 1986. The c-*myc* oncogene perturbs B lymphocyte development in Eμ-*myc* transgenic mice. *Cell*, 47, 11-18.

Leroux, D., Le Marc'hadour, F., Gressin, R., Jacob, M.-C., Keddari, E., Monteil, M., Caillot, P., Jalbert, P. and Sotto, J. J. 1991. Non-Hodgkin's lymphomas with t(11;14) (q13;q32): a subset of mantle zone/inter-mediate lymphocytic lymphoma? *Brit. J. Haematol.*, 77, 346-353.

Lovec, H., Grzeschicek, A., Kowalski, M.-B. and Moroy, T. 1994a. Cyclin D1/BCL-1 cooperates with myc genes in the generation of B-cell lymphoma in transgenic mice. *EMBO J*, In press.

Lovec, H., Sewing, A., Lucibello, F. C., Muller, R. and Moroy, T. 1994b. Oncogenic activity of cyclin D1 revealed through cooperation with Ha-*ras*: link between cell cycle control and malignant transformation. *Oncogene*, 9, 323-326.

Matsushime, H., Ewen, M. E., Strom, D. K., Kato, H.-., Hanks, S. K., Roussel, M. F. and Sherr, C. J. 1992. Identification and properties of an atypical catalytic subunit (p34^{PSK13}/CDK4) for mammalian D type G1 cyclins. *Cell*, 71, 323-334.

Matsushime, H., Roussel, M. F., Ashmun, R. A. and Sherr, C. J. 1991. Colony-stimulating factor 1 regulates novel cyclins during the G1 phase of the cell cycle. *Cell*, 65, 701-713.

Motokura, T. and Arnold, A. 1993. Cyclins and oncogenesis. *Biochimica et Biophysica Acta*, 1155, 63-78.

Motokura, T., Bloom, T., Kim, H. G., Juppner, H., Ruderman, J. V., Kronenberg, H. M. and Arnold, A. 1991. A novel cyclin encoded by a bcl1-linked candidate oncogene. *Nature*, 350, 512-515.

Norbury, C. and Nurse, P. 1992. Animal cell cycles and their control. *Annu. Rev. Biochem*, 61, 441-470.

Ohtsubu, M. and Roberts, J. M. 1993. Cyclin-dependent regulation of G$_1$ in mammalian fibroblasts. *Science*, 259, 1908-1912.

Pardee, A. B. 1989. G$_1$ events and regulation of cell proliferation. *Science*, 246, 603-608.

Quelle, D. E., Ashmun, R. A., Shurtleff, S. A., Kato, J.-y., Bar-Sagi, D., Roussel, M. F. and Sherr, C. J. 1993. Overexpression of mouse D-type cyclins accedlerates G$_1$ phase in rodent fibroblasts. *Genes & Devel.*, 7, 1559-1571.

Reed, S. I. 1992. The role of p34 kinases in the G1 to S-phase transition. *Annu. Rev. Cell Biol.*, 8, 529-561.

Rosenberg, C. L., Wong, E., Petty, E. M., Bale, A. E., Tsujimoto, Y., Harris, N. L. and Arnold, A. 1991. *PRAD1*, a candidate *BCL1* oncogene: Mapping and expression in centrocytic lymphoma. *Proc. Natl. Acad. Sci. USA*, 88, 9638-9642.

Serrano, M., Hannon, G. J. and Beach, D. 1993. A new regulatory motif in cell-cycle control causing specific inhibition of cyclin D/CDK4. *Nature*, 366, 704-707.

Sewing, A., Burger, C., Brusselbach, S., Schalk, C., Lucibello, F. C. and Muller, R. 1993. Human cyclin D1 encodes a labile nuclear protein whose synthesis is directly induced by growth factors and suppressed by cyclic AMP. *J Cell Sci*,

Sherr, C. J. 1993. Mammalian G$_1$ cyclins. *Cell*, 73, 1059-1065.

Tsujimoto, Y., Yunis, J., Onorato-Showe, L., Erikson, J., Nowell, P. C. and Croce, C. M. 1984. Molecular cloning of the chromosomal breakpoint of B-cell lymphomas and leukemias with the t(11;14) chromosome translocation. *Science*, 224, 1403-1406.

Wiman, K. G. 1993. The retinoblastoma gene: role in cell cycle control and cell differentiation. *Faseb J*, 7, 841-5.

Withers, D. A., Harvey, R. C., Faust, J. B., Melnyk, O., Carey, K. and Meeker, T. C. 1991. Characterization of a candidate *bcl*-1 gene. *Mol. Cell. Biol.,* 11, 4846-4853.
Won, K.-A., Xiong, Y., Beach, D. and Gilman, M. 1992. Growth-regulated expression of D-type cyclin genes in human diploid fibroblasts. *Proc. Natl. Acad. Sci. USA,* 89, 9910-9914.

DISCUSSION

R. Perlmutter

Jerry, can you clarify for me how the experiments on retarding lymphocyte differentiation were done. You said young mice. What were they exactly?

J. Adams

Twelve day-old spleen from transgenic and non-transgenic littermates. We did five litters of those animals; we took the spleens from all of them, determined which ones had which transgenes and then evaluated the numbers of cells. It is somewhat puzzling that in fact we have not been able to see any marked difference in cellular composition in adult animals. It is only at this age that we have been able to see this effect. But I think it could be that, in the adult animals, the drive to have an effective immune system might override a modest effect on retarding differentiation.

R. Perlmutter

Have you examined bone marrow?

J. Adams

We have not examined bone marrow. The only tissue we looked at was spleen. The difference holds for both the B and the T cells in the spleen.

R. Perlmutter

So you cannot actually distinguish between retarding development and retarding the migration of cells from the bone marrow or from the thymus into the spleen?

J. Adams

No, we cannot distinguish those possibilities at present.

L. Nadler

This model, like the bcl-2 model transgene, is very far from the human malignant phenotype. Mantle zone lymphomas is a very strange disease. It reforms the normal mantle zone which is arresting B cell and although it is supposed to be a low grade lymphoma, treatment of this malignancy is very poor. It is widely disseminated. None of the clinical phenotypes of this malignancy, either with myc or anything else, follows the phenotype of the transgene, very much like the bcl-2 transgenics.

J. Adams

I would not question that additional genes are involved in the human diseases. I think our work very clearly shows that the cyclin D1 can only be one component and I do not have any definite suggestions as to what other types of alterations might be occurring. There is the obvious possibility that some type of antigenic stimulation might have a role in particular patients. The only clue we have is the fact that we find very impressive synergy with myc. Hence, it may be useful to explore whether genes of the myc family could be involved in some of the human tumors.

L. Nadler

I think there is something very significant about the tumor remaining as a mantle zone, which would be the opposite of anything antigen driven. That would be the resting B cell, whereas the diffuse large cell lymphomas and the germinal center follicular lymphomas certainly would be antigen driven. There is no evidence that anything in the mantle zone ever is antigen driven. So I think there is something very, very special about the biology here that's yet to be unturned.

R. Dalla Favera

I have very little to add except obviously that myc and bcl-1 are never present as lesions in the same tumor. The concept of de-regulation of myc independent of structural lesion has been circulated widely so one cannot exclude that it is deregulated in transit. I have another question: The translocation involving bcl-1, the cyclin D1 gene, has always been a little bit controversial, I think, because 1) the gene is distant, and 2) the same region is very frequently involved in amplification in solid tumors yet the size of the amplicons is very large and includes a number of other genes, some very good candidates to be oncogenes. And although some have been excluded because not expressed when the region is amplified, some others have not been excluded. So one possibility is that the translocation and the amplification involve cyclin D1 and another gene and that is why maybe deregulating cyclin D1 in transgenic mice does not seem to lead to a tumorigenic phenotype as clearly as one would expect.

J. Adams

I would certainly agree that the amplification unit in the carcinomas could contain more than one relevant gene. I think that could be the case, though as you have already indicated, some of the most obvious candidate genes do not seem to be expressed, even though they are amplified. So the situation is somewhat obscure but I would not be resistant to the idea that in the carcinomas a second gene is activated by the amplification. There is less reason to suspect that a second gene is activated by the translocations in the B lymphomas, but that cannot be ruled out.

D. Metcalf

My comment is a little along the same lines. I get a little worried when a gene product that is involved in cell cycle control (a rather general process), is being linked to a very specific type, or a subset of one particular type, of leukemia. I gather from your review that D1 elevation has not been implicated in myeloid leukemias for example?

J. Adams

No, the cell-type specificity remains a puzzle.

D. Metcalf

The experiments of Sherr and his colleagues clearly show that M-CSF stimulates expression of D1 in macrophages. It is slightly worrying, therefore, that this agent with such a prominent action in this lineage, has not so far been linked with neoplasia development in that lineage. So one wonders whether there is a special function of D1 in lymphoid cells, that is separate from its role in G-1 activation? In your model, you have biased the outcome by having a promoter which is lymphocyte specific. You are talking about an agent that acts on the cell cycle presumably in every cell in the body and yet appears to be associated with leukemia development in cells of only one lineage.

J. Adams

Well, I do not know that I can give a convincing answer to that. A puzzle that all of us have to face is that there are these tight correlations at particular neoplasms with particular genes that we do not understand. Why should there be such a correlation? It is perhaps worth reiterating, however, that cyclin D1 is not normally expressed in proliferating lymphocytes. Hence, D1 could do something rather different in lymphocytes than what it does in other cell types where it normally is expressed. But since we do not know exactly what it does in any cell type, except that it helps to advance the G1 phase, we cannot really give a good answer.

D. Livingston

A couple of questions. First of all, are there multiple families with multiple integration sites on the D1 side that give rise to the bi-allelic lymphoma?

J. Adams

We only tested cooperation with one strain because of the difficulty of getting enough animals, but the strain used was arbitrarily chosen among eight strains which we examined more widely from all other aspects. We chose that particular strain simply because it had good expression in both T and B cells. I should note, however, that in very similar experiments by another group, cyclin D1 has been shown to cooperate with N-myc and L-myc (Lovec et al., EMBOJ 13: 3487, 1994).

D. Livingston

Have you tried the same kind of experiment or have you looked for any evidence of tumorigenesis with D2 transgenics or D3 transgenics.

J. Adams

No, we have not.

D. Livingston

Given the experience of Jolicoeur with activation of D2 as a result of one or another retrovirus integration event, one wonders whether D2 plays a role here.

J. Adams

Yes, there are a small number of mouse tumors in which cyclin D2 is implicated and so it could well be that D2 would have a similar effect to D1, but we have not tested that yet.

D. Livingston

And the level of the protein in these tumors?

J. Adams

Well, we believe the level of cyclin D1 protein is very good. We used as a comparison a myeloid cell line that has one of the highest levels of D1 messenger RNA and the levels of the protein in the transgenic cells are quite comparable to that.

D. Livingston

What about by comparison with the myc lymphomas for the levels of D1 protein?

J. Adams

I think that level is probably low and therefore we are not convinced that endogenous cyclin D1 plays a role in the myc tumor development. The unexpected presence of endogenous cyclin D1 may be simply telling us something about the signal transduction pathways that regulate D1 expression.

D. Livingston

Is there a net amplification in the pool of D1 molecules in the bi-allelic disease?

J. Adams

Yes, the bi-transgenic tumors have a much higher cyclin D1 level.

R. Perlmutter

This is more or less in response to David's question. We did make lck promoter driven cyclin D2 and D3 transgenics some years ago and those animals are normal. There is nothing wrong with thymocyte development, they do not develop tumors although the D1's are also normal and the only bi-transgenic experiment that we tried with an SV40 transgene and there was no augmentation of tumorigenesis with the cyclins but... there are so many reasons for an experiment to fail.

15

CLINICAL RESULTS OF ALL TRANS RETINOIC ACID

A Targeted Drug for Differentiation Treatment in Acute Promyelocytic Leukemia

L. Degos, H. Dombret, S. Castaigne, M. T. Daniel, C. Chomienne, and
P. Fenaux

Institut Universitaire d'Hematologie
Université de Paris
Hôpital Saint Louis
Paris, France

INTRODUCTION

The state of malignant cell seems to be irreversible. If one considers gene defects and chromosomal translocations, is it conceivable to find a pure differentiation agent bypassing the abnormalities (1) which could be administered to patients ?

All-trans RA is specifically active on acute promyelocytic leukemic cells. Christine Chomienne tested fresh bone marrow cells from 60 leukemic patients and disclosed a specific differentiation effect of retinoic acid in all specimens from acute promyelocytic patients using the Nitroblue Tetrazolium reduction assay. Among the derivatives, all trans retinoic acid gave better results than 13-cis isomer or 4-oxo metabolite (2). One log difference of concentration for the same magnitude in favor of ATRA was found. The ethyl-ester was not effective and we have recently tested the 9-cis which gave similar results to all-trans.

Acute promyelocytic leukemia is a rare leukemia : 10% of acute non lymphoblastic leukemia, and is characterized morphologically by the aspects of the abnormal cells called M3 subtype, clinically by a bleeding diathesis exacerbated by chemotherapy, and cytogenetically by a specific translocation t(15;17).

In 1977 Janet Rowley reported than the t(15;17) translocation was a consistent chromosomal change in APL (3). The specific RA sensitivity of APL on one hand, and the mapping of the retinoic acid receptor alpha close to the breakpoint on chromosome 17 on the other hand, prompted us to investigate the expression of the RA receptor gene in the patients cells.

The specific rearrangement of retinoic acid receptor alpha gene was first recognized by Christine Chomienne (4). In Northern blot an additional band was found in each case of APL but not in normal individuals or in other types of leukemia (M1, M2, M4, M5). The

Normal and Malignant Hematopoiesis, Edited by Enrico Mihich and Donald Metcalf
Plenum Press, New York, 1995

15;17 translocation fuses the retinoic acid receptor to a novel transcribed locus (5), which is called PML for ProMyelocytic Leukemia.

The breakpoint on chromosome 17 is constantly within the second intron of retinoic acid receptor excluding the A domain (5,6,7). On the PML gene two major known breakpoints are observed : intron 3 and intron 6 or exon 6. These two breakpoints lead to two major PML-RAR fusion transcripts, called long and short PML-RAR. The two fusion proteins both contain the functional domains for dimerization and DNA binding of retinoic acid receptor. The reciprocal small transcripts are sometimes absent.

PML-RARα hybrid protein antagonized transactivation of endogenous and transfected wild type RARα in presence of retinoic acid (8, 9, 10) which is removed with high concentration of retinoic acid. In myeloid cell lines (HL60, U937) responsive to retinoic acid for differentiation effect, PML-RARα abrogates the retinoic acid (and vitamin D3) mediated granulocytic differentiation also reversible with high concentration of retinoic acid (11). Another translocation (11;17) related to rare cases of acute promyelocytic leukemia induces the formation of a hybrid product called PLZ-RAR which also blocks the transactivation of wild type retinoic acid receptor but in an irreversible manner (12). The patients with this particular translocation are not sensitive to ATRA treatment (13).

The knowledge of the PML-RAR breakpoints allowed Sylvie Castaigne (14) to use the amplification of reversed transcribed cDNA sequence of the fusion transcript as a diagnostic laboratory test, and for the assessment of the minimal residual disease. A positive signal at diagnosis ascertained the *in vivo* retinoic acid sensitivity. A positive signal in the follow up of the patients predicts the relapse.

The results on more than 1000 patients from France and China confirmed by series of patients from USA, Japan, Australia and European countries (15), lead to three common conclusions : 1) no resistance if the diagnosis of APL is ascertained by the specific rearrangement between the chromosome 15 and 17 ; 2) no aplasia and thus no severe infection ; 3) a rapid disappearance of bleeding diathesis in the first week of treatment.

No resistance, a differentiation of malignant cells without aplasia, a rapid correction of bleeding diathesis, an easy prevention of adverse effects, a high rate of complete remission, all these spectacular results have been tarnished by a progressive acquired resistance to ATRA which provokes relapses when, ATRA is given alone.

The shortness of the duration of complete remission and the weaker activity of subsequent treatment pointed to an acquired resistance. Warrell links the resistance to a progressive decrease in plasma concentration of ATRA (16). The two reasons actually proposed result in an increased catabolytic state for ATRA. On one hand an induction by ATRA itself of the 4-hydroxylase of cytochrome P450, which plays a major role in the catabolism of the drug itself (16). On the other hand a progressive appearance of cytoplasmic binding protein II (CRABPII) in the cytoplasm of myeloid cells (17).

The appearance of CRABP could sequester free retinoic acid in the cytosol and facilitate its catabolism. CRABP appears after 2 to 3 months of treatment. The reversibility is also progressive, and patients become again sensitive to ATRA after 6 months to one year out of treatment.

Thus, two types of treatments could be proposed to the patients either ATRA which is safe, with no resistance but relapses occured in all cases, or the conventional chemotherapy which induces a high frequency of early deaths but 30% of individuals are probably cured.

Therefore our proposal for the strategy in the treatment combines ATRA to avoid early death with chemotherapy in order to sustain long term remission.

MATERIAL AND METHODS

A. Clinical Studies

1. Pilot Study (18). This study had included 26 patients aged 25 to 63 (median 46) with newly diagnosed APL and leukocytes > 10 x 10^9/l at presentation. Patients were scheduled to receive ATRA (45 mg/m2/day) until complete remission (CR), followed by an intensive "4+7" daunorubicin 'DNR, (60 mg/m2/day d1-4), Ara-C (100 mg/m2/day d1-7) course, and three "2+5" DNR + AraC courses. However the "4+7" course was administered in emergency if hyperleukocytosis rapidly developed, in order to prevent the ATRA syndrome.

2. A Randomized Trial of ATRA Followed by Chemotherapy and Chemotherapy Alone (APL 91) (19). In 1991, an European trial (APL 91) randomizing ATRA followed by chemotherapy (ATRA group) versus chemotherapy alone (chemotherapy group) in newly diagnosed APL was starded. Chemotherapy consisted of 3 DNR - Ara-C intensive courses. In the ATRA arm, chemotherapy was administered after CR achievement, or in emergency if leukocytes were above 6000/mm3 by day 5, or 10 000/mm3 by day 10, or 15 000/mm3 by day 15. The trial was ended after inclusion of 101 patients because of a significantly better event free survival in the ATRA group.

3. Design of APL 93 Trial. APL 93 trial consists of two successive randomized studies :
    - a first study comparing ATRA followed by chemotherapy and concomittant use of ATRA and chemotherapy
    - a second study assessing the role of maintenance therapy with intermittent ATRA, continuous low dose chemotherapy, both or none.
    The purpose of APL 93 trial is therefore also to test the role of intermittent maintenance treatment with ATRA (15 days every 3 months) on the relapse rate. Because, as seen above, a beneficial role has also been suggested for low dose maintenance chemotherapy with 6-mercaptopurine and methotrexate (6MP+MTX), patients will be randomized to receive maintenance with either intermittent ATRA, continuous 6 MP+MTX or both (a last control group receiving no maintenance therapy). Up to now 120 patients are enrolled.

B. Morphological Studies: Expression of PML and PML-RAR (20)

APL Samples. A total of 14 untreated APL patients were analyzed, and 4 of them were subsequently tested during RA therapy. Controls were performed on 5 aspirates, 3 patients with normal bone marrows, 1 patient with chronic granulocytic leukemia (CGL) and 1 patient with acute myeloid leukemia (AML ; M2). Blood and bone marrow samples were obtained with the patients' informed consent. Mononuclear cells were further isolated by centrifugation on Ficoll-Hypaque (Pharmacia) gradients.

Immunochemical Studies. Cells were grown in Lab-Tech micro well plates and fixed in 4 % paraformaldehyde in PBS at 4)C for 20 minutes followed or not by methanol for 5 minutes at 4°C. Cytospins of nonadherent cells were performed with 40.000 cells at 400 rpm for 10 minutes followed by air drying and fixation. The primary antibody was incubated on the cells for 1 to 3 hours at 37°C at a dilution of 1/50 in PBS. After an extensive washing in PBS, a second fluorescine-coupled antirabbit (or mouse) antibody was added for 1 hour at

37°C. Immunostaining was performed using the APAAP immuno alkaline phosphatase technique.

C. Coagulation Disorders (21)

Between November 1990 and July 1993, 27 APL patients (15 females and 12 males; median age : 40 years ranging from 14 to 68 years) were studied at presentation.

Fourteen of these 27 patients (pts. n° 1-14) were then serially tested during differentiation therapy with ATRA (9 females and 5 males median age : 46 years ; seven newly-diagnosed APL and seven APL in first relapse).

Thrombin activation was assessed on prothrombin activation fragment F1+2 (F1+2) and thrombin/antithrombin-III complexes (TAT) detection (Enzygnost Behring, France). Plasmin activation was assessed on α2PI consumption evaluated by rocket electrophoresis (Laurell Method) using monospecific rabbit antiserum (Assera Stago, France). Tissue plasminogen activator (tPA) antigen, total plasminogen activator inhibitor-1 (PAI-1) antigen and tPA/PAI-1 complexes were also tested by an enzyme immunoassay technique.

Plasma neutrophil serine protease activities (elastase and cathepsin G) were measured using a spectrophotometric assay following hydrolysis of specific substrates (N-succinyl-ala-ala-ala-p-nitroanilide and N-succinyl-ala-pro-phe-p-nitroanilide for elastase and cathepsin G respectively) measured at 410 nM.

RESULTS

Clinical Results

1. Pilot Trial. Twenty five (96 %) of the patients achieved CR. Three patients were allografted in first CR (18). Updated results showed 7 relapses, after 8, 11, 15, 27, 31, 34, 36 months, respectively, two deaths in CR, and 16 patients still in first CR, after 38+ to 54+ months (19). In the historical control group of 29 patients treated with chemotherapy alone (patients entered in our previous APL 84 trial, after excluding those who presented with hyperleukocytosis) 22 (76%) achieved CR. Three were allografted in first CR. Updated results showed 12 relapses, after 1 to 38 months, two deaths in CR, and 8 patients still in first CR after 61+ to 86+ months. Actuarial disease free interval (DFI, taking into account relapses, but censoring patients who died in CR) at 4 years was 70 ± 9 % in the pilot ATRA study and 42 ± 11 % in the control chemotherapy group. The difference was significant (p < 0.05 by the log rank test, figure 1). Four late relapses (beyond 24 months) were seen in both groups of patients. Actuarial event free survival (taking into account failure to achieve CR, relapse and death in CR as "events") and survival at 4 years were 62 ± 9% and 77 ± 8 % respectively, in the pilot ATRA (Fig. n°1).

2. Results of Randomized Trial APL 91. A CR rate of 91 % was obtained in the ATRA group as compared to 81 % with chemotherapy alone (difference not significant) (19). Event free survival was significantly superior in the ATRA group as compared to the chemotherapy group (79 % vs 50 % at one year, p = 0.001) (Fig. n°2). The difference in EFS was mainly due to a lower incidence of relapse in the ATRA group (19 % vs 40 % at one year, p = 0.005) (Fig. n°3). The (non significant) difference in CR rate resulted from a lower incidence of resistant leukemia in the ATRA group (0 % vs 11 % in the chemotherapy group) as the incidence of early death was identical in both treatment groups (9 % vs 8 %).

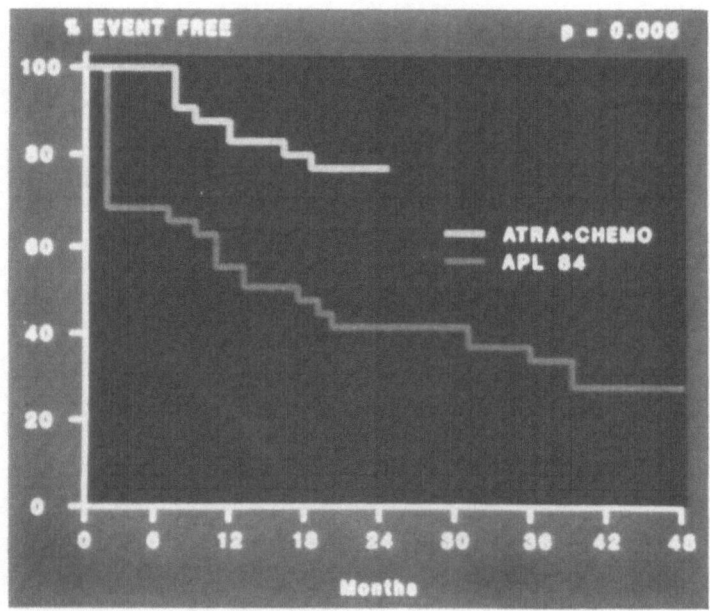

Figure 1. Event free survival curves of a pilot study of 26 patients treated by ATRA followed by a chemotherapy and of historical series (29 patients) treated by chemotherapy alone (APL 84).

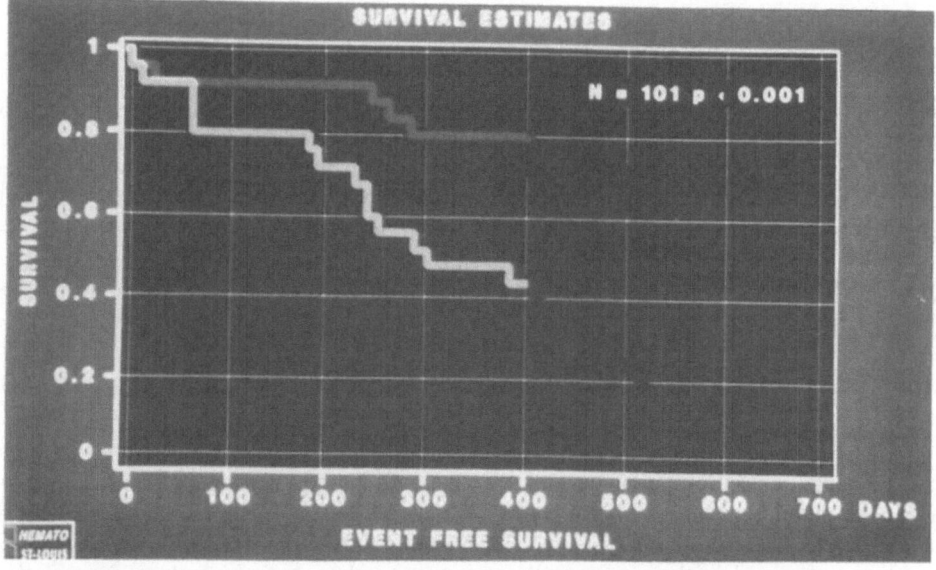

Figure 2. Event free survival curves of a multicentric randomized trial comparing conventional chemotherapy (lower curve) to ATRA treatment followed by the same conventional chemotherapy (upper curve).

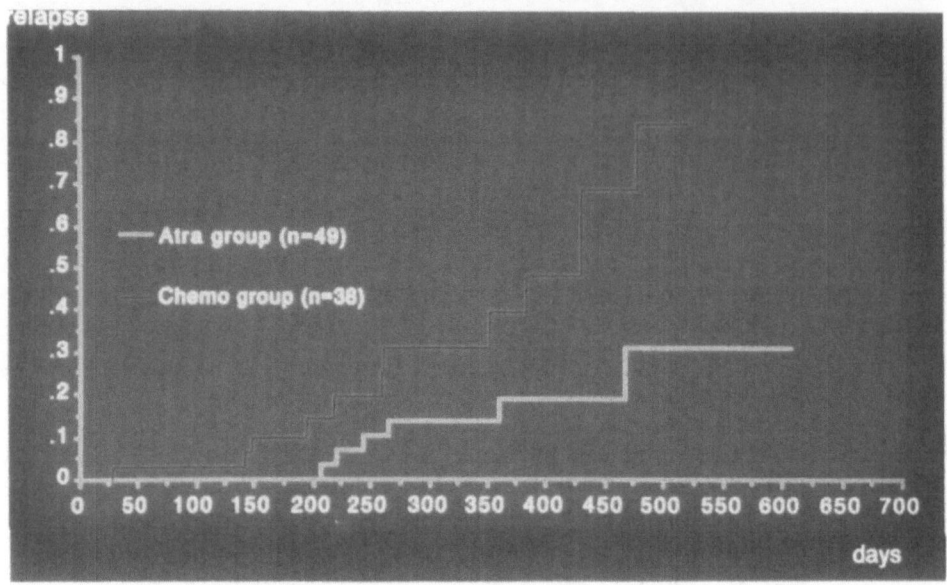

Figure 3. Incidence of relapses in the two arms of a multicentric randomized trial comparing induction by a conventional chemotherapy (upper curve) to an induction by ATRA followed by the same conventional chemotherapy (lower curve).

In the chemotherapy group (47 patients), 4 had early death, 5 had resistant leukaemia, and 38 (81 %) reached CR. 19 relapsed. 1 was salvaged by chemotherapy and 17 by ATRA. 7 of the 19 patients who relapsed died from resistant leukaemia (4 cases) or in second CR (3 cases). 5 patients died in first CR. The 5 patients who were resistant to chemotherapy achieved CR with ATRA, but 1 died in first CR and 2 died in second relapse. Thus, 19 patients died in the chemotherapy group. The 9 deaths in first or second CR occured after an allograft or autograft (6 cases) or chemotherapy (3 cases) in patients who had all been heavily pretreated.

In the ATRA group (54 patients), 5 had early death and 49 (91%) achieved CR. 10 patients relapsed, of whom 5 were salvaged by ATRA and 4 by chemotherapy. 5 of the 10 patients died in first or second relapse. Thus 10 patients died in the ATRA group.

Significant differences between the two groups were still found for event-free survival (ATRA group 83 % at one year ; chemotherapy group 50 % ; p : 0.0001, log-rank test) (figure 2) and for relapse (ATRA group 13 % at one year ; chemotherapy group 41 % ; p = 0.0006, log-rank test) (figure 3). Moreover, a significant difference in survival was also found (ATRA 91 % at one year; chemotherapy group 74 % ; p = 0,01, log-rank test) (Fig. n°4).

Morphological Findings

1. Differentiation of Acute Promyelocytic Leukemic Cells. Usual features of bone marrow acute acute promyelocytic leukemic patients include homogenous infiltration of large cells, simulating abnormal promyelocytes with several large granules and Auer rods in the cytoplasm. Maturing bone marrow cells appeared during the tratment. Auer rods were sometimes observed in the mature cells after two weeks, confirming the differentiation process of malignant cells. After 2 to 3 months treatment, bone marrow appeared normal (22) (Fig. n° 4, 5, 6).

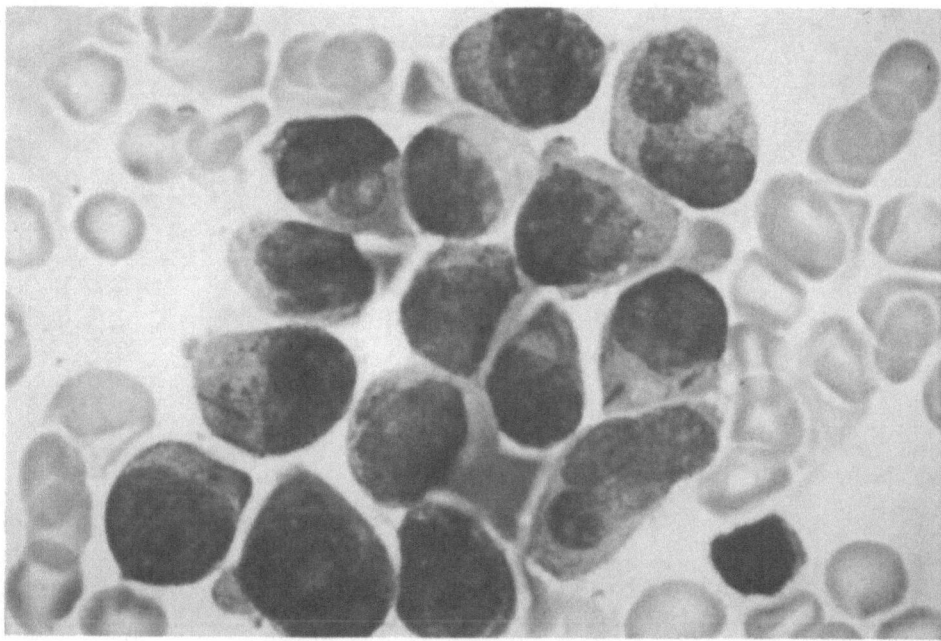

Figure 4. Morphological differentiation of acute promyelocytic leukemic cells. Bone marrow samples were taken at diagnosis (Fig n°4), after 15 days of treatment (Fig n°5) and after 2 months (Fig n°6).

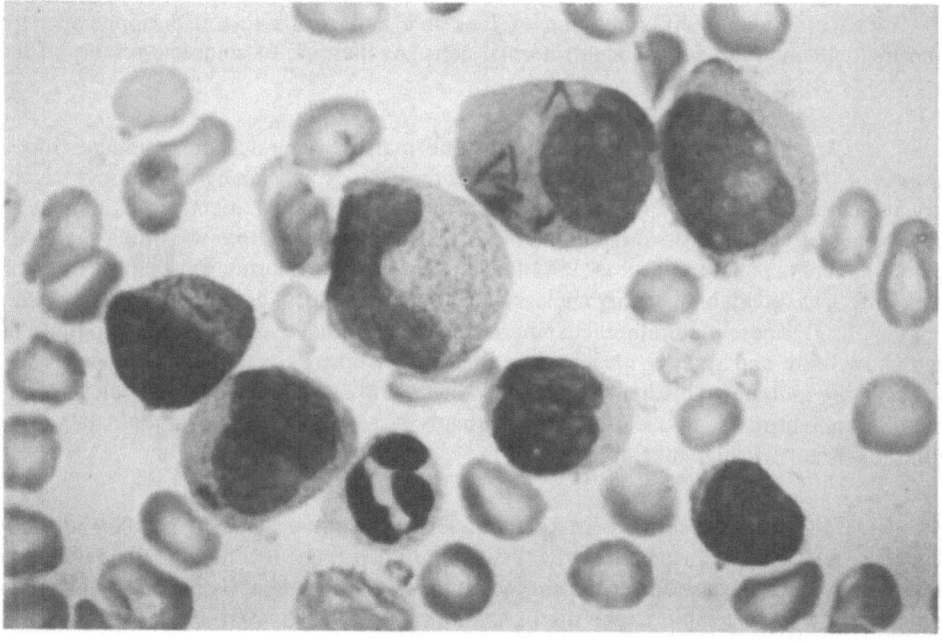

Figure 5.

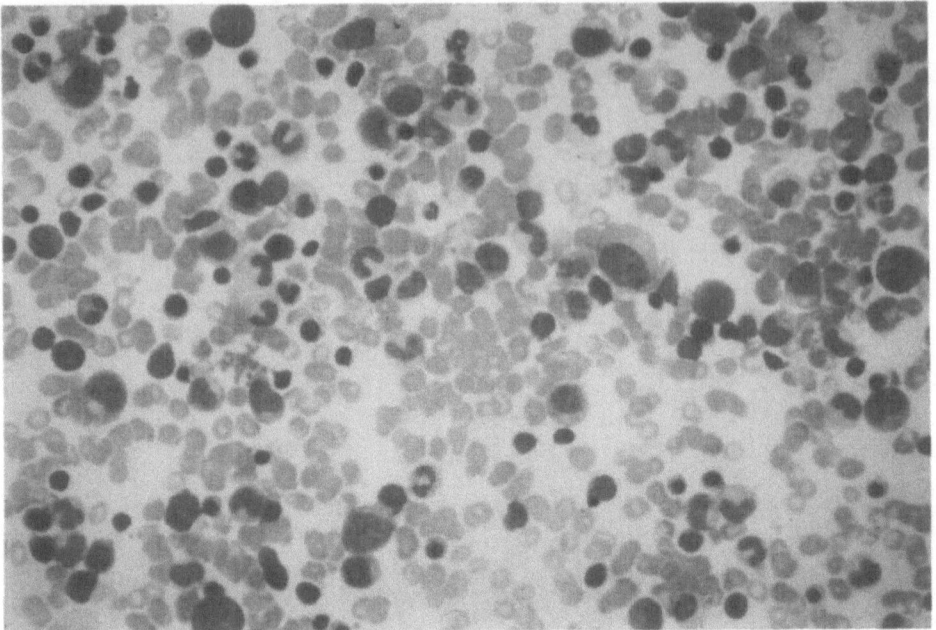

Figure 6.

The differentiation is ascertained by the absence of bone marrow hypoplasia, by the presence of Auer rods in mature cells and by the findings of an intermediate cell population with mature and immature markers.

The progressive differentiation of leukemic cells leads to the disappearance of abnormal cells and the emergence of normal cells. At the time of complete remission the caryotype becomes normal.

2. PML Expression. Cytospins from bone marrow aspirates or peripheral blood samples of 14 untreated APL patients were tested for PML expression by immunocytochemistry, the label was cytoplasmic for at least 90% of the blasts in all patients. The intensity of the labeling of the blasts is variable. Blood or marrow samples from 4 of the previously studied patients containing at least 90% blasts were analyzed during RA therapy. After 5 days, PML was found to be mainly nuclear with a speckled aspect by immunocytochemistry (Fig.n°7, 8, 9). These maturating leukemic cells appear to have an intermediate pattern as they show more and smaller nuclear dots than in the normal granulocytic lineage (20). Interestingly, in untreated patients, a few cells that could be spontaneously maturing blasts have the same profile. PML sites labeled distinctly as early as 5 days of treatment, before with the onset of visible leukemia maturation.

3. Coagulation Disorders. Markers of DIC, plasmin-dependant primary fibrinogenolysis, and diffuse unspecific proteolysis were all present in the 27 APL patients at presentation. Both elastase and cathepsin G activities were present in six patients, while only one activity was detectable in the nine remaining patients (elastase in four patients and cathepsin G in five patients). $\alpha 2PI$ deficiency was only found in patients with fibrinogen level ≤ 2.0 g/l. As it turned out, patients with fibrinogen level < 2.0 g/l did not present any

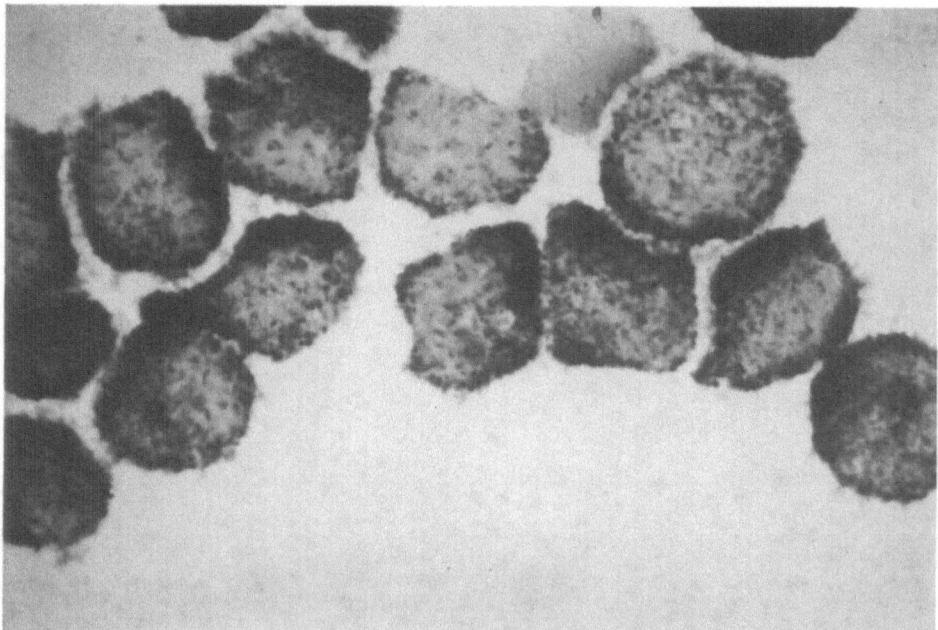

Figure 7. Immunostaining using an anti PML antibody. Bone marrow samples were taken at diagnosis (micropunctated aspect, localized in cytoplasm and nuclears, fig n° 7), after 5 days of ATRA treatment (some cells with large dots relocalized in the nucleus, fig n° 8), and after complete remission (normal aspect with a speckled pattern in the nucleus and not in the cytoplasm of each cell (Fig n° 9).

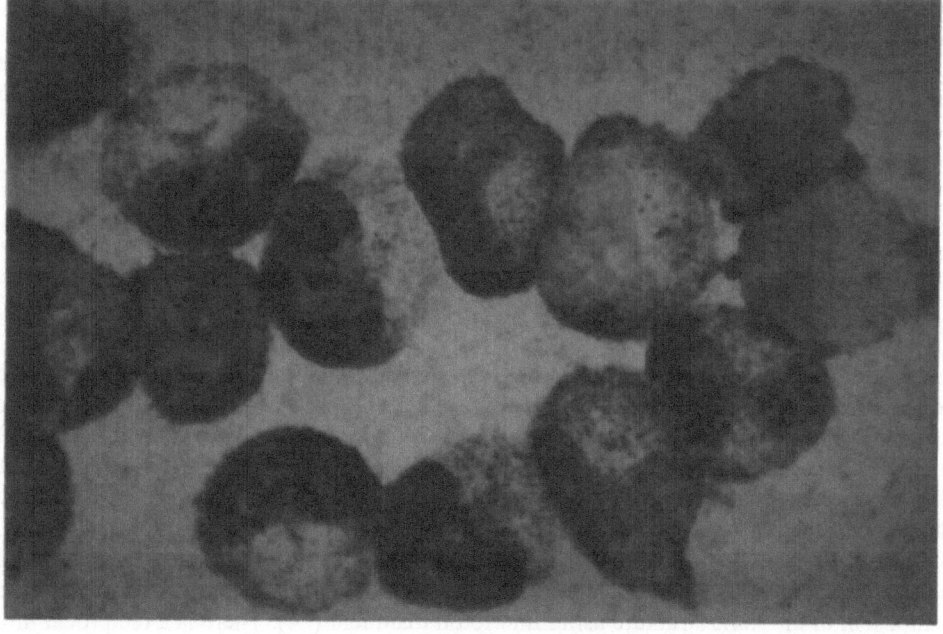

Figure 8.

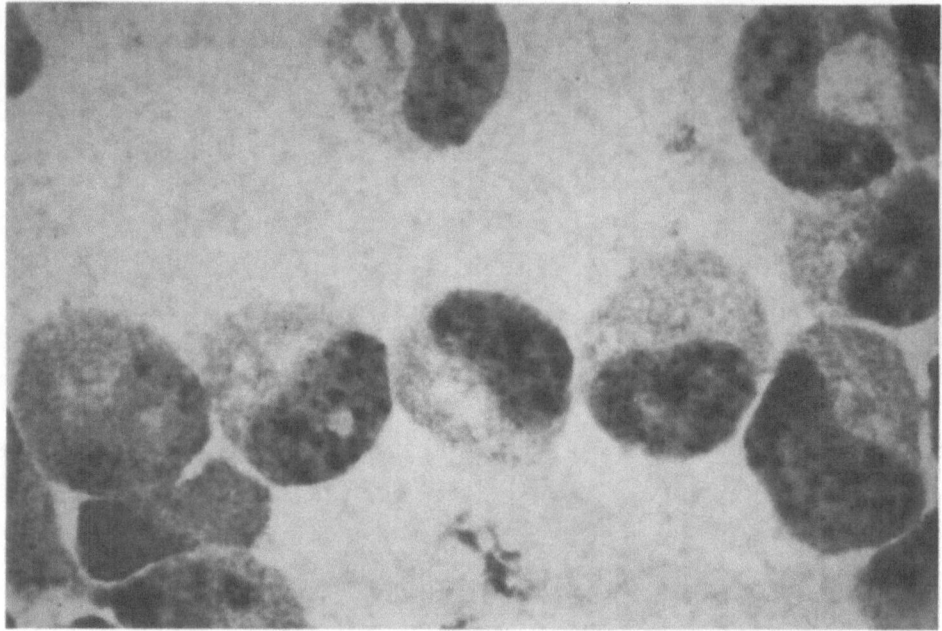

Figure 9.

significant α2PI deficiency. This positive correlation between fibrinogenopenia and α2PI deficiency is statistically significant (p<0.01).

There was a trend toward a higher incidence of clinical bleeding in the neutrophil protease positive patients (68 % versus 28 %). There was no clear correlation between the severity of fibrinogenopenia and the peaks of neutrophil protease activities. Initial fibrinogenopenia and acquired α2PI deficiency disappeared simultaneously as early as day 5 of ATRA treatment. On the contrary, there was a more prolonged persistence of DIC and neutrophil protease activities (21).

DISCUSSION

1. Differentiation of Malignant Cells and PML Associated Nuclear Body Pattern

All-trans retinoic acid (ATRA) in vivo treatment in acute promyelocytic leukemic patients induces a real differentiation of malignant cells (22,23). Leukemic cells mature, dye and and replaced by normal cells. When complete remission occurs, maturing cells are still present, explaining a positive signal detected by RT-PCR techniques for the PML-RAR fusion gene. The maturation of malignant cells is early disclosed at day 5 of treatment by the reappearance of a nuclear body of which PML product is a component. Normally, PML proteins are present on the outer shell of the nuclear body. Swelling of nuclear body is found by an overexpression of PML (gene transfer), or by an overtranscription of sp100 another composent of the same structure using Interferon treatment (24). There is a reciprocal link between the level of the two proteins. PML associated nuclear body pattern varies during the cell cycle from small dots to speckled feature. In APL cases the PML associated nuclear body is disrupted leading to a micropunctated pattern in the nucleus and the cytoplasm. The hybrid effect in the disruption is dominant (gene transfer in normal cells) (24). After ATRA treatment the structure reappears rapidly being one of the earliest marker of the differentiative effect of the drug.

2. Clinical Results

ATRA treatment induces a complete remission in all cases ascertained by the presence of PML-RAR rearrangement and gives poor response in cases with. PLZ-RAR rearrangements. However, ATRA induces the expression of catabolytic proteins, as plasmatic hydroxy-lases (4 OH hydroxylase of cytochrome P450) and cytoplasmic binding proteins (CRABPII). ATRA is thus metabolized in the plasma or sequestered in the cytosol and could not reach the nucleus where the nuclear receptors are located, and a progressive (but reversible) resistance occurs.

In order to sustain the complete remission we have proposed to add a chemotherapy. Pilot study showed a clear beneficial effect by the addition of ATRA before chemotherapy. European Multicentric randomized trial comparing conventional chemotherapy to the same treatment preceded by ATRA until complete remission demonstrated statistically significant differences in the event free survival (50 % versus 80 % at one year), in overall survival (74 % versus 91 % at one year), mainly due to a decrease of relapse rate.

An on going trial questions on the role of maintenance treatment in chances to cure patients.

3. Bleeding Disorders and Thrombotic Risk

Among the three major disorders involved in the fibrinopenia i. e. disseminated intravascular coagulation (DIC), primary fibrinolysis and lysozomial enzyme dependant proteolysis, it appears that the fibrinogen variations are correlated to the primary fibrinolysis and the hemorrhagic tendency to the plasmatic free activity lysozomial enzymes. DIC persists during ATRA treatment and the procoagulant activity could be related to the thrombotic risk during the first month of treatment.

Treatments using antifibrinolytic drugs (first week) and heparin (first month) have to be tested, since bleeding disorders are responsible of 10% early mortality and thrombotic events.

4. Hyperleucocytosis and Leukocyte Activation

Leucocyte activation is generally associated with an hyperleucocytosis. It affects 35 % of newly diagnosed and 15 % of first relapse patients, and is characterized by fever, dyspnea, respiratory distress, kidney failure and coma (20, 25). It occurs during the first month of treatment.

Hyperleucocytosis could be due to cell cycles of multiplication during the initial maturation but also to a bone marrow chase concomitant to changes in adhesion properties and in cell deformability (26). Leukocyte activation could be a consequence of secretion of cytokines (IL1β, IL6, TNFα and IL8) (27,28).

The syndrome is now well known in western countries but is almost absent in China, for unknown reasons, while more than 700 patients have been treated (15).

In order to avoid the retinoic acid syndrome we proposed to prevent it by a chemotherapy delivered when white blood cells are above 6.000 by day 5, 10.000 by day 10, and 15.000 by day 15 (19). In USA Warrell et al prefer to treat the patients by corticosteroids since only half of patients with hyperleucocytosis experienced the clinical

syndrome (25). Leukopheresis was previously proposed in USA which reduces the hyperleucocytosis but has no effect on the fatal issue of the syndrome (23).

Sylvie Castaigne, in the hope of reducing the adverse effects, proposed to treat
patients with lower doses of ATRA. She treated 30 patients with 25 mg/m2 (29) and even
12 patients with 15 mg/m2 (15). She obtained the same complete remission rate but no
reduction of the frequency of hyperleucocytosis and of retinoic acid syndrome.

CONCLUSION

ATRA is an effective and safe treatment for inducing complete remission in acyte
promyelocytic leukemia, with no resistance, without aplasia, with a rapid recovery of
bleeding diathesis. ATRA induces a differentiation of malignant cells. The proposed strategy
is a combination of ATRA and chemotherapy. So, malignancies are not always irreversible.
Differentiation and cytoreductive therapies are a partners in the battle against leukemia. The
cure at least of one type of acute leukemia by this combination of treatments being within
our reach.

The hybrid gene PML-RAR alters the transactivation of normal retinoic acid receptor
and blocks the retinoic acid induced maturation of myeloid cells, which could explain at
least in part the leukemogenesis.

PML molecules in acute promyelocyte leukemia reveal an alteration in the nuclear
body structure which is possibly linked to oncogenesis and cell cycle. Strikingly the
disruption of nuclear body is reversible under ATRA treatment.

From these investigations one can speculate that PML-RAR hybrid molecules impair
the myeloid maturation altering the normal RAR function and disrupt a nuclear body
interfering the normal PML function. ATRA restores the myeloid program and rebuilds the
nuclear body structure.

In conclusion, all trans RA in acute promyelocytic leukemia is the first model of
differentiation therapy in malignancy, the first specific treatment for a genetic defect due to
the translocation t(15;17), the first evidence for a reversible nuclear structure disruption and
greatly improves the survival of patients.

REFERENCES

1. Sachs L., 1978 : Control of normal differentiation and the phenotypic reversion of malignancy in myeloid
   leukemia cells. Nature, 274 : 535.
2. Chomienne C., Ballerini P., Balitrand N., Daniel M.T., Fenaux P., Castaigne S., Degos L., 1990 : All trans
   retinoic acid as a differentiation therapy for acute promyelocytic leukemia. I. Clinical results. Blood 76
   : 1710-1717.
3. Rowley J., Golomb H.M. , Dougherty C, 1977 : 15/17 translocation, a consistent chromosomal change
   in acute promyelocytic leukemia. Lancet 1 : 549-550.
4. Chomienne C., Ballerini P., Balitrand N., Huang M.E., Krawice I., Castaigne S., Fenaux P., Tiollais P.,
   Dejean A., Degos L. , de Thé H., 1990 : The retinoic acid receptor alpha gene is rearranged in retinoic
   acid sensitive promyelocytic leukemias. Leukemia, 4 : 802-807.
5. de Thé H., Chomienne C., Lanotte M., Degos L., Dejean A., 1990 : The t(15;17) translocation of acute
   promyelocytic leukemia fuses the retinoic acid receptor alpha gene to a novel transcribed locus. Nature,
   347 : 558-561.
6. Borrow J., Goddard A.D., Sheer D., Solomon E., 1990 : Molecular analysis of acute promyelocytic
   leukemia breakpoint cluster region on chromosome 17. Science, 249 : 1577.
7. Alcalay M., Zangrilli D., Pandolfi P.P., Longo L., Mencarrelli A., Giacomuci A., Rocchi M., Biondi A.,
   Rambaldi A., Lococo F., Diverio D., Dionti E., Grignani F., Pelicci P.G., 1991 : Translocation breakpoint

of acute promyelocytic leukemia lies within the retinoic acid receptor alpha locus. Proc. Natl. Acad. Sci. USA, 88 : 1977.

8. de Thé H., Lavau C., Marchio A., Chomienne C., Degos L., Dejean A., 1991 : The myl RAR alpha fusion mRNA generated by the t(15;17) translocation in APL encodes a functionaly altered retinoic acid receptor. Cell, 66 : 675-684

9. Kakizuka A., Miller W.H. Jr and Umesono K., 1991 : Chromosomal translocation t(15;17) in human acute promyelocytic leukemia fuses RAR alpha with a novel putative transcription factor, PML. Cell, 66 : 663-674.

10. Pandolfi P.P., Grignani F. and Alcalay M., 1991 : Structure and origin of the acute promyelocytic leukemia myl/RAR alpha cDNA and characterization of its retinoid-binding and transactivation properties. Oncogene, 6 : 1285-1292.

11. Rousselot P., Hardas B., Patel A., Guidez F., Gaken J., Castaigne S., Dejean A., de Thé H., Degos L., Farzaneh F., Chomienne C., 1994 : The PML/RAR alpha gene product of the t(15;17) translocation inhibits retinoic acid induced differentiation and mediated transactivation in human myeloid cells. Oncogene, 9 : 545-551.

12. Chen Z., Guidez F., Rousselot P., Agadir A., Chen S.J., Wang Z.Y., Degos L., Waxman S., Zelent A., Chomienne C., 1994 : PLZ/RAR fusion protein generated from the translocation t(11;17) (q23;q21) displays altered transactivation properties against the wild-type retinoic acid receptors. PNAS (in press).

13. Guidez F., Huang W., Tong J.H., Dubois C., Balitrand N., Waxman S., Michaux J.L., Martiat P., Degos L., Chen Z., Chomienne C., 1994 : Poor response to all trans retinoic acid therapy in a t(11;17) PLZF/RARα patient. Leukemia, 8 : 312-317.

14. Castaigne S., Balitrand N., de Thé H., Dejean A., Degos L., Chomienne C., 1992 : A PML/RAR alpha fusion transcript is constantly detected by RNA based polymerase chain reaction in acute promyelocytic leukemia. Blood, 79 : 3110-3115.

15. Wang Z.Y., Degos L., 1993 : Retinoic acid in hematological malignancies. Abstract. International Symposium on Retinoids in Hematologic Malignancies, Shanghaï, China. October 19-21.

16. Muindi J., Frankel S.R., Miller W.H., Jakubowski A., Sheinberg D.A., Young C.W., Dimitrowsky E., Warrell RP., 1992 : Continuous treatment with all trans retinoic causes a progressive reduction in plasma drug concentrations. Implications for relpases and retinoid resistance in patients with acute promyelocytic leukemia. Blood, 79 : 299.

17. Delva L., Cornic M., Balitrand N., Guidez F., Miclea J.M. , Delmer A., Teillet F., Castaigne S., Fenaux P., Degos L., Chomienne C., 1993 : Resistance to all trans retinoic acid (ATEA) therapy in relapsing acute promyelocytic leukemia : Study of in vitro ATRA sensitivity and cellular retinoic acid binding protein levels in leukemia cells. Blood, 82 : 2175-2181.

18. Fenaux P., Castaigne S., Chomienne C., Dombret H., Archimbaud E., Duarte M., Lamy T., Tilly H., Guerci A., Maloisel P., Bordessoule D., Sadoun A., Tiberghien P., Fegueux N., Daniel M.T., Degos L., 1992 : All trans retinoic acid followed by inten sive chemotherapy gives a high complete remission rate and prolonged remissions in newly diagnosed acute promyelocytic leukemia. Blood, 80 : 2176-2181.

19. Fenaux P., Robert M.C., Castaigne S., Archimbaud E., Chomienne C., Link H., Guerci A., Duarte M., Daniel M.T., Bowen D., Huebner G., Bauters F., Fegueux N., Fey M., Sanz M., Lowenberg B., Maloisel F., Auzanneau G., Sadoun A., Gardin C., Bastion Y., Ganser A., Dombret H., Chastang C., Degos L. and the European APL91 group., 1993 : Effect of all trans retinoic acid in newly diagnosed acute promyelocytic leukemia. Results of a multicenter randomized trial. Blood, 82 : 3241-3249.

20. Daniel M.T. , Koken M., Romagne O., Barbey S., Bazarbachi A., Stadler M., Guillemin M.C., Degos L., Chomienne C., de Thé H., 1993 : PML protein expression in haematopoietic and APL cells. Blood, 82 : 1858-1867.

21. Dombret H., Scrobohaci M.L., Renesto P., Daniel M.T., Miclea J.M., Castaigne S., Chomienne C., Sigaux F., Fenaux P., Degos L., 1994 : In vivo thrombin, plasmin and neutrophil protease activities in patients with acute promyelocytic leukemia (APL) : effect of all-trans retinoic acid (ATRA) therapy. Leukemia (soumis)

22. Castaigne S., Chomienne C., Daniel M.T., Ballerini P., Berger R., Fenaux P., Degos L., 1990 : All trans retinoic acid as a differentiation therapy for acute promyelocytic leukemia. I. Clinical results. Blood, 76 : 1704-1709.

23. Warrell R., Frenkel S.R., Miller W.H., Sheinberg D.A., Itri L., Hittelmen W.N., Vyas R., Andreef M., Taffuri A., Jakubowski A., Gabrilove J., Gordon M.S., Dmitrowski E., 1991 : Differentiation therapy of acute promyelocytic leukemia with tretinoin (all trans retinoic acid). New Engl. J. Med., 324 : 1385.

24. Koken M.H., Puvion-Dutilleul F., Guillemin M.C., Viron A., Linares-Cruz G., Stuurman N., Szostecki C., Calvo F., Chomienne C., Degos L., Puvion E., de Thé H., 1994 : The t(15;17) translocation alters a nuclear body in a retinoic acid reversible fashion. EMBO J., 13 : 1073-1083.

25. Frankel S.R., Eardley A., Lauwers G., Weiss M., Warrell R.P., 1992 : The "retinoic acid syndrome" in acute prçmyelocytic leukemia. Ann. Intern. Med., 117 : 292-296.

26. Dombret H., Geiger S., Daniel M.T., Lacombe C., Micléa J.M., Castaigne S., Degos L., 1993 : Change in micro-rhéology and deformability of acute promyelocytic leukemia (APL) cells under all trans retinoic acid (ATRA) treatment : a mechanism for ATRA-induced hyper-leukocytosis ? Abstract International Symposium : Acute Promyelocytic Leukemia : a curable disease ? Roma, November 11-12.

27. Dubois C., Schlageter M.H., de Gentile A., Guidez F., Balitrand N ., Toubert M.E., Krawice I., Fenaux P., Castaigne S., Najean Y., Degos L., Chomienne C., 1994 : Hematopoietic growth factor expression and ATRA sensitivity in acute promyelocytic blast cells. Blood, 83 : 3264-3270.

28. Dubois C., Schlageter M.H., de Gentile A., Guidez F., Balitrand N., Toubert M.E., Krawice I., Fenaux P., Castaigne S., Najean Y., Degos L., 1994 : Modulation of IL-6 and IL-1β and G-CSF secretion by all trans retinoic acid in acute promyelocytic leukemia (Leukemia accepted)

29. Castaigne S., Lefebvre P., Chomienne C., Suc E., Rigal-Huguet F., Gardin C., Delmer A., Archimbaud E., Tilly H., Janvier M., Isnard F., Travade P., Montfort L., Delannoy A., Rapp M.J., Christian G., Montastruc M., Weh H., Fenaux P., Dombret H., Gourmel B., Degos L., 1993 : Effectiveness and pharmacokinetics of low-dose all trans retinoic acid (25 mg/m2) in acute promyelocytic leukemia. Blood, 82 : 3560-3563.

DISCUSSION

C. Brugnara

What is your explanation for the persistence of the increased fibrin degradation after retinoic acid? Is it just the defect of anti-thrombin, protein S, or something else?

L. Degos

The DIC is the same during the treatment. The plasmin activation is reduced but the thrombin activation is the same. So the explanation could be the impairment in the balance between thrombin and plasmin activation.

S. Landolfo

Maybe I missed the point. When you treat the cells with retinoic acid, you have shown relocalization of nuclear bodies, is that right? Does this correspond to the disappearance of PML/RAR Alpha and the appearance of PML protein? What is the explanation? This is the first question. The second question is: Many interferon inducible genes have also a respon-sive element to retinoic acid, has this some implications for the responsiveness of these cells to interferon?

L. Degos

For the first question, we have examined the relocalization of nuclear bodies with antibodies against PML, so it is very difficult to tell you if PML/RAR or PML molecules alone are present at that time. We found that PML related nuclear bodies are disrupted before treatment and nuclear bodies are reconstructed after treatment. We do not know if it is related to the appearance of normal PML or to a disappearance of PML/RAR. It is difficult to answer your second question. PML products are coexpressed with SP100 which is interferon inducible. We do not know if interferon acts directly or indirectly because nobody has worked on the PML 5 prime region to see if there is some responsive element.

S. Landolfo

What is known about the interaction of PML with SP100?

L. Degos

I have reported preliminary results on cross expression of the two genes. I have no more data.

D. Livingston

Do HL60 cells differentiate upon introduction of PML/RAR and subsequent exposure to retinoic acid?

L. Degos

With PML/RAR in they do not differentiate.

D. Livingston

Right, so is that opposite to what you think might be happening clinically and if so, why?

L. Degos

First, HL60 is not a promyelocytic leukemic cell line. Second, it is a problem of doses. When we add retinoic acid at 10^{-8} molar, the cells become insensitive to retinoic acid but at 10^{-6} molar, the cells are again sensitive.

D. Livingston

What is deblocked?

L. Degos

At physiological concentrations of retinoic acid, there is an arrest of differentiation. If we treat with a one hundred more pharmacological concentration, they are again sensitive to retinoic acid. The mechanism of action is not actually known.

D. Livingston

Have you seen cell death? And is that the therapeutic effect?

L. Degos

No, during treatment of patients we see a cell differentiation.

D. Livingston

It appears to be differentiation but it is hard to detect a precursor to product relationship. These are not kinetic experiments. So, is there any reason to suspect that failure of differentiation is really failure of survival at 10^{-8}M RA and the end product is not only no differentiation, but no survival?

L. Degos

The number of cells does decrease during the culture which is an indirect answer.

D. Livingston

Is this a transient assay?

L. Degos

Yes, by liposome assay.

D. Livingston

So you are introducing the protein?

L. Degos

No, the gene.

D. Livingston

And what percentage of the cells take up the gene and express it?

L. Degos

It is about 40 to 50%.

D. Livingston

What if you make a stable HL60 on an inducible promoter and then induce it and do a titration?

L. Degos

It is very difficult to transfer genes into HL60. Giuseppe Pelicci has made stable transfer in U937 but not in HL60. Farzin Farzireh and Philippe Rousselor have made transient transfer with liposomes, but even with liposomes it is not easy. Results depend on the batch of liposome

J. Griffin

I am confused about the relationship of retinoic acid treatment to proliferation of the cells. One would anticipate from in vitro studies that they would arrest in G1 or G0 fairly quickly and then proceed to differentiate without growing much further once the inducing agent is added. Yet *in vivo*, there is this hyperleukocytosis syndrome which is either due to release of cells or proliferation.

L. Degos

According to previous studies reported in the literature, the cell seems to be blocked in S phase.

J. Griffin

So you are saying they would complete one cycle and then arrest?

L. Degos

They could have one cycle first and second it is very difficult to know if hyperleukocytosis is due to a release from bone marrow or if it is due to a multiplication of those cells.

J. Griffin

What happens in *vitro* if you add retinoic acid, does proliferation continue?

L. Degos

In vitro without retinoic acid, acute promyelocytic leukemic cells from patients die. If retinoic acid is added in the culture, they survive and proliferate.

J. Griffin

If you take these cells and add growth factors, such as IL-3, with or without retinoic acid, do they grow better or worse with retinoic acid?

L. Degos

They seem to grow better with GCSF. However Christine Chomienne found that after 6 days of culture malignant cells become insensitive to GCSF. Simultaneously, there is a decrease of bcl-2 protein density. Is GCSF insensitivity related to retinoic acid induced maturation or to bcl-2 modulation? Is bcl-2 modulation related to apoptosis?

C. Brugnara

What kind of system do you use to measure white cell deformability and what is that a function of?

L. Degos

We used a method measuring the viscosity in a capillary.

E. Mihich

You are using chemotherapy as a toxicity reductive treatment. But if you try to *in vitro,* initially with HL60, to treat the PML RAR cells with various concentrations of RA, which you say give a bell shaped curve of some kind, if you combine that with different doses of say Ara-C or some of the other agents that can induce differentiation by themselves in that system, what happens? Do you have some interaction among these agents? Could you translate those interactions into the clinical setting?

L. Degos

Firstly, HL60 is not a promyelocytic leukemic cell line so there is no expression of PML/RAR. So, HL60 is a M2 and not a M3 sub-type. Secondly, it is difficult to transfer PML/RAR gene in HL60. We have done experiments of transient transfer and we cannot compare various independent experiments. On the other hand, NB4 cell line is a promyelocytic leukemic cell line. I cannot answer your question applied on NB4. However, differentiation is obtained with very low concentrations of antimitotic drug and conversely we use this treatment in patients at high concentrations. So we have few data on low dose Ara-C delivered to the patient except in relapsed patients who are not informative for the real value of such combinations. So, you raised a pertinent question. I cannot answer to your question if there is an additive or synergetic *in vivo* effect of the low concentration antimitotic and differentiating agents.

D. Metcalf

I am a little bit surprised by your statement that the promyelocytic leukemic cells are blocked in the S phase of the cell cycle and I wonder how you arrived at that conclusion. I would like to see thymidine labeling data to back that up. What we find when we purify promyelocytes either from a patient with promyeloctyic leukemia or from a normal subject, is that the two cell populations behave identically *in vitro*. Both populations have a high cloning efficiency, and essentially all of the cells proliferate clonally with exactly the same concentrations of G-CSF or GM-CSF. They generate clones of identical size and the initiation of cell division is identical in the culture. So I am questioning whether they really are out of cycle in the patient. If they really are blocked *in vivo*, what do you suggest is in the culture medium that is unblocking these cells?

L. Degos

Thank you for your comments on normal and malignant promyelocytes. I have not said that we have shown that malignant promyelocytes are blocked in S phase, I have only referred to literature reports made by Drs. Raza and Preisler.

J. Griffin

In the limited studies we have done they have also not been clearly blocked in S phase whereas they are spread through the cell cycle, most of the cells are in G1. We have not studied that many, however.

16

INDUCTION OF ALLOANTIGEN SPECIFIC T CELL ANERGY BY INHIBITION OF THE B7: CD28 PATHWAY

Vassiliki A. Boussiotis, Gordon J. Freeman, John G. Gribben, and Lee M. Nadler

Division of Hematologic Malignancies
Dana-Farber Cancer Institute and
Department of Medicine
Harvard Medical School
Boston, Massachusetts 02115

INTRODUCTION

Engagement of TCR by antigen in the absence of requisite costimulation fails to induce an effective immune response, but results in a state of antigen-specific unresponsiveness, termed anergy (1, 2, 3, 4). Anergy is characterized by the inability of T cells to proliferate and produce cytokines on rechallenge with specific antigenic stimulation in the presence of costimulation. Anergy appears to be an important mechanism of peripheral T cell tolerance.

A large number of molecules expressed on Antigen Presenting Cells (APCs) and their receptors expressed on T cells are candidates for providing critical costimulatory function including B7-1 (CD80): CD28/CTLA4; B7-2 (CD86): CD28/CTLA4; ICAM-1 (CD54)-2, -3: LFA-1 (CD11a/CD18); LFA-3 (CD58): CD2; CD40: CD40L and heat stable antigen (CD24): heat stable antigen (CD24). Of the numerous costimulatory molecules, B7 mediated costimulation appears to be critical to prevent the induction of anergy (3, 5, 6, 7, 8, 9).

B7 FAMILY AND THEIR COUNTER-RECEPTORS

The B7 family is composed of at least two members, B7-1 (CD80) (10, 11) and B7-2 (CD86) (12, 13). These molecules are members of the immunoglobulin gene superfamily, and are differentially expressed either constitutively (13, 14) or following stimulation on a wide variety of APC, including monocytes, dendritic and B cells, (12, 15, 16, 17, 18, 19, 20, 21) Following activation, B7-2 is rapidly expressed, whereas B7-1 appears later (16, 22).

B7-1 and B7-2 have moderate structural similarity characterized by similar organization of their extracellular domains but markedly different cytoplasmic domains. However,

Normal and Malignant Hematopoiesis, Edited by Enrico Mihich and Donald Metcalf
Plenum Press, New York, 1995

both B7-1 and B7-2 are low affinity receptors for CD28 and high affinity receptors for CTLA4 expressed on T cells (23, 24). CD28, is constitutively expressed on 95% of resting human CD4$^+$ cells and 50% of resting CD8$^+$ peripheral blood T cells and its expression increases following activation. CTLA4 is not expressed on resting T cells, but is induced following activation (25, 26). CTLA4 shares 31% amino acid identity with CD28 with only a limited conservation between their cytoplasmic domains. However, there is 100% conservation of the cytoplasmic domains of murine and human CTLA4, suggesting a conserved signaling function. Since B7 family members have higher affinity for CTLA4 than for CD28, a CTLA4-immunoglobulin fusion protein (CTLA4-Ig) can efficiently bind B7 and totally abrogate B7 mediated costimulatory function both in vitro and in vivo (6, 27, 28, 29).

CONSEQUENCES OF SIGNALING VIA THE B7:CD28/CTLA4 PATHWAY

Following a TCR-mediated signal, ligation of CD28 by either B7-1 or B7-2 leads to increased expression of many cytokine genes (30) cytokine secretion, T cell proliferation (31, 32) and upregulation of activation induced surface molecules including IL-2Rα, β and γ (33, 34, 35), CD40 ligand (36) and CTLA4.

In comparison to CD28 signaling, little is known about the function of CTLA4 and its signaling. There is evidence that CTLA4 may mediate an antigen-specific downregulatory signal. Murine anti-CTLA4 mAbs and their Fab fragments augment T cell proliferative responses in an allogeneic mixed lymphocyte reaction, suggesting that CTLA4 molecule delivers a negative signal that is blocked by anti-CTLA4 mAb (37). Studies with human anti-CTLA4 mAbs indicated that crosslinking of CTLA4 can mediate apoptosis of previously activated human T lymphocytes. This function appears to be antigen-specific, since a concomitant TCR signal is required (38) .

THE UNIQUE ROLE OF B7/CD28 PATHWAY IN REGULATING THE INDUCTION OR PREVENTION OF T CELL ANERGY

Successful human allogeneic bone marrow and organ transplantation is presently dependent upon the chronic administration of toxic nonspecific immunosuppressive agents and the availability of a suitable histocompatible donor. If long-lasting and irreversible unresponsiveness to alloantigen could be induced, the attendant clinical toxicities might be ameliorated and the eligible donor pool would be increased. One approach to achieve this objective would be to identify and block the relevant costimulatory signals involved in the generation of the T cell immune response to alloantigen. Following primary stimulation with alloantigen in the absence of costimulation, T cells will become anergic and incapable of inducing a secondary response on rechallenge with the specific alloantigen(s).

To determine the role of adhesion or costimulatory pathway(s) in antigen recognition and prevention of anergy, we developed an in vitro model system to identify costimulatory molecules involved in the generation of the T cell immune response to alloantigen. We employed as experimental system a model of mixed lymphocyte culture in which responders were unprimed alloreactive CD4+ T cells (4) or HLA-DR7-specific T cell clones (27) and stimulators were artificial alloantigen presenting cells transfected with the specific alloantigen (HLA-DR7) alone or in combination with either B7-1 or ICAM-1. Artificial APCs expressing class II antigens but not B7-1 did not induce T cells to proliferate or secrete appreciable quantities of IL-2 but induced them into a state of anergy. Stimulators expressing class II antigen in combination with either B7-1 or ICAM-1 delivered equally effective costimulatory signals resulting in an equivalent proliferative response of alloreactive T lymphocytes. However, B7-1 but not ICAM-1 induced detectable IL-2 secretion and prevented the induction of alloantigen specific anergy on rechallenge (4). These results demonstrated that costimulatory molecules which mediate a proliferative response, do not necessarily rescue T cells from the induction of anergy.

To determine the role of adhesion and costimulatory pathways in T cell immune response mediated by professional APCs, we examined the response of HLA-DR7-specific T cell clones to stimulator EBV transformed lymphoblastoid B-cell lines (LBL), which express the specific alloantigen and all the potential candidate costimulatory molecules including ICAM-1, LFA-3, B7-1 and B7-2. To examine the contribution of B7:CD28, ICAM-1:LFA-1 and LFA-3: CD2 in the primary recognition of alloantigen and the induction of anergy, primary stimulation was performed with LBL cells in the presence or absence of monoclonal antibodies (anti-ICAM-1, anti-LFA-1 or anti-LFA-3) or fusion proteins (CTLA4-Ig) inhibitory for the pathways under study. Following primary culture, T cells were rechallenged with artificial APCs, LBL-DR7 or IL-2. Primary culture with LBL-DR7 resulted in a significant secondary proliferative response and IL-2 accumulation. Addition of each one of the mAbs or fusion proteins under study resulted in significant inhibition of primary T cell proliferative response and IL-2 secretion. However, the resulting capacity of T cells to respond on subsequent secondary antigen-specific rechallenge differed significantly. Secondary rechallenge of T cell clones primarily cultured in the presence of anti-ICAM-1, anti-LFA-1 and anti-LFA-3 mAbs induced proliferation and IL-2 accumulation (Figure 1B and C) (4, 27). In contrast, addition of CTLA4-Ig in the primary culture resulted in anergy (27) (Figure 1D).

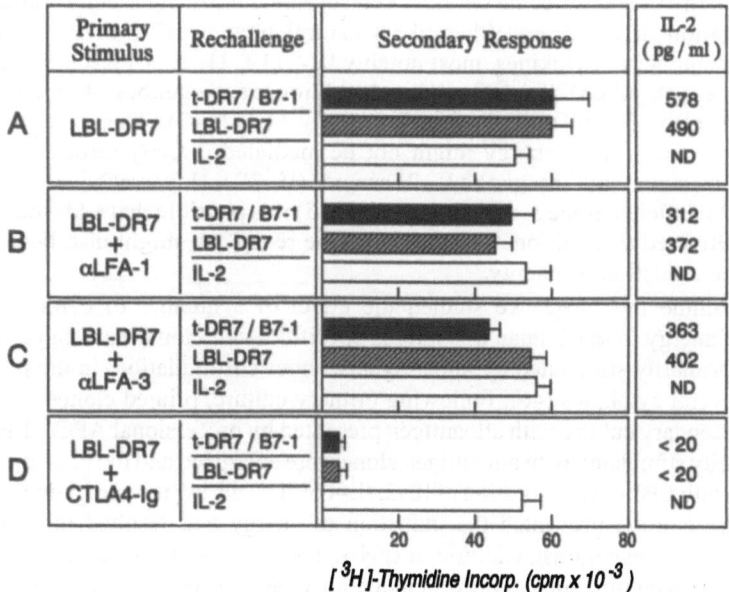

Figure 1. Lack of B7 costimulation is necessary and sufficient to induce alloantigen-specific T cell clonal anergy. Alloantigen-specific T cell clones were cultured with the indicated primary stimuli. Each population was subsequently rechallenged with a secondary stimulus as shown. IL-2 concentration was assessed in the supernatants of 24 hr of culture and thymidine incorporation was measured for the last 16 hrs of a 72 hr culture period.

These results suggest that ICAM-1, LFA-1, and LFA-3 mainly provide adhesion during primary stimulation therefore inhibition of adhesion prevents recognition of antigen and results in blockade of primary response or immunosuppression. In contrast, inhibition of the B7:CD28 costimulation results in both blockade of primary response and induction of anergy. These results also suggest that blockade of the B7 family is both necessary and sufficient to induce anergy. In summary, inhibition of adhesion results in blockade of primary response or immunosuppression, whereas inhibition of the B7:CD28 costimulation results in both inhibition of primary response and induction of anergy. It is of note that, immediately following the induction of anergy, neither B7 family members nor other non-B7 costimulatory molecules expressed on LBL-DR7 cells appear to be capable of reversing this state (4, 27) (Figure 1D).

In agreement with these *in vitro* results, several *in vivo* animal models underlined the significance of the B7/CD28 pathway in regulating the induction or prevention of anergy. CTLA4-Ig treatment resulted in long-lasting tolerance to human xenoantigens in mice (6), suppressed antibody responses to sheep red blood cells *in vivo* (39), prolonged survival of cardiac allografts in rats and mice (9, 40) and reduced the incidence of lethal graft versus host disease, without affecting hematopoetic reconstitution, following bone marrow transplantation (28)

PREVENTION OF T CELL ANERGY BY SIGNALING VIA γ_c OF THE CYTOKINE RECEPTORS

The above described studies, provide compelling evidence that one critical costimulatory signal necessary and sufficient to prevent the induction of anergy is mediated through the CD28 molecule on the T cell surface. Following T cell receptor (TCR) signaling, ligation of CD28 by either of its natural ligands B7-1 or B7-2, results in secretion of a number of cytokines, most notably IL-2 (12, 31, 32, 41). It has been shown that in the absence of CD28 costimulation, addition of exogenous IL-2 during TCR activation can also prevent the induction of anergy (4, 42, 43). Therefore, the critical signal necessary to prevent anergy might not be mediated directly through CD28, but alternatively, by signaling through the IL-2 receptor (IL-2R). However, since T cells from IL-2 and CD28 deficient mice are not anergic (44, 45) it is possible that additional signals, potentially delivered through one or more cytokine receptors, might also be capable of preventing the induction of energy.

To examine this issue, we studied the effect of a number of cytokines on the prevention of anergy in our human alloantigen specific T cell clonal model system. T cell clones were primarily stimulated by alloantigen without costimulation, in the presence or absence of several cytokines (46). Following primary culture, primed clones were rechallenged in a secondary culture with alloantigen presented by professional APCs (LBL-DR7). Clones primarily stimulated with alloantigen alone were anergized and did not respond when rechallenged with LBL-DR7. Addition of IL-2, IL-4 or IL-7 during the anergizing conditions of the primary culture, prevented the induction of anergy and resulted in a significant secondary response. In contrast, addition of IFN-γ, TNF-α, IL-6, IL-10 or IL-12 during the anergizing primary culture, did not prevent induction of anergy. Because IL-2, IL-4 and IL-7 receptors share the common gamma chain (γ_c) (47, 48, 49), we next examined whether the effect of these cytokines in preventing the induction of anergy was mediated via the γ_c. Indeed, Jak3 protein tyrosine kinase (50, 51, 52, 53) was constitutively associated with γ_c, and was activated by IL-2, IL-4 and IL-7 or by crosslinking of γ_c with anti-γ_c mAb. During antigen-specific stimulation of T cell clones under non-anergizing conditions Jak3 underwent significant phosphorylation (46). Therefore, γ_c signaling results in both activation of Jak3 and prevention of anergy .

CD28 costimulation results in upregulation of IL-2, IL-2Rα (33, 54) and IL-2R β (34). We have recently shown that CD28 costimulation results in a rapid upregulation of the α and γ chains of the IL-2R on the majority of IL-2R$\alpha^- \gamma^-$ CD4$^+$ T cells within 12 hours (35). In contrast, culture with TCR signal alone, results in upregulation of IL-2Rα and γ on a minority of cells and only after 48 hrs of culture. This observation, taken together with the above results, explains a mechanism by which CD28 costimulation may prevent the induction of anergy by hastening and increasing the production of both IL-2 and the IL-2R chains (α, β and γ_c). Induction of γ_c chain may also provide one explanation for the regulation of responsiveness to IL-4 by CD28 costimulation (55) and suggests a similar effect on responsiveness to other cytokines capable of signaling through the γ_c (56), resulting in prevention of anergy. Each one of these cytokines may provide its function in different microenvironments.

MAINTENANCE AND REVERSAL OF T CELL CLONAL ANERGY

Although inhibition of the B7 costimulatory family appears to be critical for the induction of anergy to alloantigen, a major concern for the clinical applicability of this approach, is whether it is permanent or soluble factors and costimulatory molecules expressed on APC can reverse the established state of anergy. It has been shown *in vitro* that T-cell clonal anergy can be reversed following prolonged culture of anergized cells in IL-2 and stimulation with professional APC (42, 43). The capacity of endogenous IL-2 to reverse anergy would be a significant obstacle to implementing the induction of alloantigen-specific anergy as a clinical strategy. Therefore, we examined how addition of exogenous IL-2 would reverse the state of human alloantigen specific anergy *in vitro*, and which costimulatory molecules would be involved in this response. As mentioned above, immediately following anergizing primary culture, anergized cells do not respond on alloantigen-specific rechallenge in the presence of a large number of costimulatory molecules, including B7-1, B7-2, ICAM-1, CD24, CD40, CD72 and LFA-3 (27). However, if anergizing primary culture is followed by incubation in exogenous IL-2 for at least 7 days, anergized cells can respond to alloantigen in the presence of LFA-3, but not B7-1, B7-2 or ICAM-1 costimulation (27)(Figure 2).

Because LFA-3 mediates activation of T cells via the functionally distinct epitopes of the CD2 molecule (57, 58, 59, 60, 61, 62), we examined whether induction, maintenance and reversal of anergy was associated with the expression of distinct epitopes of CD2. Indeed, prior to the induction of anergy, T cell clones express CD2 (T11.1 and T11.2), and CD2R (T11.3) epitopes, but following induction of anergy, CD2R (T11.3) is no longer detectable. The T11.3 epitope is re-expressed after 7 days of culture in IL-2, precisely when these cells regain responsiveness to alloantigen and LFA-3 costimulation, suggesting that the inability of the anergized cells to respond to LBL-DR7, is associated with the lack of CD2R expression. Challenge with alloantigen and LFA-3 costimulation, following IL-2 culture, restores responsiveness of anergized cells to alloantigen in the presence of previously insufficient costimulatory signals (27) (Figure 3).

More importantly, these results demonstrate that although anergy can be induced by blockade of B7 family costimulation, it is a temporary state which can be reversed under certain conditions in vitro and potentially *in vivo*.

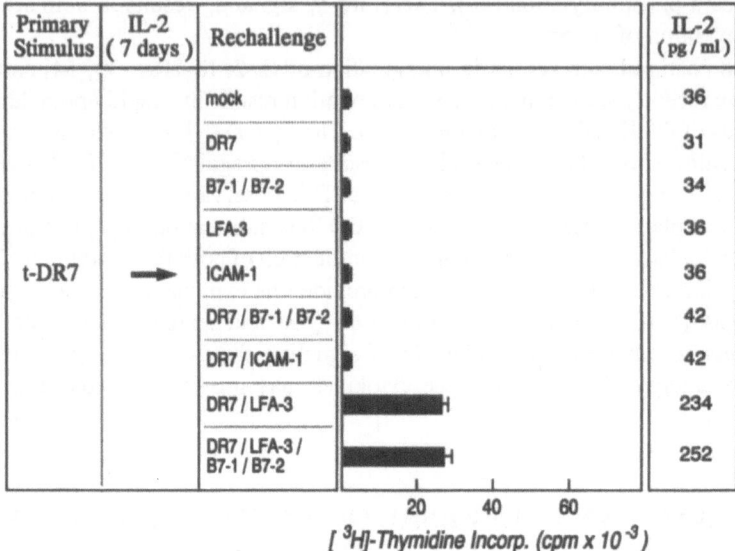

Figure 2. Alloantigen-specific anergy can be reversed by alloantigen and LFA-3 costimulation. After induction of anergy by t-DR7, T cell clones were cultured in IL-2 for 7 days and subsequently rechallenged with the indicated stimuli. Proliferation and IL-2 accumulation were measured as in Fig. 1.

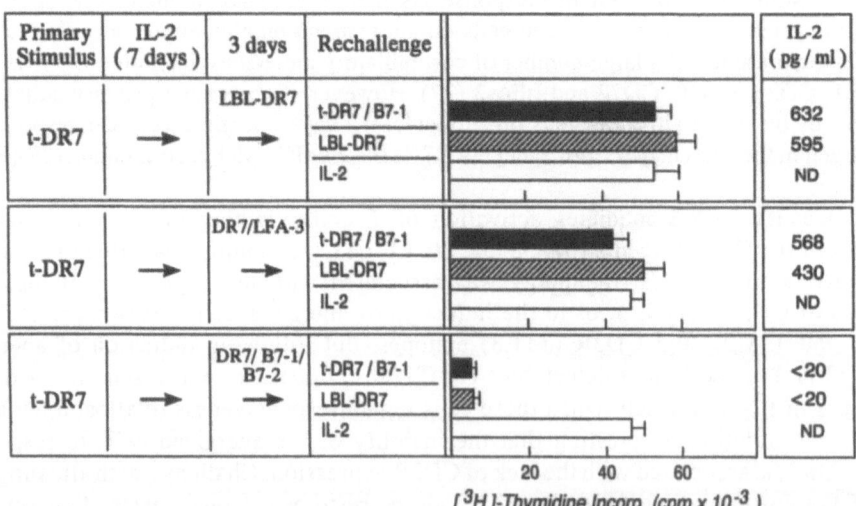

Figure 3. Alloantigen and LFA-3 costimulation restore the capacity of anergized cells to respond to subsequent rechallenge with previously insufficient costimulatory signals. After induction of anergy with t-DR7 allo-APCs and 7 days culture in IL-2, T cell clones were rechallenged for 3 days with either LBL-DR7 or COS transfected cells expressing DR7 and LFA-3 or DR7 and B7-1+B7-2. Subsequently, they were isolated and rechallenged with the indicated stimuli. Proliferation and IL-2 accumulation were measured as in Fig. 1.

INDUCTION OF ANTIGEN-SPECIFIC UNRESPONSIVENESS BY CLONAL DELETION

The potential reversal of an established state of anergy to specific alloantigen(s) in vivo, would not be a clinical concern if the responsible antigen-specific clone could be eliminated by induction of antigen specific clonal deletion. One of the major mechanisms of peripheral tolerance appears to be antigen-specific clonal deletion. However, the cellular interactions, biochemical and molecular events that mediate this effect are poorly understood. A large number of molecules known to mediate programmed cell death (apoptosis) are non-T cell restricted and do not mediate their function in an antigen-specific way.

We have recently undertaken experiments to study the nature and function of CTLA4 mediated signal on T cells. CTLA4 is not expressed on resting T cells but is only induced following activation (26). Using preactivated peripheral blood T cells, we demonstrated that at the time of the optimal CTLA4 surface expression, crosslinking of CTLA4 by specific mAbs in the presence of a concomitant TCR signal, results in downregulation of TCR mediated proliferative response (38). To precisely address the issue of antigen-specificity, we used alloantigen-specific T cell clones which were cultured with artificial alloAPCs expressing specific MHC, capable of providing a TCR signal. Addition of anti-CD28 mAb resulted in significant proliferation and IL-2 accumulation. In contrast, addition of anti-CTLA4 mAb did not stimulate proliferation but instead resulted in a significantly decreased proliferative response and no IL-2 accumulation. Under these culture conditions T-cell clones developed DNA fragmentation and underwent apoptosis (>90%) as assessed by fluorocytometry. These results suggest that CTLA4-mediated T cell clonal deletion may provide a useful approach to eliminate alloantigen-specific T cells with potential applications as a novel therapeutic strategy to induce a state of non-reversible antigen-specific unresponsiveness.

CONCLUDING REMARKS

CD28 mediated costimulation and signaling via γ_c of cytokine receptors in the presence of TCR signal, can rescue T cells from entering the anergic state. In contrast, during stimulation with professional APCs, inhibition of B7/CD28 costimulation is necessary and sufficient to induce anergy. At least one mechanism by which CD28 can prevent the induction of anergy is associated with the rapid upregulation of IL-2R (α, β and γ chains), resulting in successful signaling by cytokine receptors sharing γ_c. However, intense investigation of CD28 pathway may prove additional direct mechanism(s) via which CD28 is involved in induction or prevention of anergy. Although, anergy cannot be reversed by CD28-mediated costimulation, it is a temporary state which can be reversed under certain conditions resulting in re-expression of CD2R, which is down-regulated during induction and maintenance of the anergic state *in vitro*. Therefore, for the induction of long lasting or permanent unresponsiveness, modulation of γ_c mediated pathway and clonal deletion of alloantigen-specific T cells via CTLA4 mediated apoptosis may have great potential as a therapeutic approach in GVHD, graft rejection and autoimmune diseases.

REFERENCES

1. Jenkins, M.K., and R.H. Schwartz. 1987. Antigen presentation by chemically modified splenocytes induces antigen-specific T cell unresponsiveness in vitro and in vivo. *J. Exp. Med.* 165:302.

2. Schwartz, R.H. 1992. Costimulation of T lymphocytes: the role of CD28, CTLA-4, and B7/BB1 in interleukin-2 production and immunotherapy. *Cell* 71:1065.

3. Gimmi, C.D., G.J. Freeman, J.G. Gribben, G. Gray, and L.M. Nadler. 1993. Human T-cell clonal anergy is induced by antigen presentation in the absence of B7 costimulation. *Proc. Natl. Acad. Sci. USA.* 90:6586.

4. Boussiotis, V.A., G.J. Freeman, G. Gray, J. Gribben, and L.M. Nadler. 1993. B7 but not ICAM-1 costimulation prevents the induction of human alloantigen specific tolerance. *J. Exp. Med.* 178:1753.

5. Harding, F.A., J.G. McArthur, J.A. Gross, D.H. Raulet, and J.P. Allison. 1992. CD28-mediated signalling co-stimulates murine T cells and prevents induction of anergy in T-cell clones. *Nature.* 356:607.

6. Lenschow, D.J., Y. Zeng, J.R. Thistlethwaite, A. Montag, W. Brady, M.G. Gibson, P.S. Linsley, and J.A. Bluestone. 1992. Long-term survival of xenogeneic pancreatic islet grafts induced by CTLA4Ig. *Science* 257:789.

7. Turka, L.A., P.S. Linsley, H. Lin, W. Brady, J.M. Leiden, R.Q. Wei, M.L. Gibson, X.G. Zheng, S. Myrdal, D. Gordon, T. Bailey, S.F. Bolling, and C.B. Thompson. 1992. T-cell activation by the CD28 ligand B7 is required for cardiac allograft rejection in vivo. *Proc. Natl. Acad. Sci. USA* 89:11102.

8. Tan, P., C. Anasetti, J.A. Hansen, J. Melrose, M. Brunvard, J. Bradshaw, J.A. Ledbetter, and P. Linsley. 1993. Induction of alloantigen-specific hyporesponsiveness in human T lymphocytes by blocking interaction of CD28 with its natural ligand B7/BB1. *J. Exp. Med.* 177:165.

9. Lin, H., S.F. Bolling, P. Linsley, R.Q. Wei, D. Gordon, C.A. Thompson, and L.A. Turka. 1993. Long-term acceptance of Major Histocompatibility Complex mismatched cardiac allografts induced by CTLA4Ig plus donor-specific transfusion. *J Exp Med* 178:1801.

10. Freedman, A.S., G.J. Freeman, J.C. Horowitz, J. Daley, and L.M. Nadler. 1987. B7, a B cell restricted antigen which identifies pre-activated B cells. *J. Immunol.* 137:3260.

11. Freeman, G.J., A.S. Freedman, J.M. Segil, G. Lee, J.F. Whitman, and L.M. Nadler. 1989. B7, a new member of the Ig superfamily with unique expression on activated and neoplastic B cells. *J. Immunol.* 143:2714.

12. Freeman, G.J., J.G. Gribben, V.A. Boussiotis, J.W. Ng, V. Restivo, L. Lombard, G.S. Gray, and L.M. Nadler. 1993. Cloning of B7-2: a CTLA4 counter-receptor that costimulates human T cell proliferation. *Science.* 262:909.

13. Azuma, M., H. Yssel, J.H. Phillips, H. Spits, and L.L. Lanier. 1993. Functional expression of B7/BB1 on activated T lymphocytes. *J. Exp. Med* 177:845.

14. Nozawa, Y., E. Wachi, K. Tominaga, M. Abe, and H. Wakasa. 1993. A novel monoclonal antibody (FUN-1) identifies an activation antigen in cells of the B-cell lineage and Reed-Sternberg cells. *J. Pathol.* 169:309.

15. Freedman, A.S., G.J. Freeman, K. Rhynhart, and L.M. Nadler. 1991. Selective induction of B7/BB-1 on interferon-γ stimulated monocytes: A potential mechanism for amplification of T cell activation. *Cell. Immunol.* 137:429.

16. Boussiotis, V.A., G.J. Freeman, J.G. Gribben, J. Daley, G. Gray, and L.M. Nadler. 1993. Activated human B lymphocytes express three CTLA4 binding counter-receptors which costimulate T cell activation. *Proc. Natl. Acad. Sci. USA.* 90:11059.

17. Hathcock, K.S., G. Laszlo, C. Pucillo, P. Linsley, and R.J. Hodes. 1994. Comparative analysis of B7-1 and B7-2 costimulatory ligands: expression and function. *J. Exp. Med.* 180:631.

18. Lenschow, D.J., A.I. Sperling, M.P. Cooke, G. Freeman, L. Rhee, D.C. Decker, G. Gray, L.M. Nadler, C.C. Goodnow, and J.A. Bluestone. 1994. Differential upregulation of the B7-1 and B7-2 costimulatory molecules following immunoglobulin receptor engagement by antigen. *J Immunol.* 153:1990.

19. Caux, C., B. Vanbervliet, C. Massactier, M. Azuma, K. Okumura, L. Lanier, and J. Banchereau. 1994. B70/B7-2 is identical to CD86 and is the major functional ligand for CD28 expressed on human dendritic cells. *J. Exp. Med.* 180:1841.

20. Hart, D.N., G.C. Starling, V.L. Calder, and N.S. Fernando. 1993. B7/BB-1 is a leucocyte differentiation antigen on human dendritic cells induced by activation. *Immunology (Oxford)* 79:616.

21. Larsen, C.P., S.C. Ritchie, R. Hendrix, P.S. Linsley, K.S. Hathcock, R.J. Hodes, R.P. Lowry, and T.C. Pearson. 1994. Regulation of immunostimulatory function and costimulatory molecule (B7-1 and B7-2) expression on murine dendritic cells. *J. Immunol.* 152:5208.

22. Lenschow, D.J., G.H.-T. Su, L.A. Zuckerman, N. Nabavi, C.L. Jellis, G.S. Gray, J. Miller, and J.A. Bluestone. 1993. Expression and functional significance of an additional ligand for CTLA-4. *Proc. Natl. Acad. Sci. USA.* 90:11054.

23. Linsley, P., and J. Ledbetter. 1993. The Role of the CD28 Receptor During T Cell Responses to Antigen. *Annu. Rev. Immunol.* 11:191.

24. Linsley, P.S., J.L. Greene, W. Brady, J. Bajorath, J.A. Ledbetter, and R. Peach. 1994. Human B7-1 (CD80) and B7-2 (CD86) bind with similar avidities but distinct kinetics to CD28 and CTLA-4 receptors. *Immunity* 1:793.

25. Linsley, P.S., J.L. Greene, P. Tan, J. Bradshaw, J.A. Ledbetter, C. Anasetti, and N.K. Damle. 1992. Coexpression and functional cooperation of CTLA-4 and CD28 on activated T lymphocytes. *J. Exp. Med.* 176:1595.

26. Freeman, G.J., D.B. Lombard, C.D. Gimmi, S.A. Brod, K. Lee, J.C. Laning, D.A. Hafler, M.E. Dorf, G. Gray, H. Reiser, C.H. June, C.B. Thompson, and L.M. Nadler. 1992. CTLA-4 and CD28 mRNAs are coexpressed in most activated T cells after activation: Expression of CTLA-4 and CD28 messenger RNA does not correlate with the pattern of lymphokine production. *J. Immunol* 149:3795.

27. Boussiotis, V.A., G.J. Freeman, J.D. Griffin, G.S. Gray, J.G. Gribben, and L.M. Nadler. 1994. CD2 is involved in maintenance and reversal of human alloantigen specific clonal anergy. *J. Exp. Med.* 180:1665.

28. Blazar, B.R., P.A. Taylor, P.S. Linsley, and P.A. Vallera. 1994. In vivo blockade of CD28/CTLA4: B7/BB1 interaction with CTLA4-Ig reduces lethal murine graft-versus-host disease across the major histocompatibility complex barrier in mice. *Blood* 83:3815.

29. Wallace, P., J. Johnson, J. MacMaster, K. Kennedy, P. Gladstone, and P. Linsley. 1994. CTLA4Ig treatment ameliorates the lethality of murine graft versus host disease across major histocompatibility complex barriers. *Transplantation* 58:602.

30. Thompson, C.B., T. Lindsten, J.A. Ledbetter, S.L. Kunkel, H.A. Young, S.G. Emerson, J.M. Leiden, and C.H. June. 1989. CD28 activation pathway regulates the production of multiple T-cell-derived lymphokines/cytokines. *Proc. Natl. Acad. Sci. USA* 86:1333.

31. Linsley, P.S., W. Brady, L. Grosmaire, A. Aruffo, N.K. Damle, and J.A. Ledbetter. 1991. Binding of the B cell activation antigen B7 to CD28 costimulates T cell proliferation and interleukin 2 mRNA accumulation. *J. Exp. Med.* 173:721.

32. Gimmi, C.D., G.J. Freeman, J.G. Gribben, K. Sugita, A.S. Freedman, C. Morimoto, and L.M. Nadler. 1991. B-cell surface antigen B7 provides a costimulatory signal that induces T cells to proliferate and secrete interleukin 2. *Proc. Natl. Acad. Sci. USA.* 88:6575.

33. Cerdan, C., Y. Martin, H. Brailly, M. Courcoul, S. Flavetta, R. Costello, C. Mawas, F. Birg, and D. Olive. 1991. IL-1 alpha is produced by T lymphocytes activated via the CD2 plus CD28 pathways. *J. Immunol* 146:560.

34. Cerdan, C., Y. Martin, M. Courcoul, C. Mawas, F. Birg, and D. Olive. 1995. CD28 costimulation up-regulates long-term IL-2Rβ expression in human T cells through combined transcriptional and post-transcriptional regulation. *J. Immunol* 154:1007.

35. Freeman, G.J., V.A. Boussiotis, A. Anumanthan, G.M. Bernstein, X.-Y. Ke, P.D. Rennert, G.S. Gray, J.G. Gribben, and L.M. Nadler. in press. B7-1 and B7-2 do not deliver identical costimulatory signals since B7-2 but not B7-1 preferentially costimulates the initial production of IL-4. *Immunity*

36. de Boer, M., A. Kasran, J. Kwekkeboom, H. Walter, P. Vandenberghe, and J.L. Ceuppens. 1993. Ligation of B7 with CD28/CTLA-4 on T cells results in CD40 ligand expression, interleukin-4 secretion and efficient help for antibody production by B cells. *Eur. J. Immunol.* 23:3120.

37. Walunas, T.L., D.J. Lenschow, C.Y. Bakker, P.S. Linsley, G.J. Freeman, M. Green, C.B. Thompson, and J.A. Bluestone. 1994. CTLA-4 can function as a negative regulator of T cell activation. *Immunity.* 1:405.

38. Gribben, J.G., G.J. Freeman, V.A. Boussiotis, P. Rennert, C.L. Jellis, E. Greenfield, M. Barber, V.A.J. Restivo, X.-Y. Ke, G.S. Gray, and L.M. Nadler. 1995. CTLA4 mediates antigen-specific apoptosis of human T cells. *Proc. Natl. Acad. Sci. USA* 92:811.

39. Linsley, P.S., P.M. Wallace, J. Johnson, M.G. Gibson, J.L. Greene, J.A. Ledbetter, C. Singh, and M.A. Tepper. 1992. Immunosuppression in vivo by a soluble form of the CTLA-4 T cell activation molecule. *Science* 257:792.

40. Turka, L.A., J.A. Ledbetter, K. Lee, C.H. June, and C.B. Thompson. 1990. CD28 is an inducible T cell surface antigen that transduces a proliferation signal in CD3+ mature thymocytes. *J. Immunol.* 144:1646.

41. Freeman, G.J., F. Boriello, R.J. Hodes, H. Reiser, J.G. Gribben, J.W. Ng, J. Kim, J.M. Goldberg, K. Hathcock, L. G., L.A. Lombard, S. Wang, G.S. Gray, L.M. Nadler, and A.H. Sharpe. 1993. Murine B7-2, an alternative CTLA4 counter-receptor that costimulates T cell proliferation and interleukin-2 production. *J. Exp. Med.* 178:2185.

42. Beverly, B., S.M. Kang, M.J. Lenardo, and R.H. Schwartz. 1992. Reversal of in vitro T cell clonal anergy by IL-2 stimulation. *Int. Immunol.* 4:661.

43. Essery, G., M. Feldmann, and J. Lamp. 1988. Interleukin-2 can prevent and reverse antigen-induced unresponsiveness in cloned human T lymphocytes. *Immunology* 64:413.

44. Schorle, H., T. Holtschke, T. Hunig, A. Schimpl, and I. Horak. 1991. Development and function of T cells in mice rendered interleukin-2 deficient by gene targeting. *Nature* 352:621.

45. Shahinian, A., K. Pfeffer, K.P. Lee, T.M. Kundig, K. Kishihara, A. Wakeham, K. Kawai, P.S. Ohashi, C.B. Thompson, and T.W.. Mak. 1993. Differential T Cell Costimulatory Requirements in CD28-Deficient Mice. *Science* 261:609.

46. Boussiotis, V.A., D.L. Barber, T. Nakarai, G.J. Freeman, J.G. Gribben, G.M. Bernstein, A.L. D'Andrea, J. Ritz, and L.M. Nadler. 1994. Prevention of T cell anergy by signaling through the γc chain of the IL-2 receptor. *Science.* 266:1039.

47. Russell, S.J. 1990. Lymphokine gene therapy for cancer. *Immunology Today* 11:196.

48. Kondo, M., T. Takeshita, N. Ishii, M. Nakamura, S. Watanabe, K. Arrai, and K. Sagamura. 1993. Sharing of the interleukin-2 (IL-2) receptor γ chain between receptors for IL-2 and IL-4. *Science.* 262:1874.

49. Noguchi, M., Y. Nakamura, S.M. Russel, S.F. Ziegler, M. Tsang, X. Cao, and W. Leonard. 1993. Interleukin-2 receptor γ chain: A functional component of interleukin-7 receptor. *Science.* 262:1877.

50. Ihle, J.N., B.A. Witthuhn, F.W. Quelle, K. Yamamoto, W.E. Thierfelder, B. Kreider, and O. Silvennoinen. 1994. Signaling by the cytokine receptor superfamily: JAKs and STATs. *TIBS* 19:222.

51. Kawamura, M., D.W. McVicar, J.A. Johnston, T.B. Blake, Y. Chen, B.K. Lal, A.R. Lloyd, D.J. Kalvin, J.E. Staples, J.R. Ortaldo, and J.J. O'Shea. 1994. Molecular cloning of L-JAK, a Janus family protein-tyrosine kinase expressed in natural killer cells and activated leukocytes. *Proc Natl Acad Sci* 91:6374.

52. Johnston, J.A., M. Kawamura, R.A. Kirken, Y. Chen, T.B. Blake, K. Shibuya, R.J. Ortaldo, D.W. McVicar, and J.J. O'Shea. 1994. Phosphorylation and activation of the Jak-3 Janus kinase in response to interleukin-2. *Nature* 370:151.

53. Whitthuhn, B.A., O. Silvennoinen, O. Miura, K.S. Lai, C. Cwik, E.T. Liu, and J.N. Ihle. 1994. Involvement of the Jak-3 Janus kinase in signalling by interleukins 2 and 4 in lymphoid and myeloid cells. *Nature* 370:153.

54. Reiser, H., G.J. Freeman, Z. Razi-Wolf, C.D. Gimmi, B. Benacerraf, and L.M. Nadler. 1992. Murine B7 antigen provides an efficient costimulatory signal for activation of murine T lymphocytes via the T-cell receptor/CD3 complex. *Proc. Natl. Acad. Sci. USA.* 89:271.

55. Damle, N.K., and L.V. Doyle. 1989. Stimulation via the CD3 and CD28 molecules induces responsiveness to IL-4 in CD4+ CD29+ CD45RO memory T lymphocytes. *J. Immunol.* 143:1761.

56. Russell, S.M., A.D. Keegan, N. Harada, Y. Nakamura, M. Noguchi, P. Leland, M.C. Friedman, A. Miyajima, R.K. Puri, W.E. Paul, and W.J. Leonard. 1993. Interleukin-2 receptor γ chain: A functional component of interleukin-4 receptor. *Science.* 262:1880.

57. Moingeon, P., H. Chang, P. Sayre, L. Clayton, A. Alcover, P. Gardner, and E. Reinherz. 1989. The structural biology of CD2. *Immunological Reviews* 111:

58. Meuer, S., R. Hussey, M. Fabbi, D. Fox, O. Acuto, K. Fitgerald, J. Hodgdon, J. Protentis, S. Schlossmann, and E. Reinherz. 1984. An alternative pathway of T-cell activation: A functional role for the 50kd T11 sheep erythrocyte receptor protein. *Cell* 36:897.

59. Koyasu, S., T. Lawton, D. Novick, M.A. Recny, R.F. Siliciano, B.P. Wallner, and E.L. Reinherz. 1990. Role of interaction of CD2 molecules with lymphocyte function-associated antigen 3 in T-cell recognition of nominal antigen. *Proc. Natl. Acad. Sci. USA* 87:2603.

60. Selvaraj, P., M.L. Plunkett, M. Dustin, M.E. Sanders, S. Shaw, and T.A. Springer. 1987. The T lymphocyte glycoprotein CD2 binds the cell surface ligand LFA-3. *Nature* 326:400.

61. Bierer, B., A. Peterson, J. Gorga, S. Herrmann, and S. Burakoff. 1988. Synergistic T cell activation via the physiologic ligands for CD2 and the T cell receptor. *J Exp Med* 168:1145.

62. Meuer, S.C. 1989. T2 cluster report: CD2. *Leucocyte Typing IV, White cell differantiation anitgens, Eds* 270.

DISCUSSION

R. Perlmutter

Maybe one point of clarification would help. The experiments with respect to CTLA4, they are all blocking experiments?

L. Nadler

That is correct.

R. Perlmutter

For some of the experiments there exists apparently two sorts of sites on CTLA4 and by blocking both B71 and B72 you reveal the other category of sites, which is an apoptotic inducing site, present on the antigen presenting cell; therefore B71, B72, when ordinarily present, can override that apoptotic induction. Is that right? Whether they override directly or they override through some secondary molecule is not apparent. But in any case, crosslink is required regardless of which ligand you are talking about.

L. Nadler

Absolutely. The antigen presenting cell is a T-cell. You have a T cell receptor, the MHC class 1 and class 2. If we deliver signals here, with no second signal, what one would argue is that unless there is some kind of pathway driven here, that depletes the substrate. That is the best thought I can have right now. Ron Short has published that the mechanism of anergy is the loss of AP1. But I do not think anyone has repeated that at this point, four years later.

R. Perlmutter

There are lots of published reports on the outcome of solitary stimulation of the antigen receptor, besides the AP1 depletion. For example, Helen Quill has a long series of papers on depletion of the primary tyrosine kinases that are involved in the signal transduction pathway and of course you are familiar with the tumor induced anergy systems in which that same thing is seen. What the biochemistry of that process is, I think is unclear but the biology of it is straightforward, that is you stimulate the antigen receptor and you get anergy just as you saw.

L. Nadler

The mechanism of anergy is still unknown and there is a big black box there that we do not know. But what CD28 does is basically to co-stimulate and make IL-2 on T cells. I would certainly think that in some subsets IL-4 is produced, IL-7 is probably not produced and produced by other cells. What happens then is that IL-2 binds to the IL-2 receptor and at that point, by crosslinking gamma is what we think, because IL-4 and IL-7 do the same. And only the gamma chain crosslink does that, phosphorylates through JAK3 we think and then prevents the induction of anergy. There is a signal required to prevent something falling out. Something is going to be depleted. That is the best way we can explain it. So in the early stages of T cell activation what we think is happening is that one signals through the IL-2 pathway, or potentially, because of the knockout mouse, the IL-2 knockout mouse, or the IL-2 receptor knockout mouse or the CD28 knockout mouse, through IL-4 and IL-7, delivering the same type of signaling through gamma chain and preventing the loss of this substrate. Those cells now are intact. They proliferate. They induce effector function. And there is a point in time that we know, because when one signals through CD28, CD28 is actually downregulated. CD28 induces the CTLA4 molecule on the cell surface. B71 and B72 go down. We think B73 hangs around a lot longer and we do not know what this other ligand is. But some other ligand is either constitutively expressed or induced and since we do not have an antibody or the gene, we do not know when it comes up but when the immune response is waning. If you have IL-2, if you are signaling through B71 and B72, you are always producing co-stimulatory molecules. Your ambient environment can prevent anergy. Your clonal expansion goes on. Then, later on, when the apoptotic ligand is there and you have the ability to signal, clones are deleted. We do not know how CTLA4 mediates apoptosis. Is it associating with one of the TNF receptor family members? Is it signaling through its own domain? But what we do think is that B71 and B72 binds to CTLA4 like a giant shield and either blocks the association of this apoptotic ligand or covers up the molecule or alternatively block the association of a chain that is going to mediate apoptosis. We think that there is a balance in a pathway between the early positive signals and the late deletion signals. They are all antigen specific and they allow a T cell to enter either a state of full immunity, a state of anergy, or alternatively, to be deleted. Now how this fits into the thymus. B73 is on the thymic epithelium and whether it is involved in antigen selection in the thymus is not clear right now. If you have a way of inducing anergy, the substrate will be deleted. If you can break anergy as I showed you will reinduce that substrate. It has to come back. And so we have a pathway in and out of anergy. And so we should be able to, with the right help, to take that apart. And so I hope that in the next couple of years we can grab onto the relevant substrates that are essential to anergy.

E. Mihich

You are probably aware of our work on some of the immunostimulating effects of Adriamycin. And as you know, at least in the mouse system, Adriamycin will very selectively induce the production of IL-2, the release of IL-2 in the spleen system and also, interestingly enough, will have a selective apoptotic effect on double positive cells in the thymus and not on other cells. Having evidence that in the mouse, Adriamycin can provide a second signal to T cells as well as macrophages. I wonder what Adriamycin would do in your system, whether it would affect the stimulation of CD28, for instance, in the absence of B7.

L. Nadler

I do not know the answer to your question. What one could argue is that after one gives Adriamycin, one disorders the pathway, one may have antigen presenting cells come out that are fully co-stimulatory, making the right relevant factors.

E. Mihich

The experiments were not in a chemotherapy mode. These were conditions of stimulation rather than killing in any possible way.

L. Nadler

It is an interesting idea.

J. Griffin

Lee, I was struck by the correlation between expression of the T11-3 epitope and whether a cell is anergised or not. Two questions about T11-3. Do you have any insights about what the nature of that epitope is? It is something that comes and goes on the CD2 molecule?

L. Nadler

Ellis Reinhurst has spent, as you know, a considerable amount of time trying to look at this crystallographically, synthesizing peptides, it is not entirely clear what defines the TII-3 epitope. The alternative pathway of T cell activation does not need a first signal to stimulate the pathway; you can just crosslink the CD2 molecule on the cell surface which drives a very strong signal. LFA3 is needed to function and CD2 is needed to function, along with the TII-3 epitope. Why one needs to reverse anergy via this pathway, one can conjecture.

But one can conjecture why one needs anergy as a mechanism in the periphery in the first place. The best shot I can take at it is that the T cell repertoire is limited and if one just simply clonally deleted rapidly, one would delete one's repertoire and if one received an antigen which could not be processed and put in the context of the MHC and co-stimulatory molecules that may not be picked up by dendritic cells but presented in an environment, one might delete the immune response.

J. Ihle

So when they looked at the knockout mice, for example fyn and lyn, can you look at anergy in these mice? Does the absence of scr kinases have an effect?

L. Nadler

No, there are no animal models in which anergy is compromised.

R. Perlmutter

There is no compromise of anergy in any of those systems. In a fyn gene disruption, the cells are slightly less effective. In the thymus they are normal. In the mouse, it is a molecule in search of a function. Even the IL-7 receptor disruption, which is not affecting the common Gamma chain does the same thing. You basically get no thymocytes and no B cells.

L. Nadler

This may or may not explain the thymus phenomena whatsoever. In the periphery this appears, at least in the human clonal model and a few cracks at normal cells, appears to be active. The gamma chain knockout human is a very sick child. I will tell you how sick they are: You can reconstitute them with 5 ml of bone marrow with no immuno suppression.

J. Ihle

Just a point of clarification. With the CTLA4 inducing apoptosis, does it come up late in T cell activation, as I understand?

L. Nadler

It is not that late. It actually comes up, although there have been conflicting results, it is up there by 24 hours.

J. Ihle

And when you have an activated T cell, will triggering of that still induce apoptosis?

L. Nadler

That is the only time that it will.

J. Ihle

So it has to be in the activated stage before it will induce apoptosis?

L. Nadler

You have to give it a second T cell receptor signal. Just crosslinking it by itself, across, you will not induce it.

M. Caligiuri

Is this functioning in a comparable manner to fas? Do you have any thoughts about that?

L. Nadler

Fas is not antigen specific. You can crosslink fas without an antigen specific signal. This could associate with fas.

M. Caligiuri

How is the triggering of CTLA4 antigen specific?

L. Nadler

We cannot kill unless it has had a T cell receptor signal.

M. Caligiuri

So it itself is not antigen specific but it reacts with an antigen specific cell to induce apoptosis.

I. Weissman

You just passed by something that is interesting. You said that in the X-linked SCID child that has the common Gamma chain defect for the IL-2, 4, 7 receptor, you can transplant them with allogeneic bone marrow. Is that T cell depleted by the way?

L. Nadler

No.

I. Weissman

But how about the other kids. Can you transplant them as easily?

L. Nadler

No.

I. Weissman

So that says there is something important about that receptor. Is the Gamma receptor the functional receptor on NK cells and do these kids have NK cells? That may explain it. When we do pure stem cell transplants in mice, if you use 100 stem cells as I said in a syngeneic situation you get full recovery of all the animals but in an allogeneic situation, you have to give up to 6000 stem cells to get reconstitution and when you are at the level of 6000 stem cells you can have some rare contaminating cells. The way you get around that is to get rid of residual CD4 cells with anti CD4 and NK cells with anti NK cell antibody. That implies that cells expressing the Gamma chain, and I would say residual T cells and active NK cells, are the barriers to allogeneic transplant of stem cells.

17

NOVEL STRATEGIES FOR CANCER TREATMENT THROUGH MODULATION OF PROIMMUNE AND PROINFLAMMATORY CYTOKINES

Robert K. Hurford, Jr., Evelyne Goillot, and Robert I. Tepper

Laboratory of Tumor Biology
Massachusetts General Hospital Cancer Center
Charlestown, Massachusetts 02129

INTRODUCTION

The remarkable progress in understanding the molecular basis of the immune system has prompted new strategies for the treatment of experimental and human cancers. In particular, the cloning and identification of the biological activity of cytokine genes has allowed for the activation and expansion of various hematolymphoid populations with cytotoxic potential. Extrapolation from results obtained with cytokines in biological systems in vitro and in vivo suggest that, in general, cytokines act potently to modulate immune and inflammatory responses locally, but rarely mediate systemic activation of immune and inflammatory cells. The localized action of cytokines is also predictable from their release by helper T lymphocyte (and other cell) populations upon exposure to an antigenic challenge which, in most physiological situations, is a localized event. In terms of activating cytotoxic cells for the killing of tumors, this property of cytokine action must be considered. This report describes strategies that have been developed to direct T cell dependent and independent cytotoxicity against established cancers in experimental animals. These strategies have relied upon the identification of specific proinflammatory and proimmune properties of various cytokines with respect to tumor cells. The use of viral vectors enabling the targeting of cytokine gene products to sites of tumor cell growth have led to gene therapy strategies which take advantage of the potent localized actions of these proteins.

RESULTS AND DISCUSSION

Enhancement of Antitumor Immunity by the Localized Expression of Cytokines

We and others have postulated that the weak or absent antitumor immune response in tumor-bearing hosts may be in part a result of insufficient stimulation of cytokine release by helper T lymphocytes following antigen presentation. The notion that many tumors are non-immunogenic because they lack recognizable antigens has been challenged by experimental data establishing the existence of antigenic proteins in a variety of spontaneously-arising human and rodent malignancies (Van der Bruggen et al., 1991; Van den Eynde et al., 1991; Brichard et al., 1993). If weak immunity associated with tumor antigens results from a failure to sufficiently stimulate helper T cells, deficient cytokine production would result and, as a consequence, the activation of specific lymphocytes (cytotoxic T lymphocytes; CTL) and non-specific tumoricidal effector cells (macrophages, natural killer cells, granulocytes) would not be adequately generated. Were this the case, providing cytokines locally at the site of a tumor cell vaccination might bypass a deficient helper T cell response and allow for the generation of a potent antitumor response. To test this notion, we developed an assay, the "tumor-cytokine transplantation assay" to determine whether specific cytokines, when expressed locally at the site of tumor cell challenges in mice, would be capable of eliciting a cytotoxic immune and/or inflammatory response against transplanted tumors (Tepper et al., 1989; Tepper et al., 1992). The result of this assay, performed in our laboratory and many others, has suggested that the secretion of many cytokines, including IL-2, IL-6, IL-7, IFN-γ, TNF-α, and GM-CSF could potentiate the generation of a tumor-specific cytotoxic T lymphocyte response against mouse tumors (for a review, see Tepper and Mule, 1994). With certain other cytokines, such as IL-4 and G-CSF, the production of a massive inflammatory reaction at tumor sites can result in complete tumor cell death, with little reactivity in normal surrounding tissues (Tepper et al., 1989; Tepper, 1992; Tepper et al., 1992; Colombo et al., 1991). These inflammatory responses, however, are not effective in stimulating CTL and therefore do not promote a "memory" response against subsequent tumor challenge.

These results provide a framework for novel approaches to boost tumor immunity through the use of cytokine-containing vaccines and for targeting tumors directly with cytokine agents that can stimulate tumoricidal inflammatory reactions. With respect to the former, a number of studies have demonstrated the potential of cytokines in the augmentation of T cell-mediated immunity. Early studies involving cytokine-expressing tumor cells concluded that the localized expression of IL-2 (Fearon et al., 1990; Gansbacher et al., 1990a) or IFN-γ (Watanabe et al., 1989; Gansbacher et al., 1990b) was capable of augmenting T cell-mediated tumor immunity, as defined by the lack of tumor formation upon rechallenge of mice with viable, non-cytokine producing cells. In some of these studies, specific lysis of tumor cell targets by splenocytes from these immune animals was also demonstrated by in vitro ^{51}Cr-release cytotoxic T lymphocyte assays. Despite these data, it has been difficult to ascribe a causal role of the expressed cytokine in the augmentation of immunity, due to the lack of certain critical controls. Specifically, immunization of a control population of mice with non-cytokine expressing tumor cells (inactivated, for example, by irradiation or mitomycin C) has not accompanied the study of cytokine-expressing tumor cells in these reports. As noted, certain spontaneous tumors in man, as well as carcinogen- or virally-induced tumors of animals possess some degree of inherent immunogenicity, a point which has been recently extended by the molecular cloning of tumor-specific antigens in human melanoma (e.g. the MAGE family of proteins)(Van der Bruggen et al., 1991) and in the P815 murine plasmacytoma (Van den Eynde *et al.*, 1991). Inherent immunogenicity in tumor cell lines mandates the comparison of the immune-enhancing activities of cytokine-transduced tumor cells to that of their non-transduced counterparts, before enhanced immunogenicity upon tumor cell vaccination can be ascribed to the action of a cytokine. For example, two reports describing the enhancement of tumor immunity by the localized expression of IL-2 by inoculated tumor cells suggested that the ability to elicit protective immunity to rechallenge with viable non-cytokine-secreting tumor cells was a result of the expression of IL-2 during the time of tumor cell vaccination, since parental tumor cells grew progressively in the host (Fearon *et al.*, 1990; Gansbacher *et al.*, 1990a). However, a subsequent report by Dranoff and his colleagues (Dranoff *et al.*, 1993) has clearly shown that irradiated, non-cytokine-expressing cells of the same tumor types (CT-26 colon carcinoma, CMS-5 fibrosarcoma) as used in the original reports were as potent as immunogens as their IL-2-secreting counterparts. The use of irradiated tumor cells to effectively prime mice against a number of tumors, including methylcholanthrene-induced sarcomas (Hellstrom *et al.*, 1979), has been long-established. The putative enhancement of immunogenicity of a renal cell carcinoma line (RENCA) by IL-4 (Golumbek *et al.*, 1991) must also be re-evaluated in light of findings of the inherent immunogenicity of this line when irradiated non-cytokine secreting cells are used for vaccination (Dranoff *et al.*, 1993). Rather than a specific effect of these cytokines on the stimulation of T cell immunity, then, these control experiments suggest that the action of the cytokine served mainly to effect the localized killing of the tumor cells *in situ*; whether the expression of the cytokine provided any stimulus for enhanced cytotoxic T cell immunity beyond that achievable with inactivated tumor cells alone has not been established. A similar argument may also be made from studies involving IFN-γ-expressing tumor cells (Watanabe *et al.*, 1989; Gansbacher *et al.*, 1990b).Careful titrations of immunogenic tumors, with regard to the numbers of cells used for vaccination and rechallenge, must be performed to accurately assess the ability of a given cytokine to augment tumor immunity beyond that which is attributable to the tumor cell itself.

Such a titration has been performed for GM-CSF with a number of immunogenic tumor cell lines of distinct histologic types (Dranoff *et al.*, 1993). The findings from this study suggest that GM-CSF can augment specific immunity for several tumor types; this augmentation is dependent upon both CD4+ and CD8+ T cells. The mechanism by which this cytokine acts on T cell subsets and whether the effect is direct or indirect remains to be defined. One idea is that GM-CSF acts by stimulating the proliferation and differentiation of dendritic cells, which are potent antigen-presenting cells and, as such, play a key role in the initiation of T cell-mediated immune responses by presenting immunogenic epitopes to CD4+ T cells in the context of Class II MHC molecules, which can in turn provide help for CTL generation. Titration experiments using IL-2 and IFN-γ should address the question as to whether or not these factors can augment tumor immunity when provided locally at the site of vaccination.

The potency of immune potentiation by locally-expressed cytokines remains an important issue. The titrations necessary to demonstrate augmentation of immunity, for example, by GM-CSF, suggests that the magnitude of cytokine enhancement of immunogenicity, beyond that which can be obtained by increasing the number of vaccinating tumor cells, may not be great. As such, the efficacy of such an approach for the treatment of established human cancers is questionable. Furthermore, in some studies, rechallenge with viable tumor was carried out for no longer than four weeks after the tumor cell vaccination with cytokine-expressing cells. Thus, the long-term benefit of single, or even repeated vaccinations, has not been established. In studies using IL-2-secreting tumor cells as an immunogen, resistance to rechallenge was either lost in most mice between two and four weeks after immunization (Fearon *et al.*, 1990) or was not observed (Karp *et al.*, 1993). The tumors under study either possessed some inherent immunogenicity (CT-26 colorectal carcinoma, in the study by Fearon *et al.*) (Dranoff *et al.*, 1993) or were non-immunogenic (MCA-102 fibrosarcoma, in the study by Karp *et al.*) These observations underscore the importance of studying murine (or other animal) models of tumor immunogenicity with regard to the potency of the elicited antitumor immune response, especially if these animal studies are to serve as models for the design of human therapeutic trials. Furthermore, the need for comparing immunomodulatory strategies using the same tumor models is imperative. Finally, the importance of identifying and characterizing the tumor antigens stimulating cell-mediated antitumor immunity against various tumors will be invaluable in accurately assessing the efficacy of immune enhancement strategies.

In addition to GM-CSF, other cytokines that have been demonstrated to augment T cell-mediated tumor immunity with appropriate controls for baseline immunogenicity of the tumor cell lines investigated include IL-6 and IL-7. The antitumor activity of IL-6 has been studied using locally injected tumor cells transfected or transduced to produce IL-6 (Mullen *et al.*, 1992; Porgador *et al.*, 1992) or by the systemic administration of this cytokine (Mulé *et al.*, 1990; Mulé *et al.*, 1992). One component of the tumor inhibitory actions of IL-6 appears to be T cell-dependent. In this regard, Mulé *et al.* (1992) have demonstrated the requirement for CD4$^+$ and CD8$^+$ cells for the regression of established 3 day pulmonary micrometastases of weakly immunogenic fibrosarcomas in mice after the systemic administration of human IL-6. Similarly, Porgador and colleagues (1992) demonstrated decreased lung metastatic potential of a weakly immunogenic clone (D122) of the Lewis lung carcinoma upon vaccination with inactivated, IL-6-transfected tumor cells followed by intravenous challenge with viable D122 cells. A concomitant increase in tumor-specific CTL was demonstrated. Mice vaccinated with inactivated, non-transfected D122 also demonstrated a significant suppression of metastases and prolongation of survival compared to unimmunized animals, albeit not to the degree observed with IL-6 transfected cells. In addition to augmenting T cell immunity, IL-6 may also enhance NK activity (Luger *et al.*, 1992). For some tumors, IL-6 may have direct antiproliferative effects (Morinaga *et al.*, 1989) while, for myelomas, it may serve as an autocrine growth factor (Bergui *et al.*, 1989; Shimizu *et al.*, 1989).

IL-7 also appears to promote T cell-dependent tumor rejection (Hock *et al.*, 1991; Jicha *et al.*, 1991; Aoki *et al.*, 1992; McBride *et al.*, 1992). For two distinct tumor types, a murine fibrosarcoma and glioma, IL-7 promoted the expansion of CD8$^+$ T cells (Jicha *et al.*, 1991; Aoki *et al.*, 1992); in the case of the latter, these cells were demonstrated to be the key effector cells, as systemic administration of anti-CD8 antibody but not anti-CD4 antibody blocked tumor rejection. A third report studying the rejection of IL-7 transfected J558L plasmacytoma cells provided evidence for the requirement of CD4$^+$ T cells and macrophages, but not CD8$^+$ cells, suggesting a possible delayed-type hypersensitivity type of tumor inhibition rather than a CTL-mediated mechanism (Hock *et al.*, 1991); however a later study by the same group using the same tumor suggested that CD8$^+$ cells were required for complete tumor rejection (Hock *et al.*, 1993). The different conclusions with regard to the mechanism of action of IL-7 in inhibiting the growth of these tumors may relate to inherent differences in the distinct tumor types (e.g. the nature of the antigen (s) recognized and their MHC restrictions) or possibly by the secretion of other cytokines/ factors by the tumor cells themselves. This latter possibility is important with regard to the study of a number of tumors, including gliomas and plasmacytomas, some of which are known to endogenously express IL-6, among other cytokines (Van Meir *et al.*, 1990). Other tumors have been demonstrated to produce cytokines; for example G-CSF and TGF-β by certain murine and human sarcomas (Pekarek *et al.*, 1993) and GM-CSF by a variety of tumors of diverse histologic type (Zinzar *et al.*, 1985; Fu *et al.*, 1991). Endogenous cytokine production is an important feature to consider in the evaluation of the host response to gene-modified tumor cells, in that additive, synergistic, or opposing actions on particular effector cells may be observed as a result of the interaction among these factors. For example, it has been demonstrated that the expression (by means of gene transfection) of TGF-β in a highly immunogenic fibrosarcoma line resulted in the failure to prime tumor-specific CTL generation *in vitro* and *in vivo* (Torre-Amione *et al.*, 1990).

Cytokines may also act to alter T cell immunity by actions other than direct induction of CD4$^+$ or CD8$^+$ T cell proliferation and/or differentiation. These include the upregulation of class I or class II MHC molecules by IFN-γ, which may in turn augment antigen-presentation by tumor cells themselves (in the case of class I MHC restricted antigens) or by *bona fide* antigen-presenting cells. Cell surface molecules on antigen-presenting cell populations necessary for the costimulation of T cells for activation, such as B7 (Gimmi *et al.*, 1991; Koulova *et al.*, 1991; Linsley *et al.*, 1991) may also be regulated by certain cytokines; for example, the induction of B7 expression on macrophages by IFN-γ (Freedman *et al.*, 1991) and its downregulation by the expression of IL-10 (Ding *et al.*, 1993). The expression of this costimulatory molecule on tumor cells by means of gene transfection has demonstrated that B7 can greatly augment the ability of immunogenic tumors to stimulate the generation of tumor-specific CTL (Chen *et al.*, 1992; Townsend and Allison, 1993).

Antitumor Activity of Proinflammatory Cytokines

While some cytokines may act directly or indirectly by stimulating the generation of CTL locally, others act by promoting cytotoxicity of non-lymphoid inflammatory cells. Two examples of potent non-T cell killing of tumor cells by cytokines *in vivo* include IL-4 and G-CSF. For IL-4, the importance of tumor-infiltrating eosinophils in mediating cytotoxicity has been demonstrated by the inability of IL-4 to induce tumor killing in mice depleted of this effector cell population (Figure 1)(Tepper *et al.*, 1992). Eosinophils may also play a role in the localized killing of tumor cells in response to other cytokines, including IL-2 (Huland and Huland, 1992), although other T cell-dependent and T cell-independent mechanisms of killing have been shown to participate in the action of this cytokine (Cavallo *et al.*, 1993). Neutrophils have been shown to play a key role in the antitumor activity of G-CSF (Colombo *et al.*, 1991). For several cytokines, macrophage infiltration also is present during tumor cell necrosis; these include IL-2 (Forni *et al.*, 1987), IL-4 (Tepper *et al.*, 1989), IL-7 (McBride *et al.*, 1992), G-CSF (Colombo *et al.*, 1991), IFN-γ (Giovarelli *et al.*, 1986) and TNF-α (Blankenstein *et al.*, 1991). The extent to which macrophages play a role in cytokine-mediated tumor killing has not been clearly established, except possibly for IL-7. For this cytokine, it has been shown in one tumor model that antibodies which block macrophage infiltration interfere with tumor rejection (Hock *et al.*, 1991). Whether the major effect of macrophages in this or other tumor-cytokine systems is predominantly a function of their antigen-presenting capabilities or their effector tumoricidal actions has not been established.

Initiation of inflammatory responses by cytokines, as demonstrated for IL-4, IL-7 and G-CSF, can result in tumor cell killing by inflammatory cells whose actions are localized to the site of cytokine production. One would not predict, however, that long-lasting immunity would result from the stimulation of these effector cells types. In one study of cytokine transfectants (which included IL-2, IL-4, IL-7, TNF-α and IFN-γ transfectants) of the J558L murine plasmacytoma, long-lasting immunity was observed and in the cases of all cytokines studied required the action of CD8$^+$ T lymphocytes (Hock *et al.*, 1993). It is not necessarily the case, however, that augmented T cell immunity was causally related to cytokine production. Localized killing by a number of modalities, including irradiation, bacterial adjuvants, and cytokines may generate non-viable tumor cells and an associated inflammatory cell influx (including antigen-presenting cells, such as macrophages) which may stimulate the recognition of an immunogenic tumor. Only careful comparison of individual cytokines within a defined tumor system will allow for the determination of which cytokines are the most efficacious in stimulating T cell immunity. These studies are best performed when specific tumor antigens and their MHC restrictions have been identified for the specific tumor type under investigation.

The ability of cytokines to stimulate a localized inflammatory response resulting in tumor cell death, however, has important implications for tumor immunotherapy independent of their capacity to induce systemic immunity. For example, the eosinophilic inflammatory response induced by high localized concentrations of IL-4 is sufficient, in some cases, to induce regression of established tumor masses (see next section) (Tepper *et al.*, 1992). It has been recently established that this antitumor activity mediated by IL-4 is not limited to subcutaneous tissue sites, but can be induced at a number of other tissue sites examined, including the peritoneal cavity (inhibition of intraperitoneal growth of a mammary adenocarcinoma line; Tepper, *unpublished results*), the brain (inhibition of growth of a human glioblastoma *in situ* in *nu/nu* mice; Yu *et al.*, 1993) and experimental sarcoma metastases established in the lungs and the liver (Goillot and Tepper, *unpublished results*). Since the localized antitumor inflammatory response elicited by IL-4 is not achievable by systemic delivery of the cytokine (Tepper *et al.*, 1992), approaches which can potentially deliver IL-4 to tumor deposits, such as direct instillation of cytokine or cytokine-producing cells (e.g. in the case of brain tumors) (Yu *et al.*, 1993) or the targeted delivery of cytokine using viral vectors are worthy of investigation, given the potency of the antitumor response when high localized concentrations of IL-4 are achieved. The induction of localized inflammation and concomitant tumor killing may also be achievable, as noted, by other cytokines and cytokine combinations.

With regard to the induction of inflammation, it has been established that certain cytokines may have a direct action on the endothelium. Thus, IL-1, TNF-α, IFN-γ, and IL-4 have all been shown to induce or upregulate the expression of various adhesion molecules on endothelial cells (Bevilacqua *et al.*, 1985; Thornhill *et al.*, 1991). Through the expression of adhesion molecules, the endothelium becomes "activated" to allow binding, and subsequent transmigration, of specific inflammatory cells into tissue sites. IL-4 has been shown to specifically induce the expression of the adhesion molecule, VCAM-1, on certain endothelial cell populations (Thornhill *et al.*, 1991). The cell surface ligand for this adhesion molecule is the leukocyte integrin VLA-4. While VLA-4 is expressed on both T and B lymphocytes and monocytes, it is of interest that eosinophils and basophils, but not neutrophils, possess this integrin (Walsh *et al.*, 1991). It has also been shown that VCAM-1 activation of human endothelium *in vitro* promotes adhesion of purified eosinophils, but not neutrophils (Schleimer *et al.*, 1992). These findings suggest that the characteristic eosinophilic inflammatory infiltrate induced by IL-4 may relate to the specific induction of VCAM-1 on the endothelial surface, although this hypothesis awaits experimental verification. Cytokines also can modulate the process of angiogenesis by their proliferative action on the endothelium, as demonstrated for TNF-α, IL-8, and TGF-β. Angiogenic cytokines

Figure 1. Depletion of eosinophils reverses the *in vivo* antitumor activity of IL-4. A. Appearance of subcutaneous tumor injection sites 18 days after the inoculation of 2×10^6 IL-4-producing plasmacytoma cells (clone I2B1) in an animal (right) receiving the rat anti-murine monoclonal antibody RB6-8C5, which depletes eosinophils, or in an animal (left) receiving isotype-matched control antibody. Note the restoration of tumor formation in the absence of host eosinophils. B. Histology of tumor inoculation site from animal in A (left), following staining with Giemsa (1000X). Note the intense eosinophil infiltration and paucity of tumor cells. C. Histology of tumor inoculation site from animal in A (right), following staining with Giemsa (1000X). Note the almost complete absence of the eosinophilic inflammatory cell infiltrate and the presence of viable tumor cells. [Figs. 1B,1C from Tepper *et al.*, *Science* 257:548-551 (1992).]

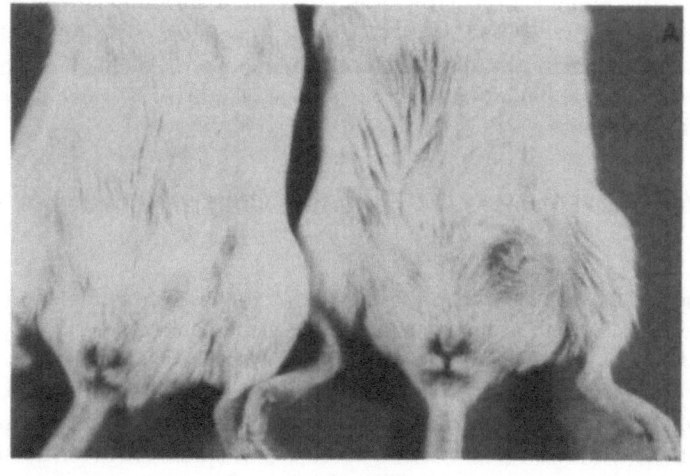

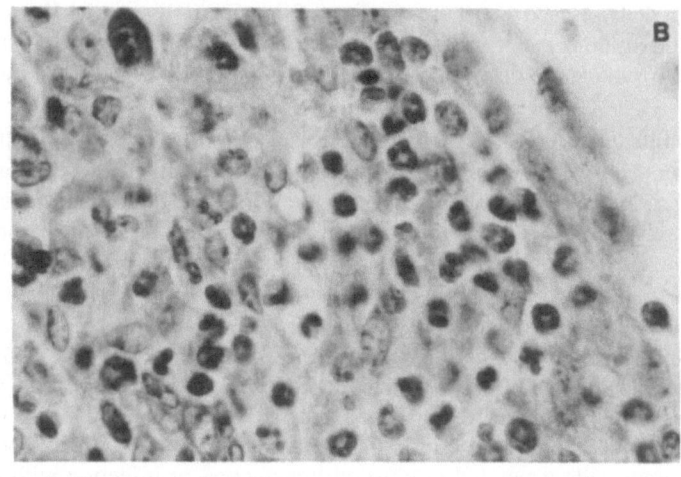

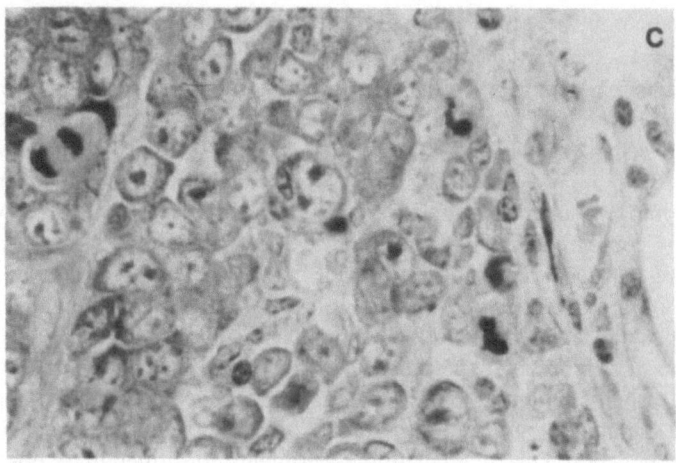

may be products of the tumor cells themselves or infiltrating T lymphocyte and macrophage populations (Leibovich *et al.,*1987; Koch *et al.,* 1992; Roberts *et al.,* 1986). In this regard, the inhibition of certain locally-expressed factors by antibodies or soluble forms of cytokine receptors may be capable of inhibiting the process of neovascularization which is central to malignant tumor growth.

The Use of Retroviral Vectors for Targeting Gene Expression to Tumor Metastases

Given the potency of locally-expressed cytokines in promoting immune and inflammatory responses which result in tumor cytotoxicity, the targeted delivery of cytokines to sites of metastatic tumor growth by the use of viral vectors has been an area of active interest in our laboratory. This approach has been utilized to direct gene transfer into focal tumor deposits, and thereby potentially applicable to the treatment of localized malignancy but not disseminated tumor (Culver *et al.,* 1992). In those experiments, retroviral producer cells were directly injected into tumor nodules. The selectivity for gene transfer into tumor cells and not surrounding tissue derives from the observation that retroviruses will only infect actively dividing cells. Based on these studies of in situ delivery, we have designed experiments to determine if the *intravascular* delivery of retroviral producer cells could be used to target gene products to disseminated metastases in an end organ (liver) supplied by the vasculature.

After establishment of three day micrometastases in the liver, CRE MFG LACZ retroviral producer cells or their supernatant, capable of transferring the LACZ marker gene, were delivered through the portal circulation. For two different tumor cell lines that were evaluated separately, producer cells injections resulted in the efficient transduction of metastases *in vivo* . However, injection of supernatant containing retroviral particles was not effective (Table 1). This *in vivo* targeting of metastases by producer cells was specific since the surrounding hepatocytes were not transduced.

For the sarcoma line, 4JKneo/mp, a single injection of CRE MFG LACZ producer cells 3 days after injection of tumor cells gave marked *in vivo* targeting results. 92% of all 4JKneo/mp metastases were targeted, and greater than 5% of tumor cells were transduced in 61% of the total metastases (Table 1 and Figure 2). Given the finding that a population of LACZ positive tumor cells generated *in vitro* by transfection or infection with the LACZ gene do not remain uniformly LACZ positive *in vivo* following the establishment of metastases, the measurement of X-gal-positive cells after *in vivo* lac Z gene targeting may be an underestimate of the percentage of tumor cells transduced.

To demonstrate that the observed LACZ positive cells resulted from *in vivo* gene transduction of tumor cells and not from marker gene expression in residual producer cells, the fate of these producer cells was established. Over several days, the immune rejection of CRE MFG LACZ cells is anticipated because they are NIH3T3 derivatives and therefore allogeneic in C57BL/6 mice. When CRE MFG LACZ producer cells were injected via the portal circulation into normal mice, the number of X-Gal positive cells in the liver was greatly reduced at 6 days compared to 1 day (not shown). The number of X-gal-positive cells obtained following *in vivo* transduction of tumor cells was 30-70 fold greater that that observed with retroviral producer cell injection alone, providing strong evidence that the observed X-gal-positive cells were not a result of a residual producer cell population. Moreover, in the *in vivo* targeting experiments described above (Fig. 2), LACZ transduced tumor cells are clearly distinguishable from residual producer cells by morphology and the presence of an inflammatory infiltrate around the producer cells that is consistent with immune mediated rejection. Dual cell labeling experiments confirmed these observations.

Table 1. *In vivo* targeting of tumor cells with CRE MFG LACZ retroviral producer cells

Tumor cell line day 0	Retroviral injection day 3	Number of animals examined at day 9	Number of metastases counted	Percentage of metastases targeted	Percentage of metastases with >5% targeting
4JK neo/mp	CRE MFG LACZ producer cells	13	668	92%	61%
4JK neo/mp	CRE MFG LACZ supernatant	2	49	0%	0%
M477	CRE MFG LACZ producer cells	6	1346	52%	28%
M477	CRE MFG LACZ supernatant	2	323	2%	0%

The CRE MFG LACZ producer cell line was generated by transfection of the retroviral LACZ expression vector, MFG LACZ, into the ecotropic packaging cell line CRE (Dranoff *et al.*, 1993). Tumor cells were injected at day 0 into the spleen—1×10^6 4JKneo/mp (sarcoma) cells or 4×10^5 M477 (breast carcinoma) cells in 0.5 ml media without serum. At day 3, splenic injections contained either 1×10^6 CRE MFG LACZ producer cells in 0.5 ml media without serum or 0.5 ml of 24 hr. CRE MFG LACZ supernatant with 20 mg/ml polybrene. Mice were sacrificed at day 9, and liver samples were fixed with 2% paraformaldehyde solution. Adjacent, 10 μm frozen sections were stained with hematoxylin and eosin or X-Gal.

The specificity of tumor transduction after retroviral producer cell injection was demonstrated by examining other dividing cell populations in the host. Following intraperitoneal injection with CRE MFG LACZ producer cells, X-Gal staining of frozen sections or cell preparations did not provide any evidence for infection of cells in the small intestine, large intestine, lung, or bone marrow. As a positive control, reconstitution experiments

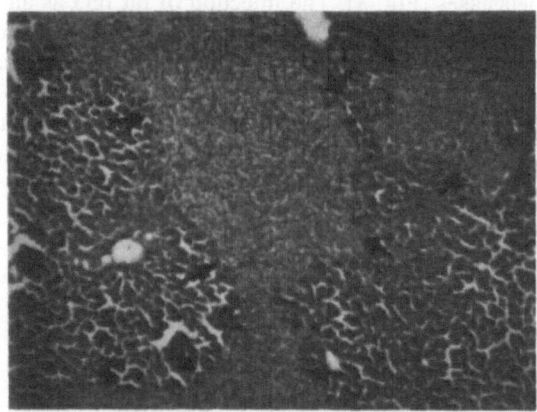

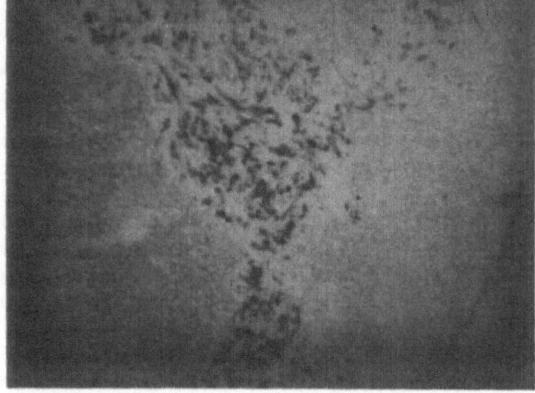

Figure 2. Intravascular Delivery of Retroviral Producer Cells Specifically Targets Hepatic Metastases. Hematoxylin & eosin (H&E; top) and X-gal (bottom)-stained sections of experimentally-induced hepatic metastasis of the murine fibrosarcoma line 4JK neo/mp. Three days following intrasplenic injection of tumor cells, 1×10^6 CRE retroviral packaging cells expressing the retroviral vector, MFG LACZ, were injected by the same route. Animals were sacrificed six days later, liver tissue was harvested, frozen in OCT compound and sectioned. Multiple arrows outline the tumor mass from the hepatic parenchyma on the H&E section.

detected the presence of the LACZ gene by PCR down to a 1:200,000 dilution of LAC Z
DNA into wild-type genomic DNA.

As previously noted, experiments have documented that the expression of either
murine IL-4 or murine IL-2 elicits a host-mediated response which is cytotoxic for a wide
variety of tumors. The observed anti-tumor effects are non-cell autonomous and depend
solely on local concentration of the cytokine (Tepper *et al.,* 1989; Fearon *et al.,* 1990;
Gansbacher *et al.,* 1990a). Thus, cytokine expression should cause the regression of established metastases after this expression is achieved in the local tumor environment. From
these previous results and in vitro infection experiments, we estimated that 5% or higher
targeting of metastases with either IL-4 or IL-2 would produce sufficient cytokine to cause
a greater than 90% reduction in tumor burden. As shown in Table 1, this targeting frequency
is achievable by a single retroviral producer cell injection for a significant proportion of
established micrometastases (e.g. 61% of 4JKneo/mp metastases).

The observed tumoricidal effects upon retroviral targeting correspond well to these
predictions. The intraportal delivery of retroviral producer cells with a vector encoding IL-4
or IL-2 (CRE MFG IL-4 and CRE MFG IL-2 respectively) caused a marked reduction in the
total number of metastases (Figure 3) when compared to animals that received control
retroviral producer cells (CRE MFG Tac). More than half of the residual metastases in
animals injected with CRE MFG IL-4 producer cells contained an extensive inflammatory
infiltration characteristic of IL-4 expression. The number of tumor cells in these infiltrated
metastases was greatly reduced due to IL-4 mediated regression. A neutralizing monoclonal
antibody against IL-4, previously shown to inhibit the anti-tumor effect of IL-4 *in vivo*
(Tepper *et al.,* 1989; Tepper *et al.,* 1992), reversed the effect of IL-4 gene targeting to
metastases (Fig. 3). The injection of rat IgG as an antibody control did not hinder tumor
reduction by CRE MFG IL-4 producer cells (not shown).

Since CRE MFG IL-4 and CRE MFG IL-2 cells directly secrete IL-4 or IL-2, it was
necessary to establish that cytokine gene transfer and not local cytokine expression by
producer cells was responsible for the observed reduction in metastases. Because CRE cells
are NIH3T3 derivatives, we isolated NIH3T3 clones infected in vitro with MFG IL-4 or

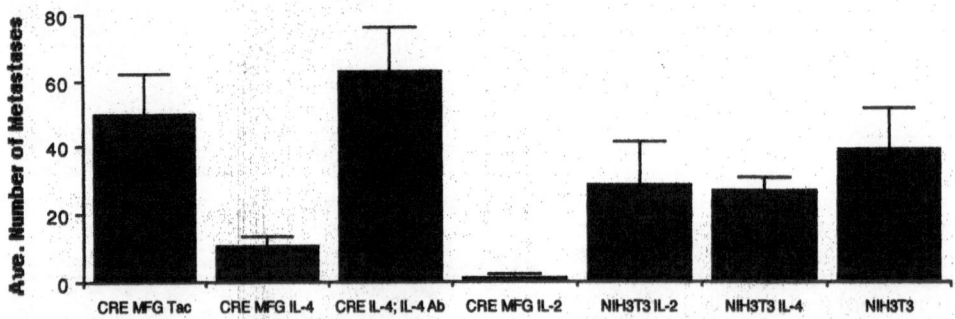

Figure 3. Reduction in number of metastases by retroviral producer cell targeting with IL-4 or IL-2. In this
series of experiments, the tumor cell population used to form metastases was a mix of six clones marked by
in vitro infection with MFG LACZ. Thus, metastases targeted with cytokines were identified both by
morphological characteristics and X-Gal staining of tumor cells in adjacent sections cut through the center of
all 5 liver lobes. A. Column graph that illustrates the average number of liver metastases in different group of
animals (3-6 mice per group) with standard error included. At day 0, 2X10^6 4JKneo/LACZ cells were injected
into the spleen, and at day 3, 1x10^6 of 6 different cell lines were then injected. Column 1 shows injections with
CRE MFG Tac as control retroviral producer cells. Tac is the low affinity human IL-2 receptor and does not
bind murine IL-2. In column 3, animals were given two injections of a MAb against IL-4, which has been
previously shown to block the anti-tumor effect of IL-4 *in vivo* (Tepper *et al.,* 1989).

Figure 4. Reduction of liver tumor mass at 21 days by retroviral producer cell targeting. Mice were injected with 4×10^5 4JKneo/mp cells at day 0 and 1×10^6 retroviral producer cells at day 3 (4-6 animals per group). The mice were sacrificed at day 21 and the entire liver, including tumor mass that grew in the liver, was resected and weighed. The control animals are littermates that were not injected with either tumor cells or retroviral producer cells.

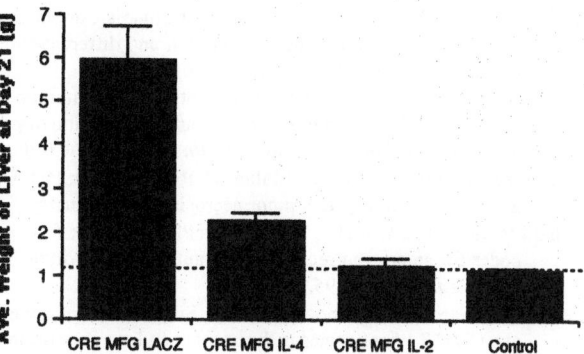

MFG IL-2. These clones express comparable levels of IL-4 or IL-2 as the producer cells themselves, but they do not make retroviral particles since they lack the packaging functions of CRE. Neither of these two cytokine expressing clones significantly reduced the number of metastases (Fig. 3), confirming the importance of *in vivo* gene transfer for the observed anti-metastatic effects. This requirement for *in vivo* transduction probably involves amplification of cytokine production in the tumor microenvironment after retroviral infection compared to the level achieved by a single injection of cytokine expressing cells.

Targeting metastases with cytokine genes by delivery of retroviral producer cells through the portal circulation also caused significant long term reduction in tumor burden. Three days after the establishment of micrometastases, mice were injected with CRE MFG IL-2, CRE MFG IL-4, or CRE MFG LACZ producer cells. These mice were sacrificed after three weeks when evidence of tumor growth was apparent in the control animals that received CRE MFG LACZ cells. The results (Figure 4) demonstrate a marked reduction in tumor mass by single injection of CRE MFG IL-2 or CRE MFG IL-4.

CONCLUSION

The ability to redirect the immune response to target invasive tumors by locally expressing cytokine gene products provides a novel approach for the treatment of malignant disorders. In experimental systems, the expression of cytokines locally *in vivo* has provided much information as to their mechanism of action in terms of the specific hematolymphoid populations they mobilize and activate. The results of these studies has suggested that many cytokines can trigger the cytotoxicity of T lymphocytes and/or non-lymphoid cellular effectors (e.g. granulocyte and macrophage populations) to kill tumor targets in the vicinity of cytokine release. These potent actions are typically not observed when cytokines are administered systemically, as localized secretion recapitulate more precisely the normal release of these gene products upon antigen induction of an immune or inflammatory response. These observations, coupled with the development of gene targeting technologies to direct the expression of genes *in vivo* at the site of tumor deposits, should enable the design of biotherapeutic strategies for metastatic cancer.

REFERENCES

Aoki T, Tashiro K, Miyatake S, Kinash T, Nakano T, Oda Y, Kikuchi H, Honjo T (1992). Expression of murine interleukin 7 in a murine glioma cell line results in reduced tumorigenicity in vivo. *Proc. Natl. Acad. Sci. USA* 89:3850-3854.

Bergui L, Schena M, Gaidano G, Riva M, Caligaris-Cappio F (1989). Interleukin 3 and interleukin 6 synergisticaly promote the proliferation and differentiation of malignant plasma cell precursors in multiple myeloma. *J. Exp. Med.* 170:613-618.

Bevilacqua M P, Pober J S, Wheeler M E, Cotran RS, Gimbrone MJ (1985). Interleukin 1 acts on cultured human vascular endothelium to increase the adhesion of polymorphonuclear leukocytes, monocytes, and related leukocyte cell lines. *J Clin Invest* 76:2003-2009.

Blankenstein T, Qin Z, Uberla K, Muller W, Rosen H, Vok HD, Diamanstein T (1991). Tumor suppression after tumor cell-targeted tumor necrosis α gene transfer. *J. Exp. Med.* 173:1047-1052.

Brichard V, Van Pel A, Wolfel T, Wolfel C, DePlaen E, Lethe B, Coulie P, Boon T (1993). The tyrosinase gene codes for an antigen recognized by autologous cytolytic T lymphocytes on HLA-A2 melanomas. *J. Exp. Med.* 178:489-495.

Cavallo F, Giovarelli M, Gulino A, Vacca A, Stopacciaro A, Modesti A, Forni G (1992). Role of neutrophils and CD4[+] T lymphocytes in the primary and memory response to nonimmunogenic murine mammary adenocarcinoma made immunogenic by IL-2 gene transfection. *J. Immunol.* 149:3627-3635.

Chen L, Ashe S, Brady WA, Hellström I, Hellström KE, Ledbetter JA, McGowan P, Linsley PS (1992). Costimulation of antitumor immunity by the B7 counterreceptor for the T lymphocyte molecules CD28 and CTLA-4. *Cell* 71:1093-1102.

Colombo MP, Ferrari G, Stoppacciaro A, Parenza M, Rodolfo M, Mavilio F, Parmiani G (1991) Granulocyte colony-stimulating factor gene suppresses tumorigenicity of a murine adenocarcinoma in vivo. *J.Exp. Med.* 173:889-894.

Culver K, Ram Z, Wallbridge S, Ishii H, Oldfield EH, Blaese RM (1992). In vivo gene transfer with retroviral vector-producer cells for treatment of experimental brain tumors. *Science* 256:1550-1552.

Ding L, Linsley PS, Huang L, Germain RN, Shevach EM (1993) IL-10 inhibits macrophage costimulatory activity by selectively inhibiting the up-regulation of B7 expression. *J. Immunol.* 151:1224-1234.

Dranoff G, Jaffee E, Lazenby A, Golumbek P, Levitsky H, Brose K, Jackson V, Hamada H, Pardoll D, Mulligan RC (1993). Vaccination with irradiated tumor cells engineered to secrete murine granulocyte-macrophage colony-stimulating factor stimulates potent, specific, and long-lasting anti-tumor immunity. *Proc. Natl. Acad. Sci. USA* 90:3539-3543.

Fearon ER, Pardoll DM, Itaya T, Golumbek P, Levitsky HI, Simons JW, Karasuyama H, Vogelstein B, Frost P (1990). Interleukin-2 production by tumor cells bypasses T helper function in the generation of an antitumor response. *Cell* 60:397-403.

Forni G, Giovarelli M, Santoni A, Modesti A, Forni M (1987). Interleukin-2 activated tumor inhibition *in vivo* depends on the systemic involvement of host immunoreactivity. *J. Immunol.* 138:4033-4041.

Freedman AS, Freeman GJ, Rhynhart K, Nadler LM (1991). Selective induction of B7/BB1 on interferon-γ stimulated moncytes: a potential mechanism for amplification of T cell activation through the CD28 pathway. *Cell. Immunol.* 137:429-437.

Fu YX, Watson GA, Kasahara M, Lopez DM (1991). The role of tumor-derived cytokines on the immune system of mice bearing a mammary adenocarcinoma. I. Induction of regulatory macrophages in normal mice by the in vivo administration of rGM-CSF. *J. Immunol.* 146:783-789.

Gansbacher B, Zier K, Daniels B, Cronin K, Bannerji R, Gilboa E (1990a). Interleukin 2 gene transfer into tumor cells abrogates tumorigenicity and induces protective immunity. *J.Exp. Med.* 172:1217-1224.

Gansbacher B, Bannerji R, Daniels B, Zier K, Cronin K, Gilboa E (1990b). Retroviral vector-mediated gamma-interferon gene transfer into tumor cells generates potent and long lasting antitumor immunity. *Cancer Res.* 50:7820-7825.

Gimmi CD, Freeman GJ, Gribben JG, Sugita K, Freedman AS, Morimoto C, Nadler LM (1991). B7 provides a costimulatory signal which induces T cells to proliferate and secrete interleukin-2. *Proc. Natl. Acad Sci. USA* 88:6575-6579.

Giovarelli M, Cofano F, Vecchi A, Forni M, Landolfo S, Forni G (1986). Interferon-activated tumor inhibition *in vivo*. *Int. J. Cancer* 37:141-147.

Golumbek P, Lazenby A, Levitsky HI, Jaffee LM, Karasuyama H, Baker M, Pardoll DM (1991). Treatment of established renal cancer by tumor cells engineered to secrete interleukin-4. *Science* 254:713-716.

Hellstrom I, Hellstrom KE, Bernstein ID (1979). Tumor enhancing suppressor activator T cells in spleens and thymuses of tumor immune mice. *Proc. Natl. Acad. Sci. USA* 76:5294-5298 .

Hock H, Dorsch M, Diamantstein T, Blankenstein T (1991). Interleukin 7 induces CD4[+] T cell-dependent tumor rejection. *J.Exp. Med.* 174:1291-1298.

Hock H, Dorsch M, Kunzendorf U, Qin Z, Diamantstein T, Blankenstein T (1993). Mechanisms of rejection induced by tumor cell-targeted gene transfer of interleukin 2, interleukin 4, interleukin 7 tumor necrosis factor, or interferon γ. *Proc. Natl. Acad Sci. USA* 90:2774-2778.

Huland E and Huland H (1992). Tumor-associated eosinophilia in interleukin-2-treated patients: evidence of toxic eosinophil degranulation on bladder cancer cells. *J. Cancer Res. Clin. Oncol.* 118:463-467.

Jicha DL, Mulé JJ, Rosenberg SA (1991). Interleukin 7 generates antitumor cytotoxic T lymphocytes against murine sarcomas with efficacy in cellular adoptive immunotherapy. *J. Exp. Med.* 174:1511-1515.

Karp SE, Farber A, Salo JC, Hwu P, Jaffe G, Asher A, Shiloni E, Restifo, NP, Mulé JJ, Rosenberg SA (1993). Cytokine secretion by genetically modified nonimmunogenic murine fibrosarcoma. Tumor inhibition by IL-2 but not by tumor necrosis factor. *J. Immunol.*150:896-908.

Koch AE, Polverini PJ, Kunkel SL, Harlow LA, DiPietro LA, Elner VM, Elner SG, Strieter RM (1992). Interleukin-8 as a macrophage-derived mediator of angiogenesis. *Science* 258:1798-1801.

Koulova L, Clark EA, Shu G, Dupont B (1991). The CD28 ligand B7/BB1 provides a costimulatory signal for coactivation of CD4+ T cells. *J. Exp. Med.* 173:759-762.

Leibovich SJ, Polverini PJ, Shepard HM, Wiseman DM, Shively V, Neuseir N (1987). Macrophage-induced angiogenesis is mediated by tumour necrois factor-alpha. *Nature 329-630-632.*

Linsley PS, Brady W, Grosmaire L, Aruffo A, Damle NK, Ledbetter JA (1991). Binding of the B cell activation antigen B7 to CD28 constimulates T cell proliferation and interleukin 2 mRNA accumulation. *J. Exp. Med.* 173:721-730.

Luger TA, Schwarz T, Krutmann J, Kirnbauer R, Neuner P, Kock A, Urbanski A, Borth W, Schauer E (1989). Interleukin-6 is produced by epidermal cells and plays an important role in the activation of human T-lymphocytes and natural killer cells. *Ann. N.Y. Acad. Sci.* 557:405-414.

McBride WH, Thacker JD, Comora S, Economou JS, Kelley D, Hogge D, Dubinett SM, Dougherty GJ (1992). Genetic modification of a murine fibrosarcoma to produce interleukin 7 stimulates host cell infiltration and tumor immunity. *Cancer Res.* 52:3931-3937.

Morinaga Y, Suzuki H, Takatsuki F, Akiyama Y, Taniyama T, Matsushima K, Onozaki K (1989). Contribution of IL-6 to the antiproliferative effect of IL-1 and tumor necrosis factor on tumor cell lines. *J. Immunol.* 143:3538-3542.

Mulé JJ, McIntosh JK, Jablons DM, Rosenberg SA (1990). Antitumor activity of recombinant interleukin 6 in mice. *J. Exp. Med.* 171:629-636.

Mulé JJ, Custer MC, Travis WD, Rosenberg SA (1992). Cellular mechanisms of the antitumor activity of recombinant IL-6 in mice. *J. Immunol.* 148:2622-2629.

Mulé JJ, Custer MC, Averbook B, Weber J, Goeddel D, Rosenberg SA, Schall TJ (1994). RANTES secretion by gene-modified tumor cells results in loss of tumorigenicity *in vivo*: Role of immune cell subpopulations. (submitted)

Mullen CA, Coale M, Levy AT, Stetler-Stevenson WG, Liotta LA, Brandt S, Blaese RM (1992). Fibrosarcoma cells transduced with the IL-6 gene exhibit reduced tumorigenicity, increased immunogenicity and decreased metastatic potential. *Cancer Res.* 52:6020-6024.

Pekarek LA, Weichselbaum RR, Beckett MA, Nachman J, Schreiber H (1993). Footprinting of individual tumors and their variants by constitutive cytokine expression patterns. *Cancer Res.* 53:1978-1981.

Porgador A, Tzehoval E, Katz A, Vadai E, Revel M, Feldman M, Eisenbach L (1992). Interleukin 6 gene transfection into Lewis lung carcinoma tumor cells suppresses the malignant phenotype and confers immunotherapeutic competence against parental metastic cells. *Cancer Res.* 52:3679-3686.

Roberts AB, Sporn MB, Assoian RK, Smith JM, Roche NS, Wakefield LM, Heine UI, Liotta LA, Falanga V, Kehrl JH (1986). Transforming growth factor type beta: rapid induction of fibrosis and angiogenesis in vivo and stimulation of collagen formation in vitro. *Proc. Natl. Acad. Sci. USA* 83:4167-4171.

Schleimer RP, Sterbinsky SA, Kaiser J, Bickel CA, Klunk DA, Tomioka K, Newman W, Luscinskas FW, Gimbrone MA, McIntyre BW, Bochner BS (1992). IL-4 induces adherence of human eosinophils and basophils but not neutrophils to endothelium. *J. Immunol.* 148:1086-1092.

Shimizu S, Yoshioka R, Hirose Y, Sugai S, Tachibana J, Konda S (1989). Establishment of two interleukin 6 (B cell stimulatory factor 2/interferon beta 2) -dependent human bone marrow-derived myeloma cell lines. *J. Exp. Med.* 169:339-344.

Tepper RI, Pattengale PK, Leder P (1989). Murine interleukin-4 displays potent antitumor activity in vivo. *Cell* 57:503-512.

Tepper RI (1992). The tumor-cytokine transplantation assay and the antitumor activity of interleukin-4. *Bone Marrow Transpl.* 9 (Suppl. 1):177-181.

Tepper RI, Coffman RL, Leder P (1992). An eosinophil-dependent mechanism for the antitumor effect of IL-4. *Science* 257:548-551.

Tepper RI and Mule JJ (1994). Experimental and clinical studies of cytokine gene-modified tumor cells. *Human Gene Therapy* 5:153-164.

Thornhill MH, Wellicome SM, Mahiouz DL, Lanchbury JSS, Kyan-Aung U, Haskard DO (1991). Tumor necrosis factor combines with IL-4 or IFN-γ to selectively enhance endothelial cell adhesiveness for

T cells. The contribution of vascular cell adhesion molecule-1-dependent and -independent binding mechanisms. *J. Immunol.* 146:592-98.

Torre-Amione G, Beauchamp RD, Koeppen H, Park BH, Schreiber H, Moses HL, Rowley DA (1990). A highly immunogenic tumor transfected with a murine transforming growth factor type β1 cDNA escapes immune surveillance. *Proc. Natl. Acad. Sci. USA* 87:1486-1490.

Townsend SE, Allison JP (1993). Tumor rejection after direct costiumlation of CD8+T cells by B7-transfected melanoma cells. *Science* 259:368-370.

Van der Bruggen P, Traversari C, Chomez P, Lurquin C, DePlaen E, Van den Eynde B, Knuth A, Boon T (1991). A gene encoding an antigen recognized by cytolytic T lymphocytes on a human melanoma. *Science* 254:1643-1647.

Van den Eynde B, Lethe B, Van Pel A, De Plaen E, Boon T (1991). The gene coding for a major tumor rejection antigen of tumor P815 is identical to the normal gene of syngeneic DBA/2 mice. *J Exp Med* 173:1373-1384.

Van Meir E, Sawamura Y (1990). Sawamura Y, Diserens AC, Hamou MF, de Tribolet N (1990). Human glioblastoma cells release interleukin 6 in vivo and in vitro. *Cancer Res.* 50:6683-6688.

Walsh GM, Mermod JJ, Hartnell A, Kay AB, Wardlaw AJ (1991). Human eosinophil, but not neutrophil, adherence to IL-1-stimulated human umbilical vascular endothelial cells is $\alpha_4\beta_1$ (very late antigen-4) dependent. *J. Immunol.* 146:3419-23.

Watanabe Y, Kuribayashi K, Miyatake S, Nishihara K, Nakayama E, Taniyama T, Sakata T (1989). Exogenous expression of mouse interferon γ cDNA in mouse neuroblastoma C1300 cells results in reduced tumorigenicity by augmented anti-tumor immunity. *Proc. Natl. Acad. Sci. USA.* 86:9456-9460.

Yu JS, Wei MX, Chiocca EA, Martuza RL, Tepper RI (1993). Treatment of glioma by engineered interleukin-4 secreting cells. *Cancer Res.* 53:3125-3128.

Zinzar SN, Svet-Moldavsky GJ, Fogh J, Mann PE, Arlin Z, Iliescu K, Holland JF (1985). Elaboration of granulocyte-macrophage colon-stimulating factor by human tumor cell lines and normal urothelium. *Exp. Hematol.* 13:574-580.

DISCUSSION

M. Caligiuri

Why do you think it is that systemic delivery of low doses of cytokines do not mimic the effects you see with the local transfection?

R. Tepper

Well, I oversimplified the IL-4 story a bit. Low doses of IL-4 delivered into the lymphatics of tumors, as shown by Guido Forni from Torino, can actually stimulate, in certain tumors, a T cell response. What is known about systemic delivery of many cytokines is that there are circulating soluble forms of cytokine receptors that are encoded by normally expressed mRNAs that result in binding of cytokine activity. At least one hypothesis is that these molecules may actually inactivate cytokines that enter the circulation following local expression . Other investigators postulate that these soluble receptors could be a transport mechanism for the cytokine but, by and large, if you look at systemically-delivered IL-2 or gamma interferon or TNF in this type of tumor system, you rarely get the response you get when you deliver the cytokine locally. I think localized delivery recapitulate to a much greater degree the normal process of cytokine release following antigen stimulation.

A. D'Andrea

Have you ever taken the same mouse and injected it in one location with the tumor cells expressing say IL-4 but in another location with tumor cells that were mock transfected and seen regression of both tumors?

R. Tepper

When you do that experiment you observe tumor growth on the non cytokine injected side, in the case of IL-4. Again, this is the case with most tumors. However, there may be a systemic, T-cell mediated response with certain immunogenic tumors. In the case of other cytokines, again assuming that you have an immunogenic tumor and you immunize the animals in the naive state, you can certainly show regression and therefore the presence of circulating T cells that are capable of inhibiting distant lesions. That has been shown for IL-2, for Gamma interferon and for GM-CSF. Interestingly, at the site of the GMCSF expression, the tumor actually grows but if you remove the tumor and then re-challenge the animal with the same tumor after a period of up to several weeks, you do not get tumor take. Regression of micrometasteses that are established for several days may also be observed. The mechanism by which GMCSF acts is not clear. It may be by the stimulation of dendritic cells. Almost all these tumor immunization experiments have been done in naive animals who are then subsequently challenged with viable tumor cells. The whole concept of presentation of antigen in the absence of co-stimulators may not apply in those situations where you are providing cytokines that can, in certain cases, have some co-stimulatory functions or induce the expression of co-stimulators on antigen presenting cells. These experimental conditions may not apply directly to human vaccination trials in patients who already have established tumors. In that regard targeting inflammatory responses by local-ized cytokine deliver is more attractive in that you do not obviously require a T cell response.

A. D'Andrea

Have you identified any colon cancer lines which are resistant to fas induced apoptosis but which do not overexpress bcl-2? It looked like there was a pretty perfect correlation there.

R. Tepper

No, there is not a perfect correlation. Other bcl-2 related molecules, such as bcl-x, may be substituting for bcl-2. These are some of the studies that we are doing now. I can tell you that there are certain oncogenes that can also interrupt the ability to induce apoptosis through fas in these tumors. Another interesting point is, at least in the case of several human tumors that we have looked at, the induction of apoptosis by fas does not require p53. We have looked at several tumor lines which are p53 null by a functional assay. So at least in some cases, fas mediated apoptosis does not depend upon p53 expression.

J. Adams

Since many would argue that the real clinical problem in cancer is micrometastases, I cannot quite picture how you would effectively use this protocol. In most cases one can only infect a fraction of the tumor cells. Since the metastases will be clonal, presumably many of them will not derive from infected cells, and therefore will escape this treatment.

R. Tepper

Well it is a non cell autonomous effect so all tumor cells need not be targeted. There are two potential problems, however. One problem is widely disseminated metastases and I do not think this is a therapy for that condition. In terms of a potential therapy, one application would be tumors that are predominantly or exclusively metastasizing to a single organ, such as colon cancer metastasizing to the liver, which has a vasculative supplying the tumor that one can access via cannulation. You do not have to target each and every tumor cell as long as you get a sufficiently high level at a given metastatic site and as long as you target the great majority of metastases. This is a highly experimental model, with rapidly dividing tumors that are therefore capable of a high rate of retroviral integration. You can target, with a single injection, over 90% of the metastases. In humans it may be possible actually to do multiple cannulations and you can think potentially about this as a therapy. But I have a lot of reservations about that for other reasons. I think we need a viral vector where you are not that close to the threshold level of cytokine required for killing. You need one perhaps that can express tenfold more cytokine activity. Adenoviral vectors may be potentially useful in that regard but then you have non integration and immunogenicity issues.

J. Griffin

Some human tumors, particularly lung carcinomas, already make cytokines like GM-CSF or G-CSF and sometimes to such an extent that there is a systemic syndrome called a leukemoid reaction. Do you think that those tumors escape this type of killing or is there some other explanation?

R. Tepper

I think there is another explanation because we reproduced some of those experiments earlier on in mice. We took an IL-3 transfected plasmacytoma and injected it subcutaneously. A large tumor results. The animals actually turn erythematous if you do this in nude mice. If you bleed the animal and look at the buffy coat it is expanded markedly. So the tumor grows and you get all the systemic effects, including histamine release and leukocytosis. So with certain cytokines there is no local cytotoxic effect; IL-3 is one of them.

18

CONCLUDING COMMENTS

Donald Metcalf

The Walter and Eliza Hall Institute of Medical Research
Melbourne, Australia

This meeting has reviewed current knowledge of the organization and regulation of normal hemopoiesis, the molecular mechanisms responsible for the development of leukemia and some newer approaches to the management of leukemia.

These subjects were linked deliberately in the program on the general grounds that, if we have a clear understanding of the manner in which normal hemopoiesis is regulated, we shall have gone a long way to understanding the nature of the cellular abnormalities responsible for leukemic transformation. The rationale behind this view is that leukemia appears to be based on errors in cell proliferation and in the process of self-renewal versus differentiative cell division - cellular events that occur continuously during normal hemopoiesis.

Our knowledge of the nature and regulation of hemopoiesis has increased remarkably in the last decade although information on certain key questions remains incomplete. The pattern emerging from the control of hemopoiesis is one in which specific regulatory factors, usually glycoproteins, regulate the formation, maturation and functional activity of cells in each of the eight major hemopoietic lineages. At least 25 of these specific regulators have now been defined, with the production of biologically-active recombinant material that is potentially available for clinical use in appropriate circumstances.

None of these regulators is restricted in its action to cells of a single lineage and the consequence of this arrangement is, typically, that six to eight regulators have demonstrable actions on cells of each lineage. This results in a control system that appears to be extremely redundant. However for each of the regulators so far analyzed, inactivation of the relevant gene by homologous recombination results in an abnormal phenotype, indicating that each regulator has unique functions that cannot be completely compensated for by other regulators. This indicates that the control systems, while exhibiting considerable overlap, are highly complex and subtle, an arrangement presumably designed to cope with situations ranging from the precise control of basal hemopoiesis to rapid expansion in response to a variety of emergency situations.

Under basal conditions, these regulators are produced in low concentrations and usually by a wide variety of cell types. Complex signaling networks have been identified that allow regulator production to be rapidly amplified either by initiation of transcription or by stabilization of mRNA. The required lability of regulator levels is achieved by a combination of the lability of mRNA transcription and the short half-lifes both of the mRNA and protein products.

Normal and Malignant Hematopoiesis, Edited by Enrico Mihich and Donald Metcalf
Plenum Press, New York, 1995

An unexpected feature emerging for all regulators is their polyfunctionality. Typically, such regulators not only control cell division but also at least some aspects of differentiation commitment, maturation induction and the functional activation of the resulting mature cells. The membrane receptors mediating these actions of individual regulators are usually of a single type but are homo- or heterodimers with specific regions in their intracytoplasmic domains that seem likely to initiate distinct signaling cascades. Key molecular components of these signaling cascades have been identified but current information is not quite complete enough to allow identification of particular intermediates as initiating specific responses in the stimulating cells. Ultimately those signaling cascades needing to achieve changes in transcriptional activity in the nucleus do so by modulating the production and functional activity of nuclear transcription factors, but in no case has the complete set of transcription factors been identified for any specific response such as mitosis or differentiation commitment. The increasing use of animal models in which genes encoding these signaling molecules or transcription factors have been inactivated should allow these pathways to be characterized in more detail.

It has been commented during the meeting that our knowledge of what regulates the processes of lineage or differentiative commitment remains incomplete. Some evidence suggests that these processes are stochastic and not subject to extrinsic regulation. I believe this view to be incorrect but some of the important regulators involved probably have yet to be detected. However, even with the known regulators, there are clear examples of regulator-modulated commitment for both normal and certain leukemic cells.

While the information on regulatory factors appears relatively complete for the more mature members of certain hemopoietic families such as the erythroid, granulocytic, moncytic and megakaryocytic pathways, information appears quite incomplete on regulators of T- and B-lymphoid cells and for stem cells. In particular, no regulator or regulator combination can achieve self-generation by stem cells, despite the obvious capacity of such cells to exhibit this property in vivo. Difficulty is also still being experienced in readily growing primary human or murine lymphoid leukemias in vitro - difficulties probably ascribable to our incomplete knowledge of the regulators of these cell populations.

Two aspects of the biology of leukemic cells that were not commented upon in the discussions are the clonality of most leukemic populations and the fact that, typically, leukemic cells divide more slowly than corresponding normal cells. In analyzing the regulatory mechanisms controlling hemopoietic cells, we need to address these two questions. We need to devise experimental systems that can explain how a more slowly dividing cell can gain clonal dominance over a pre-existing population of hemopoietic cells. The slower cell cycle times of leukemic cells are due usually to a longer than normal G1 period and with our improving understanding of the role played by cyclins in regulating transit of cells through G1, it may be possible to identify the nature of the abnormality common in leukemic cells.

As best documented for the myeloid leukemias, analysis has shown that leukemic cells are not genuinely autonomous but exhibit dependency on and responsiveness to normal regulatory factors. This is a curious and somewhat anomalous situation since there is evidence that an acquired autocrine capacity to produce relevant growth factors is one of the changes leading to leukemic transformation in experimental model systems and autocrine growth factor production is often demonstrable in human leukemic cells. Whatever the explanation for this situation, one consequence is that normal hemopoietic regulators, even when present in normal concentrations, must play an important role in stimulating the emergence of a leukemic clone and its expansion to result in clinical disease.

From model systems, two distinct types of change need to be acquired by cells during leukemic transformation (a) an acquired capacity for autocrine growth stimulation and (b) acquisition of an abnormal capacity for self-renewal divisions. Both of these abnormalities can be acquired by a variety of mechanisms. Autocrine growth stimulation may be achieved by the acquired capacity to produce one or more relevant growth factors but can also be achieved by expression of aberrant receptors that are functionally activated even in the absence of ligand binding. Alternatively, various protooncogene products may become produced in a dysregulated manner and achieve cell proliferation either because they are normal components of the mitotic signaling cascade or can achieve mitotic signaling by some alternative pathway.

Whether to self-renew or to form more differentiated progeny is a choice cells must make during each cell division but the mechanisms controlling this choice have not yet been characterized. It must be presumed that a variety of nuclear transcription factors can be involved in such decisions and that many of the oncogenes or protooncogenes identified as being involved in leukemogenesis operate by influencing such decisions. This appears to be the case for the Hox 2.4 gene, shown to be coleukemogenic with an acquired autocrine CSF production, and to enhance self-renewal divisions. It seems reasonable to speculate that the PML-RARa fusion protein in acute promyelocytic leukemia might also act by impairing the capacity of cells to differentiate during cell division.

One of the major problems concerning leukemogenesis is the difficulty in linking the above body of information on normal hemopoiesis or leukemia development in model systems with the rapidly increasing body of information on molecular abnormalities in human leukemic cells revealed by analysis of the specific translocations associated with various types of leukemia.

In general, the leukemias differ from solid tumors in the less frequent occurrence of random chromosomal abnormalities and the very common occurrence of precise translocations or inversions. Thus translocations such as the 9:22 translocation are virtually universal in chronic myeloid leukemia and the 15:17 translocation in acute promyelocytic leukemia. That translocations of this type can induce leukemic transformation has been documented in at least some animal models.

The problem arises because none of these translocations involves genes encoding any known hemopoietic growth factors or their receptors. The same situation is evident in the mouse where all murine myeloid leukemias exhibit deletions of chromosome 2 yet no genes for potentially relevant growth factors or growth factor receptors appear to be located on this chromosome.

Thus, two bodies of internally consistent data have arisen from growth factor studies in normal and leukemic systems on the one hand and from translocation-identified protooncogenes that must somehow be combined. It may be that the translocation-derived fusion proteins or aberrantly-expressed proteins can be categorized as influencing one or other of the two general abnormalities needing to be acquired by transforming cells. It is not possible on present information to advance specific proposals for most of these products although the bcr-abl product of the Philadelphia translocation seems able to deliver a proliferative signal when introduced into cells and this would provide a partial explanation of the role of bcr-abl in chronic myeloid leukemia. Some evidence was presented during this meeting that oncogene products may, on occasion, deliver a proliferative signal to a cell through an alternative signaling pathway. This is an intriguing possibility that warrants a careful search in leukemic cells for such alternate signaling pathways.

It has to be admitted that advances in knowledge of the biology of hemopoiesis so far have had only a minor impact on the clinical management of leukemia. Elimination of leukemic cells by chemotherapy still remains as the sole method for achieving a cure in leukemia. A significant exception has been the use of all-trans retinoic acid to achieve spectacular initial remissions in acute promyelocytic leukemia where a precise defect involving the retinoic acid receptor is demonstrable. Even here, sustained remissions still depend on early chemotherapy. There is some role for growth factors to enhance hemopoietic regeneration following marrow transplantation and a possible role for such agents in achieving increased responses of appropriate leukemic cells to cycle-specific chemotherapeutic agents.

What has emerged as a disappointing aspect of current studies on leukemia is the identification of specific translocations as being involved in a high proportion of leukemias. Why should this be disappointing? The reason is that no likely mechanisms have been identified that would be responsible for inducing such translocations. Moreover, few investigators now study such possible mechanisms and there is a real likelihood that these are random abnormalities resulting from no specific initiating cause. If the latter is in fact the situation, then we are faced with the admission that leukemia arises because of random chromosomal rearrangements and will continue to occur as a series of disease states for which no prevention is possible.

The identification of specific translocations in leukemic cells has provided a sensitive method, using the polymerase chain reaction with appropriate probes, for the detection of the very small numbers of residual leukemic cells. This is proving to be a useful guide for the effectiveness of chemotherapy or the need for further chemotherapy. In principle, the same approach could be used for the very early detection of leukemic clones but this is not a practicable method for application to a general population, even if combined probes were used to screen for the presence of multiple types of leukemic cells. The approach may have some merit when applied to persons at high risk of leukemia such as myelodysplastic patients or patients receiving intensive chemotherapy for other types of cancer.

With these possible exceptions, nothing so far learned provides obvious opportunities for the prevention of cancer. The newer knowledge of the molecular abnormalities in leukemic cells and of regulators controlling hemopoietic cells can only be foreshadowed as potentially providing improved methods for the treatment of diseases that will continue to develop.

On this mixed note, I will conclude this discussion and would like to thank all the speakers for the fair and clear manner with which they have reviewed their assigned subjects and for their thoughtful comments during the various discussion sessions. My discussion is itself now open for comments and discussion.

DISCUSSION

L. Nadler

One of the questions I have comes from your proposed model of leukemogenesis. A neoplasm is always internally selecting. We always think of neoplasms as clonal malignancy because they are clonal in their genetic abnormality. But the tumor cells are clearly not clonal in their biological phenotype. There is always selecting to win. If there is an immune system it is selecting to win. The tumor cells are selecting to win in the microenvironment in which they live. So any model has to factor in a continuing genetic change by somatic mutation, additional oncogenic events, or expression of molecules allowing cells to exit into tissues etc., to keep the malignant clone dominant. We so rarely cure patients with tumors and this is certainly due to the tumor evolving. We see this well in follicular lymphoma, involving the bcl-2 translocation that Riccardo talked about. There was a clinical study at NIH where they randomised people to treat them early or treat them late. When they treated them early with the same amount of bulk tumor, they responded much better. This is a malignancy under such driven somatic mutation, that the late malignant populations are very different from those present earlier in the disease. Do you have a comment on this?

D. Metcalf

Yes, I agree. More tumors are clonal by some criteria yet the populations of cells within the tumor are certainly heterogeneous. I do not agree that tumors always generate more potent, more vicious progeny. In the FDC-P1 leukemogenesis model we have been studying lately, it has been quite striking that in most leukemic populations, certified autonomous clonogenic cells continually generate progeny, most of which have reverted to a factor-dependent, presumaly less malignant, state. A similar sequence is presumaly responsible for the fact that most cells in a chronic myeloid leukemic clone are fully factor-dependent. Similarly, in early studies on Friend virus-induced erythroleukemia, a large proportion of the progeny leukemic cells died during tumor progression. It does remain a familiar clinical reality, best seen with the lymphomas, that there is usually a progression in the course of the disease to populations of greater and greater malignancy.

L. Nadler

Just one additional point. It is remarkable to me that, even in the hematologic malignancies as in the solid tumors, populations emerge with Class I or Class II histocompatibility antigens and with changes in their adhesion receptors. Thus with tumor progression, cells emerge with very different properties and clinicians annot regard tumors as being composed of cells of a fixed type.

D. Metcalf

Absolutely not, tumor cell populations are highly dynamic.

L. Nadler

And if you think about them as a fixed target you will lose the battle.

H. Beug

You alluded to the fact that the experimental models of leukemogenesis, which are mainly murine, often do not quite agree with what is seen in humans. I was wondering whether this may be due to the fact that the mouse may be much more genetically unstable than the human. There appear to be fewer control mechanisms to prevent mutations and cells undergoe immortalisation even in naturally-occurring tumors, which I understand is not frequent in human tumors. It may be much easier with murine cells to achieve immortalisation and self-renewal than with human cells where you have, first by p53, to generate genetic instability.

D. Metcalf

I think that is probably a valid comment. However, even with murine cells it can be hard to generate immortalized or aberrant cells. We have been studying recently a max 41 transgenic mouse that forms an enormous excess of granulocitic cells, yet these cells appear to retain a normal phenotype and myeloid leukemia does not occur.

J. Adams

In view of the great advances in hematopoietic growth factors, why is it that it is so difficult to grow cells from humans that have a rampant leukemia? Is this just a technical problem, or does it mean that the growth control of the cells *in vivo* in some way is greatly different in kind from what we can achieve in cultures?

D. Metcalf

In the case of a disease like chronic myeloid leukemia there is absolutely no difficulty in growing leukemic cells in primary culture. What is worrying however is that the clonogenic cells grown are not capable of self generation. It may be that the growth factors used have in fact truncated this property by differentiation commitment in the act of generating the colonies. Alternatively, it may be that none of us have yet been able to culture the true ancestral clonogenic cells in these populations. There has been a recent study in the SCID mice trying to estimate by limit dilution assays, the frequency of true clonogenic cells in acute myeloid leukemic populations, as defined by the ability to generate a transplanted tumor in the SCID mice (Lapidot et al., Nature 1994, 367:645-648). They estimated that there was only one engrafment unit (?stem cell) in every quarter of a million leukemic cells. However, such data need to take into account the fact that even with a uniform cloned leukemic cell population it usually requires far more than one clonogenic cell to produce a transplanted tumor even in syngeneic recipients. So the frequency of clonogenic cells in AML may be much higher than one in 250,000. The situation is more difficult for human lymphoid leukemias and lymphomas because culture methods used to date have produced very poor results with primary tumor tissue.

L. Nadler

I would like to comment on this. Everyone has assumed that you need soluble factors to obtain lymphoid tumor cell growth. In fact, with the B-cell malignancies including multiple myeloma, follicular lymphoma, chronic lymphocytic leukemia, what seems to be needed are cell contact molecules. We can now grow these primary tumors as well as we can grow EB virus-transformed B-cell lines, but they need to be maintained using the right cell contact signal. To improve our ability to grow epithelial tumors and solid hemopoietic tumors, we really need to clone the genes, encoding these contact signals and growth factors.

D. Metcalf

I was making a comment about soluble growth factors. Membrane-displayed growth factors may be what are needed for some tumor cell types but we need to see the data first. Even if you use a Dexter type (cell contact) long-term culture with CML or AMLcells, the leukemic population still terminates. Simply the provision of a cell contact underlayer that works well for a time is not really what can sustain that population in the way we assume it can be maintained *in vivo*. So we are still missing something and cell contact or membrane-displayed molecules certainly may be part of the answer.

D. Livingston

Even with well-established immortal cell lines, not all of the cells are clonogenic. So there is more to the story than being immortal, having rich tissue culture medium and an attachment surface on which a cell can grow. We do not know how yet to make cells grow at 100% efficiency. Some of this may be due to genome instability. There is no such thing as a true clone once the genome becomes unstable because then no two cells are exactly the same genetically and there may be the growth requirements for one cell that is not relevant for the others. I think there are still many things to learn about what it takes to be able to be established as a clone, even on an attached surface.

D. Metcalf

You do not expect 100% of cells of a tumor population to exhibit any one property because of the known heterogeneity even in a cloned tumor population.

D. Livingston

Well, you say you have to inject a thousand cells, or ten thousand, or a hundred thousand to produce a transplanted tumor. There must be many physiological factors like vascular supply that can influence this process.

D. Metcalf

Yes that is quite correct. Even using cells of a uniform cloned line, some cells on transplantation will become trapped in vessels or lodge in tissues where, for various reasons, continued cell proliferation is not possible.

INDEX